MW00824134

About Island Press

Island Press is the only nonprofit organization in the United States whose principal purpose is the publication of books on environmental issues and natural resource management. We provide solutions-oriented information to professionals, public officials, business and community leaders, and concerned citizens who are shaping responses to environmental problems.

In 2004, Island Press celebrates its twentieth anniversary as the leading provider of timely and practical books that take a multidisciplinary approach to critical environmental concerns. Our growing list of titles reflects our commitment to bringing the best of an expanding body of literature to the environmental community throughout North America and the world.

Support for Island Press is provided by the Agua Fund, Brainerd Foundation, Geraldine R. Dodge Foundation, Doris Duke Charitable Foundation, Educational Foundation of America, The Ford Foundation, The George Gund Foundation, The William and Flora Hewlett Foundation, Henry Luce Foundation, The John D. and Catherine T. MacArthur Foundation, The Andrew W. Mellon Foundation, The Curtis and Edith Munson Foundation, National Environmental Trust, The New-Land Foundation, Oak Foundation, The Overbrook Foundation, The David and Lucile Packard Foundation, The Pew Charitable Trusts, The Rockefeller Foundation, The Winslow Foundation, and other generous donors.

The opinions expressed in this book are those of the author(s) and do not necessarily reflect the views of these foundations.

SAMPLING RARE *or* ELUSIVE SPECIES

SAMPLING RARE *or* ELUSIVE SPECIES

Concepts, Designs, and Techniques for Estimating Population Parameters

Edited by
William L. Thompson

Foreword by
Kenneth P. Burnham

Island Press
Washington . Covelo . London

Library of Congress Cataloging-in-Publication Data.

Sampling rare or elusive species : concepts, designs, and techniques for estimating population parameters / edited by William L. Thompson ; foreword by Kenneth P. Burnham.
p. cm.
Includes bibliographical references and indexes.
ISBN 1-55963-450-2 (cloth : alk. paper) — ISBN 1-55963-451-0 (pbk. : alk. paper)
1. Rare animals—Monitoring. 2. Rare plants—Monitoring. I. Thompson, William L. (William Lawrence), 1962-
QL82.S26 2004
591.68—dc22

2004008006

British Cataloguing-in-Publication Data available.

Book design by Teresa Bonner

Printed on recycled, acid-free paper 

Manufactured in the United States of America
10 9 8 7 6 5 4 3 2 1

Contents

Part Five The Future

Foreword

> The starship settles into a standard orbit around the M-class planet while the science officer studies the sensor readouts.
>
> "Captain, this planet has an amazing diversity of life; there are 16,783 species of mammal-like animals alone. The most rare of these has only 28 individuals left, correction, 27 now; the most common species numbers close to 7 billion individuals. With our sensors we can determine the abundance and diversity on every hectare down there."
>
> "That is more than I need to know, Mr. Spock."

I do not remotely expect this scenario ever to be achieved. However, I have been active in statistical aspects of population estimation since 1969 (34 years), and I have witnessed advances in methodology that truly were not envisioned in 1969 (or in 1965, the benchmark year for capture-recapture).

The phenomenal progress in wildlife population estimation in the past third of a century has been driven by the practical need to estimate population abundance and related parameters and to monitor changes over time in animal abundance. Visionary development of both new models and new technologies have played big roles in these advances. It is easier to note some of the new technologies: radio telemetry, satellite telemetry, passive integrated transponder (PIT) tags, coded wire tags, DNA as the "tag," acoustical methods, global positioning systems (GPS), geographic information systems (GIS), photographic identification, camera traps, affordable personal computers, and comprehensive free software for the required specialized analyses of the data, which arise from a plethora of new statistical models (and applications) and inference methods like Markov chain Monte Carlo (MCMC), not to mention many new innovative field sampling designs.

Another important reason for the outpouring of new and refined methods in population estimation is the now substantial and ever-increasing number of statisticians working on the subject, often cooperatively with a growing number of well-trained quantitative wildlife biologists and ecologists. Increasingly, there is a meaningful blending of statistics, ecology, and field methods within a spirit of rigorous theory yet practical applicability.

Methodology for estimating the abundance of biological populations continues to be a very challenging yet active and productive area of biometrical research. However, an underemphasized yet critical aspect of this subject area is the topic of this book: population estimation when the subject populations are sparse, rare, or elusive (or all three—these are three different concepts). Most work in population estimation does not particularly focus on the problem of the very-difficult-to-sample species and situations, wherein field methods may be as important as statistical models (hence methods such as capturing not the animal but rather its photograph, or its DNA, or its acoustical signature).

The uniqueness and strength of this book is that it focuses on what its title says and implies: sampling rare or elusive species to estimate abundance and other biologically meaningful parameters. The chapter authors are persons with real and substantial experience in their subject material, which is often some aspect of capture-recapture or distance sampling. They are experienced, often substantially so, with the general theory and tools of abundance estimation, with the species and methods they write about, with real data, real studies, and often with management and legal issues that arise from the Endangered Species Act.

There are many biases and problems in sampling of, and inference about, rare or elusive animals or plants. (The book concentrates on animals, but plants are not ignored.) This book presents in one place comprehensive information about these biases and how we can reduce them. Many of these issues involve how data are collected: design and field protocols. This is not a book just about abstract statistical models and data analysis.

Another feature of the book I like is the wide taxonomic range of problems and examples considered. It is, thus, cross-disciplinary as regards applications (birds, fish, herpetofauna, mammals, plants). In fact, the foundations of population sampling and abundance estimation are quite general, and it is important to know this. A person considering estimating abundance of a given taxon must realize there is much of value in the general literature on the subject. To be knowledgeable about abundance estimation, a person must read a wide range of the literature, not just the literature on the taxon of special interest. A reader may well find the information needed in literature that nominally is about a different taxon. Thus, I regard it as advantageous to have a book on a common theme that includes examples across many taxa and sampling applications.

This book is a practical blending of sound theory and useful applied and applicable state-of-the-art methods in the expanding subject of sampling rare and elusive species. The authors and editor are to be thanked for their contributions herein.

KENNETH P. BURNHAM
November 2003

Acknowledgments

The idea for this volume arose from a half-day symposium I organized and chaired, "Sampling Rare or Elusive Species: Challenges and Choices," which was held at the Ninth Annual Conference of The Wildlife Society in Bismarck, North Dakota, on September 26, 2002. I am grateful to The Wildlife Society and to the Biometrics Working Group of The Wildlife Society for sponsoring this symposium.

The strength of any edited volume is derived from the quality of its chapters. This volume benefited greatly from the breadth of experience and expertise of its chapter authors (see "Contributors" at the end of this volume). I thank them for their time and efforts. I also am deeply indebted to the many external reviewers for their editorial assistance (see "Reviewers" at the end of this volume). Their insightful comments generated a much-improved version of this book.

Finally, I offer my sincere thanks to the editorial staff of Island Press, especially Barbara Dean, Barbara Youngblood, and Cecilia González. This project never would have been successful without their patience, support, and guidance.

1

Introduction

William L. Thompson

Natural resource professionals are typically faced with multiple sources of uncertainty when attempting to manage plant and animal populations in ways that ensure their persistence at acceptable levels into the foreseeable future. These sources of uncertainty can be broadly categorized as environmental variation, structural uncertainty, partial observability, and partial controllability (Nichols et al. 1995; Williams 1997). Environmental variation refers to inherent differences in the physical setting, such as those induced by habitat and weather, that may influence population status and trend. Structural uncertainty arises from the lack of knowledge about the underlying processes influencing population dynamics and how these processes relate to management practices. An example is the uncertainty related to whether hunting has an additive or a compensatory effect on waterfowl populations (Williams 1997). Partial observability concerns the incomplete knowledge of population status and trend, whereas partial controllability is the inability to strictly regulate the implementation of management actions.

Sources of uncertainty affecting management of plant and animal populations are not mutually exclusive. For instance, complex habitats (environmental variation) could lead to an incomplete and variable assessment of population status or trend (partial observability), which in turn could mask demographic responses to management practices (structural uncertainty). Therefore, effective management of plant and animal populations is predicated upon both recognizing and properly accounting for (hopefully, minimizing when possible) these various sources of uncertainty (Williams 2001). The optimum approach is to incorporate management options within an adaptive resource management framework (Holling 1978; Walters 1986). This is sometimes referred to as management by experiment (MacNab 1983), which recognizes uncertainty inherent within natural systems and incorporates learning as a fundamental objective so that management can improve with knowledge gained by previous actions. Competing models or hypotheses encapsulate this uncertainty and are evaluated as part of the learning process.

A critical element of adaptive resource management is monitoring responses of the population(s) under study to the management action(s) (Morrison et al. 2001). The monitoring program typically is based on obtaining an estimate of occupancy (spatial distribution or proportion of area occupied), abundance, density, or survival recorded at specified time intervals within the area of interest. However, estimation of these population parameters often is confounded by detection probability, that is, probability of correctly noting the presence of an individual (or species, for species-level surveys) within some area and time period. Failure to properly account for detection probability leads to biased estimators and, perhaps, misleading estimates of population status and trend (see Chapter 4). This could undermine the learning process in adaptive resource management and erode empirical support for management decisions and policies.

Development of methods and models to properly account for detection probability, especially for mobile species, continues to be one of the hotter areas in biometric research. This is reflected in the recent proliferation of books describing these approaches (e.g., Thompson et al. 1998; Elzinga et al. 2001; Borchers et al. 2002; Williams et al. 2002). However, sampling designs and counting methods described in these books are geared primarily toward moderately abundant to abundant species; many do not translate well to rare, elusive, or otherwise difficult-to-detect species (see Chapter 2). Unfortunately, these species are usually the ones of management concern because of their low numbers, limited geographical distribution, and/or our lack of reliable information on their population status and trend.

General Sampling Framework

The purpose of this volume is to describe the latest sampling designs and counting (estimation) techniques for reliably estimating occupancy, abundance, and other population parameters of rare or elusive plants and animals. Rare refers to low abundance or restricted geographical distribution (clustered or not) or both, whereas elusive refers to low probability of detection for whatever reason(s) (see Chapter 2). This demographic information in turn may be used as a basis for informed management and policymaking. Because mobile species typically present greater sampling challenges, the majority of this volume focuses on free-ranging animals. However, many of the designs and counting or estimation techniques also apply to sampling sessile organisms such as plants.

Throughout this volume, population or species surveys usually follow a sampling framework whereby a geographical area containing the population or species of interest is partitioned into sampling units, such as plots, quadrats, and transects. Then a subsample of these units is chosen based on some form of probability-based selection process, and some type of count or measurement is applied within chosen units. Such a scenario may be viewed as a two-stage sampling design (after Hankin 1984 and Skalski 1994), where the random selection of units is the first stage and the count or measurement is the second stage. This usage deviates from that in sample survey texts (e.g., Cochran 1977), which define a two-stage design as a random sample of subunits within randomly selected sampling units (e.g., subareas within a chosen plot). However, the traditional definition of a two-stage sample assumes a complete enumeration or measurement within selected subunits, which is usually not possible in plant or animal surveys. Therefore, this volume follows the definition of two-stage designs (or, generally, multistage designs) that treats the count as the last stage. Note that additional stages are possible, depending on the spatial extent of the area of interest.

This volume generally follows statistical notation suggested by Thompson et al. (1998) for plant and animal surveys, such that U is the total sampling units used for inference, u is the number of sampling units in a sample, and N is the true population size of individuals in the sampled population. However, notation in Chapter 13 deviates from this convention because it uses previously published notation in its equations; hence, in this chapter only, M is the total sampling units used for inference, m and n denote number of sampling units in a sample, and T is the true population size (total).

Scope of this Volume

Because of the extreme paucity of published books on sampling rare or elusive species, this volume contains a mixture of both theory and application, with a strong emphasis on application. This volume is organized into five sections: Overview and Basic Concepts; Sampling Designs for Rare Species and Populations; Estimating Occupancy; Estimating Abundance, Density, and Other Parameters; and The Future.

The first section provides an overview of sampling rare or elusive species and populations. In Chapter 2, McDonald discusses the results of his informal survey of biologists and biometricians in which he solicited their definitions of "rare" and asked them to recount their successes and failures when sampling rare species. This chapter serves as both an overview of designs commonly used to sample rare species and as a guide to their practicality and usefulness based on experiences of biologists and biometricians. The next two chapters in Part I deal with the issue of detection probability and its effects on parameter estimators. Pollock et al. (Chapter 3) discuss the components of detection probability and present applied examples of how to estimate these components. In Chapter 4, Conn et al. use simulation to rigorously evaluate implications of using counts (captures, in this case) uncorrected for incomplete detectability as surrogates for abundance in monitoring population trends of low-abundance species. Topics discussed in Chapters 3 and 4 are relevant to moderately abundant or abundant species as well as to rare ones.

Part II of this volume explores sampling designs for efficiently estimating abundance of rare species, with a greater emphasis on the first stage of a two-stage design, that is, probability-based selection of sampling units. Smith et al. provide a thorough review and practical evaluation of adaptive cluster sampling in Chapter 5. Manly follows with a description of an alternative adaptive approach to estimate abundance of rare species, based on a two-phase, stratified sampling regime. In Chapter 7, Christman reviews the sequential sampling design, another form of adaptive sampling, and applies it to estimating abundance of a waterfowl population.

The objective of a monitoring program may be to track changes in presence or occupancy of a species within the area of interest. Such information tends to be less costly to collect than data on abundance, density, or survival (MacKenzie et al. 2001, 2003). However, just like other population estimators, occupancy estimators are susceptible to bias from incomplete detectability. Therefore, Chapters 8 and 9 describe alternative approaches to estimating occupancy through correction of incomplete detection probability. MacKenzie et al. (Chapter 8) provide a review of the topic, present their latest work on developing a reliable estimator of occupancy, and apply their estimator to sampling rare populations. In the next chapter, Peterson and Bayley offer a Bayesian alternative for estimating probability of presence for rare or difficult-to-detect species. Poon and Margules complete this section by describing a practical sampling method for locating populations of rare plants in remote areas, which is essentially a form of species-presence survey.

Because abundance and density are commonly used in population monitoring, a section of this volume is devoted to methods for estimating these and other population parameters for rare or elusive species or populations. Chapters in Part IV generally have a greater emphasis on the second stage of a two-stage design (counting methods), as compared to chapters in Part II. Nonetheless, both stages are important to obtaining reliable parameter estimates.

The first four chapters in Part IV describe noninvasive methods for estimating abundance of rare or elusive species. Waits (Chapter 11) reviews the latest literature on genetic sampling methods for estimating abundance of rare animals and then critically appraises the current state-of-the-art for using these techniques. Karanth et al. follow with an overview of photographic sampling methods. They provide practical advice for applying this approach to sampling elusive mammals in tropical forests. At the other end of the climatic spectrum, Becker et al. (Chapter 13) describe their work with using tracks in the snow to estimate abundance of rare and elusive carnivores. We move to the marine environment in Chapter 14, where Hanselman and Quinn evaluate the use of adaptive sampling and double sampling, in conjunction with trawls (invasive) and hydroacoustics (noninvasive), to estimate abundance of rare and clustered, bottom-dwelling fish species.

In Chapter 15, O'Shea et al. tackle the difficult problem of sampling bats. They conclude that reliable survival estimates of bats are much more obtainable than meaningful abundance estimates. Therefore, they discuss the pros and cons of banding bats, provide an extensive review of survival studies of bats, critically evaluate these studies, and offer examples of their own research in estimating survival rates of bats. Ganey et al. conclude Part IV with a comprehensive evaluation of the effectiveness of a monitoring protocol for detecting trends in abundance and in finite rate of population growth (λ) of Mexican spotted owls (*Strix occidentalis lucida*). All other chapters could be viewed, and perhaps should be viewed, within this monitoring context. As with other chapters in this volume, concepts discussed by Ganey et al. are not limited in application to one species or taxon.

The final part and chapter discuss future avenues of research and methodological development for estimating abundance of rare or elusive species. Chapter 17 is presented within a two-stage sampling framework discussed earlier but could be applied to larger-stage contexts. Nonetheless, the focus is both on probability-based selection of plots and on counting or estimation techniques that properly correct for incomplete detectability of animals or plants within these plots.

Because relatively little has been published on sampling rare or elusive species, the scope of this volume is necessarily broad, with examples from plants and animals within terrestrial, aquatic, and marine environments. This breadth of application extends to geography as well; chapter authors hail from the United States, India, New Zealand, and Australia. I hope this volume will serve as a source of information for biologists and resource managers and as a source of motivation for biometricians to engage in further research and methodological development in this area of biological sampling. When possible, biologists and resource managers should try to incorporate designs and survey methods discussed in this volume to population monitoring programs, and in turn, conduct these programs within the adaptive management framework that was discussed earlier in this chapter.

References

Borchers, D. L., S. T. Buckland, and W. Zucchini. 2002. *Estimating Animal Abundance: Closed Populations.* Springer-Verlag, London.

Cochran, W. G. 1977. *Sampling Techniques.* 3rd ed. Wiley, New York.

Elzinga, C. L., D. W. Salzer, J. W. Willoughby, and J. P. Gibbs. 2001. *Monitoring Plant and Animal Populations.* Blackwell Science, Malden, Massachusetts.

Hankin, D. G. 1984. Multistage sampling designs in fisheries research: Applications in small streams. *Canadian Journal of Fisheries and Aquatic Sciences* 41:1575–1591.

Holling, C. S., ed. 1978. *Adaptive Environmental Assessment and Management.* Wiley, New York.

MacKenzie, D. I., J. D. Nichols, J. E. Hines, M. G. Knutson, and A. B. Franklin. 2003. Estimating site occupancy, colonization, and local extinction when a species is detected imperfectly. *Ecology* 84:2200–2207.

MacKenzie, D. I., J. D. Nichols, G. B. Lachman, S. Droege, J. A. Royle, and C. A. Langtimm. 2001. Estimating site occupancy rates when detection probabilities are less than one. *Ecology* 83:2248–2255.

MacNab, J. 1983. Wildlife management as scientific experimentation. *Wildlife Society Bulletin* 11:397–401.

Morrison, M. L., W. M. Block, M. D. Strickland, and W. L. Kendall. 2001. *Wildlife Study Design.* Springer-Verlag, New York.

Nichols, J. D., F. A. Johnson, and B. K. Williams. 1995. Managing North American waterfowl in the face of uncertainty. *Annual Review of Ecology and Systematics* 26:177–199.

Skalski, J. R. 1994. Estimating wildlife populations based on incomplete area surveys. *Wildlife Society Bulletin* 22:192–203.
Thompson, W. L., G. C. White, and C. Gowan. 1998. *Monitoring Vertebrate Populations.* Academic Press, San Diego.
Walters, C. J. 1986. *Adaptive Management of Renewable Resources.* MacMillan, New York.
Williams, B. K. 1997. Approaches to the management of waterfowl under uncertainty. *Wildlife Society Bulletin* 25:714–720.
_____. 2001. Uncertainty, learning, and the optimal management of wildlife. *Environmental and Ecological Statistics* 8:269–288.
Williams, B. K., J. D. Nichols, and M. J. Conroy. 2002. *Analysis and Management of Animal Populations.* Academic Press, San Diego.

Part I

OVERVIEW AND BASIC CONCEPTS

Part I of this volume lays the foundation for the rest of the book with an introduction and overview of methods and concepts for sampling rare or elusive species. In Chapter 2, McDonald presents results of an informal survey of biologists and biometricians in which he solicited their definitions of "rare" and asked them to recount their successes and failures when sampling rare or elusive species. These real-world experiences serve as a practical guide for biologists and biometricians who are sampling rare species in a variety of habitats and contexts.

The next two chapters in Part I deal with the important issue of detection probability and its effects on parameter estimators. In Chapter 3, Pollock and his colleagues provide a detailed introduction to this topic. They then discuss the components of detection probability and present applied examples of how to estimate these components for different species and environments. In the following chapter, Conn et al. use simulation to rigorously evaluate implications of failing to properly account for detection probability (capture probability, in this case) when monitoring population trends of low-abundance species. Topics discussed in Chapters 3 and 4 are relevant to moderately abundant or abundant species as well as to rare ones.

2

Sampling Rare Populations

Lyman L. McDonald

A rare population sometimes is defined as one with a low number of individuals. However, even large populations can appear to be rare either because the species practices elusive behavior or because the population is sparsely distributed over large ranges. Populations also may appear to be rare because of ineffective survey procedures. In any case, it is difficult both to detect and to estimate abundance or distribution of rare populations.

In this chapter, I briefly review definitions of a rare population and procedures typically recommended for detection and study of rare populations of animals and plants. I also report on my personal experiences, successes and failures, in attempts to detect, study, or monitor the abundance or distribution of rare populations. In addition, I report the results of a survey of statisticians and biologists in which they reported their successes and failures while attempting to study rare populations.

Survey Form

I had been planning to prepare an overview on "Sampling Rare Populations" for some time. I had my own experiences on which to call, but these were limited primarily to terrestrial wildlife studies that I have worked on in the western United States, with some work on marine mammals off the coast of Alaska. To broaden the scope of the chapter, I sent a request via e-mail to a number of statisticians and biologists. Frankly, this request was a convenience survey of my associates for whom I had current e-mail addresses, a direct violation of one of the basic conclusions of this chapter, namely, the need to have some type of randomness in the study and to spread the sample over, in this case, the population of statisticians and biologists. Regardless of the use of this nonrandom sample, the responses were informative, and I trust they will be useful and interesting to the reader. I also have woven my own experiences into the discussion.

I have enough experience with surveys to avoid open-ended questions, however, this was an exception. I have chosen to paraphrase or quote from the responses, hoping to capture the essential elements. I apologize in advance if I misrepresent any facts or opinions. Also, there was overlap in the rather long responses, and I apologize if I fail to give credit to similar comments. The text of the e-mail message read:

I am planning to prepare a paper on Sampling Rare Populations. I have attached a draft abstract to give you an idea of what I have in mind. Please do me a favor and provide some information concerning your personal experiences in attempting to sample rare populations. I intend to summarize and report on my (biased) sample of statisticians and biologists. Published work may be reviewed and included in the literature cited. Please let me know if I can quote you and cite personal communication on unpublished material.

First, write a definition for a "rare population" or modify mine: A rare population is one where it is difficult to find individuals because of small numbers, secretive and/or nocturnal behavior, or clumped distribution over large ranges, that is, a lot of zeros occur in the data.

Second, list those sampling procedures that come to mind if you are faced with designing a study to detect, investigate, determine abundance, or monitor a rare population. A couple of mine include distance sampling, adaptive cluster sampling, etc.

Third, please describe your most successful experience (including a brief discussion of the study design) to detect, study, determine abundance, or monitor a rare population. Give references to any published material or send gray literature if available. My most successful/satisfying experience was in design of a study to determine densities of clams (including Saxidomus gigantean, Protothaca staminea, *and* Serripes groenlandicus*) in an area of Prince William Sound following the* Exxon Valdez *oil spill. There was lower abundance of clams in the subjective sample of sites in the "preferred habitat" (stratum 1) than in the systematic sample of shoreline sites in the rest of the area (stratum 2).*

Finally, please describe your biggest failure (including a brief discus-

*sion of the study design) in an attempt to detect, study, determine abundance, or monitor a rare population. What was wrong? My biggest failure was in helping to design surveys to estimate the abundance of black-footed ferrets (*Mustela nigripes*) in the colony in the Big Horn Basin of Wyoming following their discovery in 1981. Again, please provide references to published literature or send copies of gray literature.*

Any other comments or suggestions?

Survey Results

This section presents results of my informal survey of statisticians and biologists regarding their definitions of a rare population and procedures they recommended for detection and study of rare populations of animals and plants. This section ends with both the respondents and me recounting our successes and failures in these efforts.

Definitions and Objectives Should Be Made Concise

My definition of rare is relative to detectability. A rare population is one in which it is difficult to find individuals because of small numbers, secretive and/or nocturnal behavior, or clumped distribution over large ranges.

I was quickly reminded that the word "population" should be defined and plays a major role in the objectives and methods of an associated study. Both Mike Morrison (wildlife ecologist, University of Nevada) and Jim Jenniges (environmental specialist, Nebraska Public Power Disrtict, Kearney) appropriately reminded me that this was a widely misunderstood and misused term (personal communication; this reference to personal communication will be dropped throughout the rest of this chapter due to the nature of the source material). Before one can define "rare," one must first define "population." Was a "clumped distribution over large ranges" even a population, or was the study concerning a metapopulation structure? A statistical definition of a population (or sampling frame) was one thing, but here, the word "population" has biological meaning and the objectives of a study should be carefully documented.

Definition of the word "rare" elicited even more responses because it did not simultaneously fit everyone's concept of common species (that may be difficult to detect), small populations (e.g., threatened and endangered species under the Endangered Species Act), and everything in between. "Perhaps the rarity of a species originates mainly from the small abundance of appropriate habitat in the study area and not detectability, in which case it does not fit your definition of rare populations" (Nick Markov, wildlife biologist, Russian Academy of Science). Bill Thompson (biometrician, USDI National Park Service) provided a good reference on the issues entitled, "What is rarity?" by Gaston (1997). Most biologists and some statisticians objected to defining rarity relative to detectability and not relative to an absolute. For instance, Bill Gould (biometrician, New Mexico State University) commented that a survey method could be considered effective if detectability was high and ineffective if detectability was low, thus making an abundant population appear to be "rare." Several individuals commented that seasonality and species life stages should be considered. Many animals are easily observable only during short periods of their life history, and that life history must be understood before adequate survey procedures can be devised. Charles Bomar (professor of biology, University of Wisconsin-Stout) considered rare a poor term in his field of entomology because the general consensus was that it had no meaning; he offered an opinion with which I agreed: " . . . the reality of rare is we don't know anything about the particular species."

David R. Smith (biometrician, USGS Leetown Science Center) provided the observation that a population small in numbers was rare, a population that was secretive and/or nocturnal was elusive, and a population that was clumped was spatially clustered. These characteristics did not necessarily go together, although they could. Smith has been working on sampling of freshwater mussels (*Unionoida*), which can be rare, clustered, and elusive—the last because some proportion of the population is buried in substrate and not easily detected (Smith et al. 2001). When all three of these characteristics are present, they have a multiplicative effect on the difficulty of designing an effective sampling procedure.

I was involved in designing similar surveys for clams in the shallow subtidal zones in Prince William Sound, Alaska (Holland-Bartels 2001). Actually, this turned out to be one of my more satisfying attempts to design a survey of rare-clustered-elusive populations, but not because we learned a great deal about the population of clams on the first survey attempt (see below).

In my defense, many individuals had little problem with defining rare in terms of detectability—up to a point. Ken Burnham (statistician, USGS Colorado Cooperative Fish and Wildlife Research Unit) wrote that it might not be the population that is rare, but rather that individuals are difficult to detect relative to effort in the survey methods. Lowell Diller (senior biologist, Green Diamond Resource Company) expanded my definition, "A rare population is one where it is difficult to find individuals, utilizing known sampling techniques, either because of small numbers, secretive and/or nocturnal behavior, or because of clumped distribution over large ranges, that is, a lot of zeros occur in the data. Therefore, a rare population is often conditional on the sampling techniques available."

Diller illustrated his point based on surveys for snakes in the Snake River Birds of Prey Area in southern Idaho in the mid-1970s. At the time, he indicated that night snakes (*Hypsiglena torquata*) were thought to be one of the rarer snakes in Idaho, with only four known records for the state. Using standard collecting techniques at the time (turning rocks and such along some transect or driving roads at night), he too came to the conclusion that night snakes were very rare. In the second year of his study, he experimented with drift fences and funnel traps and suddenly began capturing many night snakes. They turned out to be the most common species of snake in certain habitats and were the third most commonly captured of these species within the entire study area. Diller concluded that some species may appear to be very rare based on current survey procedures, but can in fact be found to be quite abundant when the proper technique is applied.

Distinguishing between truly functionally rare populations/species and operationally rare populations is an issue of considerable societal and ecological importance (Mike Marcus, senior biologist, Tetra Tech EM Inc.). In some cases, rarity may be only a consequence of incompetent sampling, that is, a population may have intermittent, geographic rarity or behaviorally induced rarity, operational factors that make the population very difficult to sample effectively or efficiently. I agree with Marcus that rarity may be only an artifact of our operational ignorance on how to adequately sample certain populations.

Dan Goodman (professor of ecology, Montana State University) commented that it was interesting that the sampling problems were really most intense with sparse and/or clumped populations, regardless of the total population numbers. Bryan Manly (senior biometrician, Western EcoSystems Technology, Inc.) addressed this issue with the statement that "Pacific walrus (*Odobenus rosmarus divergens*) are rare because although there are a lot of them, they are distributed over a huge area at very low density." This, combined with dangerous, logistically difficult, or expensive survey procedures, produces some difficult situations. In my experience with flights in Alaska for survey of Dall sheep (*Ovis dalli dalli*) (McDonald et al. 1990; Kern et al. 1994) and flights off the northwest coast of Alaska for survey of Pacific walrus (Gilbert 1999) and polar bear (*Ursus maritimus*) (McDonald et al. 1999), the animals were at times extremely sparse and difficult to detect. However, at unpredictable times, the animals occurred in huge concentrations too numerous to count accurately (e.g., herds of sheep in rugged habitat or clusters of walrus hauled out on ice flows). These efforts were marginally successful at best. None of the three species currently qualify as rare in the sense of being threatened or endangered, but walrus and polar bear were certainly rare when one was looking out of the window of an airplane. New technology, new procedures (perhaps presence-absence surveys described below), or new objectives for study and monitoring are needed before these widely and unevenly dispersed species living in forbidding habitats will be anything but rare from an operational point of view.

Whether a population is rare in the sense of small numbers apparently has very little to do with design of a study to detect, estimate abundance of, or monitor the population. Many of the same general principles apply to study of sparse, elusive, and/or spatially clustered populations. In the following, I review and investigate principles for designing or improving surveys for such populations through interspersion of my own experiences with responses provided by respondents to my survey. I use the word "rare" primarily to mean low detectability, that is, rare from a statistical sampling point of view.

Critical Sampling Issues

To detect, estimate abundance of, and/or monitor rare (i.e., rare, elusive, or spatially clustered) populations, Jim Nichols (wildlife biologist, USGS Patuxent Wildlife Research Center) commented that two critical sampling issues existed and referenced his paper with Lancia et al. (1994), an excellent summary of the most common procedures available for estimating the number of animals in wildlife populations. The first issue was that frequently the population was suspected to inhabit an area that was too large to survey completely; that is, we cannot conduct our survey methods over the entire area of interest. This issue requires that we sample space in a manner that permits inductive (statistical) inference about sampling units that we do not visit, that is, we must adopt a probabilistic sampling procedure. This is a classical sampling problem dealt with in texts such as Cochran (1977) and Thompson (1992).

The second issue that Nichols commented on was that survey methods for animal populations hardly ever permit censuses (complete counts). Instead, our methods yield counts of animals or indirect evidence (e.g., tracks or fecal material), and these counts represent some unknown fraction of the population of interest. To draw inferences about abundance or distribution, we require some inference about detection probability (i.e., p = the probability that a member of the population of interest appears in our count on a surveyed area, if present). Nichols emphasized the importance of increasing p and improving the precision and accuracy in estimation of p. The basic correction in the estimation of abundance when p is less than 1.0 is the ratio of the count to p (i.e., $\frac{\text{Count}}{p}$). This ratio can be misleading if the estimator of the denominator, p, has poor precision or accuracy (particularly when p is low).

Three additional references that cover these issues specifically as they relate to the design of wildlife studies are the second edition of Seber's (1982) classic text *The Estimation of Animal Abundance and Related Parameters* and the recent books by Morrison et al. (2001) and Williams et al. (2002). One of my points is that the methods for study of biological populations are applicable for common or rare populations. As mentioned previously, rare populations have a way of becoming "common," or at least easier to study, when proper methods are developed that provide improved estimates of detection probability. Successful study in the case of rare, elusive, and spatially clustered populations is just more difficult to achieve than for common species. Success requires the merging of appropriate techniques to increase the probability of detection and, I firmly believe, a sampling procedure to spread the effort out over the entire study area, perhaps with less effort in areas that are predicted to have lower density. Not all of the respondents' success stories had a strict probabilistic sampling procedure, but all had at least ad hoc procedures to spread the sampling effort over the entire area of interest.

Designs for Sampling Area

This section covers designs for sampling areas that are potentially useful for surveys of rare populations. This design component is associated with the first stage of a two-stage design, as defined by Thompson (Chapter 1).

Stratification of Study Area

Most respondents mentioned the potential advantages of stratification of the study area based on subjective information and previous information (e.g., Morrison et al. 2001). I agree that stratification of the area with some type of probabilistic sampling within strata seems likely to be useful for rare habitat specialists. However, systematic sampling should be given serious consideration in design of studies for rare populations (see below for some success stories) before introducing the extra complexity of stratification and the accompanying unequally weighted data among strata (i.e., unequal probability sampling on the entire study area). Also, systematic sampling is common and appropriate for most long-term environmental monitoring programs.

I agreed with Robert S. Cochran (professor of statistics, University of Wyoming) that domain estimation (Cochran 1977) was an alternative to stratification in dealing with subareas, strata in this case, that cannot be defined in advance to correlate with densities of rare species (i.e., there was not a readily available sampling frame). It may be advantageous to obtain a "large" sample, random or systematic, from the overall study area to be mined for the domain of interest, habitat, and presence of rare species in this case. In my experience, the concept of domain estimation is not part of the mainstream of ideas that typically go into design of biological surveys. It deserves to play a larger role.

Adaptive Sampling of Study Area

Again, most respondents mentioned the potential application of adaptively sampling the study area for rare species (Thompson and Seber 1996; Chapter 5, this volume). In particular, most respondents felt that Thompson and Seber's procedures for study of clustered populations seem likely to be useful for populations in which individuals were rare, yet clustered. My own experience with Thompson and Seber's procedures for estimation of the abundance of spatially clustered species has been somewhat negative. For example, an attempt was made to use the procedures in an aerial survey for walrus in 1990 (Gilbert 1999), but logistical problems with fuel and flight time made use difficult. My judgment is that the procedures are most likely to be useful in survey of relatively small areas where the logistics can be more easily handled. Modifications have been made to the procedures to make the logistics more feasible (e.g., Brown and Manly 1998); these modifications should be considered in use of Thompson and Seber's methods.

Greg Linder (ecologist-field biologist, USGS Columbia Environmental Research Center) indicated that he has had good luck with Thompson and Seber's (1996) adaptive cluster sampling approach in his work with soil invertebrates and soil fungi in what I suspect are relatively small and easily accessible areas. Other respondents indicated plans for use of adaptive cluster sampling (e.g., Mike Morrison for monitoring the endangered southwestern willow flycatcher (*Empidonax traillii extimus*) in the Sierra Nevada of California).

Adaptive sampling procedures have, of course, been developed for many situations besides spatially clustered populations. Bryan Manly has found an adaptive two-phase, stratified random survey to be very useful in the study of abundance of three shellfish species in New Zealand (Manly et al. 2002a; Chapter 6, this volume).

Two individuals suggested snowball sampling (Kalton and Anderson 1986) as a potential method for surveying rare populations. This procedure is a form of adaptive sampling used in detection and study of rare human populations. Humans might be interviewed to identify locations of rare species and asked to identify other humans who probably know of other occurrences. There are no adjustments for bias or probability of detection (as in Thompson and Seber's adaptive cluster sampling). The technique seems suited for initial exploratory and qualitative investigations, but possibly not for quantitative studies (Kalton and Anderson 1986; see also Chapter 7, this volume).

Unequal Probability Sampling and Resource Selection Functions

I was pleasantly surprised to see the use of resource selection functions (Manly et al. 2002b) mentioned for design of unequal probability sampling schemes. I was surprised because it had never occurred to me to use the information from resource selection functions directly in the design of biological surveys in similar habitat. Jennifer Brown (statistician, University of Canterbury) explained it this way, "The idea is that you have information on the habitat variables for sites where there are koalas (*Phascolarctos cinereus*), and sites where there are not koalas and use logistic regression to come up with a function that describes the likelihood that the habitat is used." The function could be used to predict relative probability of use of units in similar habitat and to select units for survey using the unequal probabilities. An initial sample over the study area might be used not only to estimate abundance, but to estimate a resource selection function. The resource selection function might then be used to select additional units for survey with probability proportional to the probability of occurrence of the species in the sense of adaptive sampling. Alternatively, a similar area might be stratified based on the estimated relative probabilities of use.

Ray Czaplewski (mathematical statistician, USDA Forest Service) commented that in the mid-1990s, he worked on designing monitoring systems for forests in the Pacific Northwest involving surveys for hundreds of species of lichens, fungi, mollusks, etc. Some of the fungi could be identified only by their fruiting bodies, which might appear for a week or two once every few years. These species were thought to be rare and dependent upon old-growth forest habitats, but no one really knew. Some species might be common, and some might not be dependent upon old-growth forests, but they could be difficult to detect. The simple idea behind the monitoring plan was to use habitat models in a geographical information system (GIS) to produce a "probability of detection" surface for large study areas, then select sample sites using those probabilities (Tolle and Czaplewski 1995). Over time, the habitat models were expected to improve and the sampling frame to become more efficient.

Michael Samuel (wildlife biologist, USGS Wisconsin Cooperative Wildlife Research Unit) borrowed the case-control method from the medical epidemiology literature to evaluate wetland environmental factors associated with the occurrence of rare avian botulism (*Clostridium botulinum*) outbreaks (Rocke and Samuel 1999). This approach is useful when the cases (e.g., wetlands with disease outbreaks) are uncommon, but they can be paired with a control (wetland without a disease outbreak). Samuel commented that they have found this approach much more effective in contrast to the usual plan that might sample random wetlands in hopes of a disease outbreak occurring.

Other Sampling Plans

There are several plans not promoted by the respondents that I would judge to be useful for sampling areas in the study of rare populations. In particular, systematic sampling (e.g., Morrison et al. 2001) was in fact used in some of the references provided (Max et al. 1996) and in some of the success stories below. Morrison et al. (2001) provided an excellent introduction to alternative sampling plans.

Patch-Occupancy Models and Presence/Absence Surveys

Jim Nichols pointed out the relatively new patch-occupancy models in his work with MacKenzie et al. (2002, 2003), where the objective is not to estimate abundance but the probability that a patch (habitat unit) is occupied (Chapter 8, this volume). These methods are based on repeated visits to selected units but require only information on species detection or nondetection. Because it is not necessary to mark or identify individuals, the authors suspect that it may be logistically feasible to cover larger areas with such surveys than with other approaches. Such surveys can be based on animal sign (e.g., tracks or scats) in some cases (Karanth and Nichols 2002). I agree with Nichols that surveys of this sort have great potential for rare populations and believe that we are likely to see increased use of them for this purpose, especially for long-term monitoring programs. Gary White (professor of fishery and wildlife biology, Colorado State University) pointed out that the MacKenzie et al. (2002) procedures are included in the computer software package MARK (White and Burnham 1999).

Glen Sargeant's (wildlife biologist, USGS Northern Prairie Wildlife Research Center) success story (see below) involved a presence/absence survey of swift fox (*Vulpes velox*) in western Kansas (Sargeant et al. 2002). They interpreted the resulting map of observations as a degraded representation of the true distribution and used Markov chain Monte Carlo methods for image reconstruction to estimate probabilities of swift fox occurrence for townships in western Kansas.

Resource selection functions (habitat models) also can be used to predict relative probability that a patch (unit) is occupied, and in some cases, to estimate absolute probabilities (Manly et al. 2002b). Bill Gould commented that he had been involved in three projects in New Mexico where the individuals were rare (Jemez Mountain salamanders (*Plethodon neomexicanus*), Mexican spotted owls (*Strix occidentalis lucida*) and Colorado chipmunks (*Tamias quadrivittatus*) on the White Sands Missile Range). He has developed logistic regression models for the latter two species to identify those habitat features that best explained presence/absence. However, in his case, interval estimates of the predictions of occurrence resulted in disappointingly large intervals because of small sample sizes.

The Green and Young (1993) procedure for sampling to detect rare species has been integrated into study and monitoring protocols (e.g.,

Bonar et al. 1997; Smith et al. 2001). The procedure also is currently being used to examine the extent of the whirling disease host (*Tubifex tubifex*) in the San Juan River, New Mexico (Bill Gould). Green and Young based their model on the assumption that when populations are sparse, a Poisson model can adequately describe the distribution of animals per unit. Bonar et al. (1997) applied the Green and Young procedures to be confident that a rare species occurred in densities less than a certain level (e.g., less than 0.60 fish per 100 m reach in their case).

Jim Peterson (fisheries biologist, USGS Georgia Cooperative Fish and Wildlife Research Unit) and Jason Dunham (research fishery biologist, USDA Forest Service) commented on their work to develop a protocol for estimation of probability of presence/absence of bull trout (*Salvelinus confluentus*), a species of concern under the federal Endangered Species Act in the Pacific Northwest (Peterson et al. 2001). The presence/absence protocol built on previous approaches (Bonar et al. 1997) and was recommended as a model for other presence/absence surveys. Peterson and Dunham also integrated this sampling protocol with prior expectations of occurrence from a habitat model to provide updated estimates of presence when bull trout were not detected in sampling surveys (e.g., Chapter 9, this volume).

Methods for Estimation of Detection Probability

This section is devoted to counting methods that adjust for incomplete detectability of rare or elusive individuals. Thompson (Chapter 1, this volume) considered this component to be the second stage of a two-stage design for sampling plant and animal populations.

Capture-Recapture Methods

Several respondents felt that capture-recapture procedures (Seber 1982; Otis et al. 1978; Williams et al. 2002) can be useful if capture probabilities are sufficiently high (e.g., camera-trapping tigers (*Panthera tigris*); Karanth and Nichols 1998; Chapter 12, this volume). High capture probabilities for rare species arise either when the animals can be strongly attracted to some sort of bait or when the investigator knows the species and area well enough to focus capture efforts at areas of high animal activity. I agree with Ken Burnham that DNA "fingerprinting" for evidence of capture and recapture of individuals in rare populations is a developing area to watch (see Chapter 11). Bryan Manly and I have worked with the U.S. Fish and Wildlife Service to evaluate the potential of this procedure for estimating the abundance of Pacific walrus.

Distance Sampling

Similarly, respondents mentioned the successful use of distance sampling (Buckland et al. 2001) in surveys for rare populations (e.g., sightings of spotted dolphins (*Stenella attenuate*) from commercial fishing vessels). My own experience in using distance methods in large-scale surveys (e.g., polar bears, McDonald et al. 1999; walrus, Gilbert 1999) is mixed, primarily because of the logistical difficulties and limited number of sightings on which to estimate detection probabilities. Regardless of the difficulties, distance sampling will remain one of the primary tools in design of large and small-scale studies for rare populations.

Intercept Sampling

Earl Becker (biometrician, Alaska Department of Fish and Game) has developed estimation procedures based on interception of animal tracks (Becker 1991; Becker et al. 1998; Chapter 13, this volume). Becker has used probability sampling based on line-intercept sampling (reviewed by Morrison et al. 2001) and a sampling-unit probability estimator based on network sampling (Thompson 1992) to estimate abundance of elusive furbearer and large carnivore populations. These techniques merit consideration when tracks can reliably be followed after, say, a fresh snowfall.

Sighting Probability Models

Again, many respondents mentioned the potential use of sighting probability models in study of rare populations. Often, sighting probability models are developed from surveys in which known numbers of radio-marked animals are present to permit direct estimation of detection probability, possibly as a function of covariates (Samuel et al. 1987). These detection probability functions then are applied to counts from other areas to adjust for less than 100% detection. The method can work for rare species because the known presence of radio-marked animals yields more precise estimates of detection probability than standard methods using less direct estimation approaches. In some situations, two survey procedures on the same area have been combined to produce sighting probability models (reported in Seber 1982; see McDonald et al. 1990 for an example). One of the procedures is viewed as marking a sample of animals, plants, or animal sign known to be present and the other survey "recaptures" some of the individuals marked by the first. I cannot say that the Dall sheep surveys reported in McDonald et al. (1990) and Kern et al. (1994) were particularly successful attempts to estimate abundance of spatially clustered populations, but they certainly were some of the more interesting—filled with beautiful scenery and lots of sheep and potential biases.

Successes in Sampling Rare Populations

A brief discussion of the concept of rare populations from an operational or detectability point of view and brief review of some of the procedures used to detect, study, or monitor rare populations seemed to be necessary for this chapter. However, my major goal was to record some opinions, including my own, from both statisticians and biologists on the basic issues behind their most successful and unsuccessful attempts in surveys of rare populations.

Success Due to Spreading the Sample over the Study Area

I begin with one of my successes. I was a team member in the design, conduct, and analysis of the Nearshore Vertebrate Predator project in Prince William Sound, Alaska, funded in 1995 by the *Exxon Valdez* Oil Spill Trustees. In the simplest sense, the goal was to determine if food or residual oil was limiting the recovery of certain species in the areas affected by the 1989 *Exxon Valdez* oil spill. The study included surveys of clams as food for sea otters (*Enhydra lutris*) in affected and relatively unaffected areas of the Sound. As previously mentioned, clams can be sparse, elusive, and spatially clustered.

The first year's design was one of the more successful (or satisfying) of my career for two reasons. First, I was able to convince team members to sample at some of the survey sites for intertidal clams, shallow subtidal clams, saltwater mussels, and sea otters. This task was important and difficult, but unrelated to the rarity of clams. Second, the subtidal clam survey team insisted on focusing their effort (limited scuba diving, underwater dredging, etc.) in "preferred" clam habitat during the first year. I insisted on some probabilistic sampling to spread the sample out. The compromise was to expend half of the effort in surveying sites in "preferred" habitat and half in systematic location of sites along the coastline. Luckily perhaps, a higher density of clams was detected on the systematically located sites than on the "preferred" sites. It probably did not appear in any of the reports (e.g., Holland-Bartels 2001), but I remember it clearly. The results provided some new information on clam habitat and had a positive effect on the design and results of the remaining years of the study.

One of the team members on the 1995 Nearshore Vertebrate Predator project reported on a recent successful survey of intertidal clams in Glacier Bay, Alaska (Bodkin et al. 2001). The team systematically sampled 48 shoreline segments from a random origin and 12 segments from a "high-density" stratum based on the prevalence of shell litter or siphon squirts observed at low tides. Clam density in the high-density stratum was about three times greater than on the other segments. With this experience, I expect that they will be able to design more economical surveys in the future using: (1) stratification based on visual inspection for presence of shell fragments or siphon squirts, and (2) more sampling effort in the strata with the higher density of shell litter or siphon squirts.

Steve Buckland (professor of statistics, University of St. Andrews) was involved in designing surveys of *Primula scotica*, the very rare Scottish primrose. Past survey efforts were concentrated on quadrat surveys of known colonies. He was able to persuade the survey team to conduct line transect surveys in suitable habitat surrounding the known colonies. The results indicated that there were as many plants outside of the known colonies as in them because there was a large area of suitable habitat outside the small colonies, so the low density multiplied by the large area gave a high abundance.

Buckland's experience reminded me of my own on the Nowitna National Wildlife Refuge in Alaska while working with the U.S. Fish and Wildlife Service to estimate duck production in the late 1980s. The density of ducks on the muskeg pothole area was extremely low and the ducks were rare, but in the end, we estimated more production from that huge area than in the more highly productive oxbow and riverine habitat.

The basic point of Steve Buckland's experience and my own on the Nowitna Refuge is important when dealing with sparsely distributed individuals over large areas. There may well be a larger proportion of a population in the low-density stratum than in the higher density strata.

Dan Monson (wildlife biologist, USGS Alaska Science Center) responded that within the sea otter project with Jim Bodkin (wildlife biologist, USGS Alaska Science Center) and others, their biggest success has probably been with estimation of abundance of sea otters (Bodkin and Udevitz 1999). Sea otters may or may not be abundant in a particular area, but when abundant, they often have a very clumped distribution. Individuals on the surface are visible to count while others are underwater and unavailable (see Chapter 3 in this volume for a further discussion of this issue). The method relied on stratified survey of strip transects systematically placed generally perpendicular to shore with a random start point. Intensive search efforts were conducted on some subset of otter sightings to estimate the probability of detection to adjust the counts for otters that were missed (e.g., underwater).

Monson commented on a recent survey of an offshore area in lower Cook Inlet, Alaska, where no previous surveys had been conducted. Many of the transects were empty, but occasionally otters were spotted at relatively unpredictable places at great distances from shore; these few observations represented a fairly significant number of otters due to the large amount of offshore area. He concluded that it was worthwhile to sample the low-density habitat, but the highest proportion of the population was found in the high-density stratum.

Melinda Knutson (wildlife biologist, USGS Upper Midwest Environmental Sciences Center) reported that her most successful experience was in a farm-pond study (Knutson et al. 2004). They randomly placed a 10-km grid over the study area and selected intersection points as anchor points for the study sites, selecting three constructed farm ponds and a natural wetland in proximity to each anchor point. They identified 10 species of amphibians in the study ponds, including the locally rare wood frog (*Rana sylvatica*) and a new record for the county, the blue-spotted salamander (*Ambystoma laterale*). They failed to find the cricket frog (*Acris crepitans*), a species assumed to be extirpated from the study area.

Jim Nichols has been involved with a project that involves rare populations, specifically, tiger populations within nature reserves in India (Karanth and Nichols 1998; Chapter 12, this volume). He characterized the work as having been successful, especially in comparison with other attempts to estimate abundance for this species. Within each refuge, "camera-trap" sites were spread out over large areas, so that a large portion of each study area was covered by the sampling effort. The work was directed at abundance estimation and used camera-trap data in conjunction with closed-population, capture-recapture models. Because of the a priori knowledge of tiger habits and movement patterns based on extensive radio telemetry work, the project leader, Ullas Karanth (senior scientist, Wildlife

Conservation Society, India), was able to obtain fair numbers of captures and fairly good capture probabilities. Nichols concluded with the comment that other than perhaps DNA sampling, he could not think of an approach that was likely to work better than this sort of camera-trapping for these populations.

Carl Schwarz (professor of statistics, Simon Fraser University) reported on success with Paul Starr (fisheries biologist, independent consultant) in design of surveys for estimation of abundance of groundfish species off the coast of British Columbia, Canada (Starr and Schwarz 2000). They settled on a stratified (by depth and area) design with random trawl sampling in each stratum. The fishers in the industry "knew" where fish were and did not like dropping their nets where they had not fished before. In addition, they wished to avoid areas where the bottom topography was not known. The fishers were eventually persuaded to participate in the study, and it was quite successful. The results were of great interest for one species in particular, because it appears that this species was in relatively low abundance, but fairly evenly distributed.

Charles Bomar's greatest success story with rare species occurred in Wyoming. He went hunting for snow scorpion flies on a particular road east of Laramie and found the species he was looking for, *Boreus coloradensis*, which had never be collected in Wyoming before. He also collected a new species, recently named *Boreus bomari* (Byers and Shaw 1999). His moral for the story: look in suitable habitats that no one else has surveyed.

Dick Whitney's (retired fisheries biologist, University of Washington) problem was one of attempting to document newly spawned fish as direct evidence of successful introduction of any of 35 species after adults were stocked in the 880-kilometer-square Salton Sea in California. Stocking occurred over the period 1929–1956, by which time only five successful introductions had been confirmed. Beginning in 1954, he organized a substantial effort of regular monthly sampling around the shores wherever access could be gained with a beach seine. This sampling brought plenty of specimens of previously documented species, but no new species. An unexpected species, threadfin shad (*Dorosoma petenense*), was first collected in 1955, evidently having found its own way into the sea through irrigation canals from the Colorado River.

However, of most interest was determining whether the orangemouth corvina (*Cynoscion xanthulus*) or other potential sport fishes might have become established. Whitney continued the routine sampling with a constant level of effort monthly at 16 locations around the perimeter of the sea throughout the year. This sampling initially brought a large number of zero catches of *C. xanthulus* and no unusual fish. Even the catches of the abundant species dropped to zero at times in the winter samples. As a result, he wrote, there were many questions asked about his competence, and there was grumbling in the lab and the field. Finally, in 1956, the first and only one-year old *C. xanthulus* was taken in the beach seines. In 1957, there were a total of 24 one-year olds and young-of-the-year caught at six locations. It was concluded that spawning had been successful, and the project was terminated. Whitney concluded that success came from the regular sampling in time and space combined with persistence and willingness to look like a fool to people who were not accustomed to thinking in terms of probability sampling.

Glen Sargeant's success story dealt with swift fox surveys that usually were difficult because the animals occurred at a low density and were secretive (Sargeant et al. 2002). The key to success was to first admit they could not estimate abundance and probably would frequently fail to detect swift foxes where they occurred. The team addressed these limitations by developing a systematic track survey that enabled them to search large areas very quickly and achieve rates of detection that were relatively high for carnivore surveys. They interpreted the resulting map of observations as a degraded representation of the true distribution and used Markov chain Monte Carlo methods for image reconstruction to estimate detection rates and smoothing parameters and, ultimately, probabilities of swift fox occurrence for townships in western Kansas.

Linda Kerley (wildlife biologist, Lazovsky State Nature Zapovednik, Russia) reported on the annual Siberian tiger (*Panthera tigris altaica*) survey to estimate numbers using snow tracking in the far east of Russia (Smirnov and Miquelle 1999). The survey was performed twice a year just after a fresh snowfall. A large number of people were employed to walk virtually every drainage in an area to count fresh tracks. In that way, they limited double counting of the same tiger in different places.

Success Due to Stratification and Efficient Procedures

Doug Crowe (professor of biology, Casper College) indicated that his most successful experience was with white rhino (*Ceratotherium simum*) capture operations in Botswana, Africa. The team was attempting to locate and capture all rhino close to the northern border of Botswana and move them to the interior of the country. There were very few rhino left in the area, and those that remained had been harassed by poachers to the extent that they were very wary and stayed in dense cover. The team was able to stratify the search area based upon the availability of water and use a forward-looking infrared (FLIR) unit from the military to detect body heat early or late in the day. They were able to detect about one rhino for every 2 days of search. Crowe concluded with the comment that if they located an animal in the evening and there was no time to dart and remove it before dark, they dropped a bushman on the track and returned in the morning. This worked pretty well, except that the poachers shot one of their trackers. However, the FLIR worked pretty well on poachers too " . . . and we soon discouraged them from shooting our personnel!!"

David R. Smith commented that he has been involved with a number of approaches to survey freshwater mussels, including adaptive cluster sampling (see Chapter 5), systematic sampling, distance sampling, double sampling, and closed-population mark-recapture (Strayer and Smith 2003). He has found double sampling to be quite useful, both for within site sampling to correct for mussels' elusive (buried) behavior and for multiple site sampling for allocating effort to sites where mussels were at higher density and away from sites where mussels were absent or at negligible density. The application of double sampling for selecting sites was "double sampling for stratification," in which the first phase of sampling was qualitative and was used to assign a site to a density stratum (Strayer and Smith 2003).

Bryan Manly described a team project to estimate the size of three shellfish species populations on beaches near Auckland, New Zealand (Chapter 6, this volume). Initially they lacked information on which to stratify the relatively large area and therefore talked to local people who gather the shellfish for food. Based on the resulting stratification and use of an adaptive two-stage sampling procedure to locate units (Manly et al. 2002a), they were able to estimate abundance of the species with acceptable precision.

Success Due to Strong Government and Industry Support

Ken Burnham reported that one of his more successful projects was working with the National Marine Fisheries Service (NMFS) in use of distance sampling to monitor porpoise populations in the eastern tropical Pacific (e.g., see citations in Buckland et al. 2001). The design used semirandom lines surveyed by ship or plane. NMFS and the U.S. Congress wanted the surveys to be accurate and precise and were willing to pay. This support occurred at all levels of government administration and included good support from interested industries. Burnham regarded this enlightened support as the critical aspect in success of the program. This support allowed the NMFS to conduct ancillary studies on methodological issues, including adjustments for evasive movement of porpoise and evaluation of accuracy of distance and angle estimation. Burnham concluded that this project was not just a program to survey porpoise but included improvement of the design and field protocols in every possible way. Huge strides were made to make distance sampling work for monitoring these marine mammals.

Successes Following Failures: Development of Proper Techniques Leads to Success

Mark Udevitz (statistician, USGS Alaska Science Center) discussed a couple of "failures" related to their sea otter surveys (see Dan Monson above; Bodkin and Udevitz 1999). Sea otters are considered rare in certain habitats. A failure in one of the earlier surveys was not to plan for the effect of observers on detectability. They used two observers to make the survey easier to complete, but an analysis of the data revealed that the observers had quite different abilities to detect sea otters from the air. The data for estimation of separate correction factors were inadequate, resulting in population estimates that had very low precision. From that experience they learned to use one observer with enough survey effort so that adequate data could be obtained to estimate the probability of detection.

One other "failure" reported by Udevitz was associated with the existence of a few very large groups of sea otters (rafts of hundreds of otters). One or more of these groups was detected on early surveys, and the variance of estimated abundance became too large for the estimates to be useful. This is, of course, a very common problem in sampling rare populations with clumped distributions. Their approach to solve this problem involved estimating population size for two different subpopulations: (1) those that occurred in small groups; and (2) those that occurred in large groups at the time of the survey. By post-stratifying the population in this way, the variance of abundance for the large groups was very high, but because a relatively small proportion of the total estimated population occurred in large groups, it did not contribute much to the variance of the overall population estimate. Udevitz noted that this post-stratification also was useful because of the need to estimate detectability differently for large and small groups. He concluded that after overcoming the various "failures," he viewed the sea otter surveys as a successful experience. Use of this approach and its adjustment for detectability has allowed the U.S. Fish and Wildlife Service to detect declines in Alaskan sea otter populations with improved accuracy over prior survey methods.

Lowell Diller responded that his biggest "failure" was an initial attempt to survey an adult population of tailed frogs (*Ascaphus truei*). The adults are only active at night, and the initial searches with headlamps indicated that the frogs were quite rare and elusive, a general conclusion from other studies as well. After the surveys were repeated for several years without collecting enough data to generate meaningful estimates, he accompanied another herpetologist looking for the terrestrial black salamander (*Aneides flavipunctatus*). He was shown that he could pick up the eye shine of the salamanders if he took the light off and looked directly down the beam. Diller commented that suddenly the world came alive with eyes shining back. He tried the same technique on tailed frogs and there were adult frogs all over the place. Nothing is published yet, but initial ideas about the density of adult tailed frogs on the study area are certainly going to change.

Earl Becker responded that his "worst" sampling experience was a study to estimate the size of a population of wolves (*Canis lupus*) in Alaska. However, on the second attempt, he had his "best" experience and obtained precise estimates (Becker et al. 1998). In the first survey attempt, the team agreed to sample most of the good areas (high-density stratum) and use a design based on network sampling (Thompson 1992) for areas where wolves were less likely to be (medium- and low-density strata). They detected six wolf packs of similar sizes, all associated with the high-density stratum. No packs associated with only low- or medium-density strata were detected. The standard error of the estimate was very small and the confidence interval was so tight the team would not report the results. Becker concluded that in retrospect, they needed to shift more effort into sampling units within low- and medium-density strata.

Tim Gerrodette (marine ecologist, NOAA National Marine Fisheries Service) said that one of his experiences with rare populations was with a small porpoise endemic to the northern Gulf of California and known by its Spanish name, vaquita (*Phocoena sinus*) (Jaramillo-Legorreta et al. 1999). There were two issues. First, it took time to figure out the right sampling tool. It takes a fairly large ship with sufficient height to see far ahead of the ship, and these animals avoid a smaller boat before they come into sighting range. The research team was able to estimate the relative efficiency of searches from the larger ship compared to previous attempts to find vaquitas. One unexpected result was that they found indications of a significant part of the population, about 25%, in shallower water than initially expected.

Ed Kline (aquatic biologist, Kline Environmental Research, LLC) said his biggest "failure" was the first period of a survey to estimate the number of dolly varden (*Salvelinus malma*), a char, in a lake close to Juneau, Alaska. The solution was simple. "It is critical to use fresh salmon roe to bait nets rather than jarred roe from a bait store. . . . In other words, just because you have trouble catching them, doesn't mean they are rare."

Jack Griffith (fisheries biologist and professor emeritus, Idaho State University) stated, "Regarding successes and failures [with rare populations], I have one story and don't know in which category it falls." It was an effort to learn more about a fish called the Shoshone sculpin (*Cottidae: Cottus greenei*) that was endemic to springs flowing into the mid-Snake River in the canyon near Twin Falls, Idaho. The sculpin is small (less than 80 mm in length), cryptic, benthic, and sympatric with the abundant and larger species, the mottled sculpin (*C. bairdi*). No systematic survey had ever been performed, and the habitat was disappearing quickly. In 1978 Griffith began surveying with a low level of effort (electrofishing and snorkeling) that proved unsuccessful for capturing an individual. Then one of his former students reported small sculpins in and around the commercial trout hatchery he was managing. Griffith commented that just to be polite he stifled a smirk and visited the hatchery. They pulled a dip net up through the thick vegetation in a cement raceway holding thousands of catchable-sized rainbow trout, and, lo and behold, there were a handful of fat and happy Shoshone sculpin. The moral that I would add to his story is "Do not avoid surveying in all potential habitat."

Failures in Attempts to Survey Rare Populations

First, I will discuss one of my own failures in attempts to survey rare populations. The black-footed ferret was thought to be extinct in the United States until a pet dog brought a dead one into the yard of a rancher near Meeteetse, Wyoming, in 1981. The animal was identified by a local conservation officer, which set off a massive effort to "save the ferrets." The U.S. Fish and Wildlife Service and a private scientist quickly initiated surveys of the ferrets and their suitable habitat, which was white-tailed prairie dog (*Cynomys leucurus*) colonies, the primary and almost exclusive prey of the ferrets. The Wyoming State Game and Fish Department appointed me as a member of the "Black-Footed Ferret Advisory Team," and we met semiannually to review the research and survey efforts. There were other political and social problems that I will not go into (see Clark 1997).

The first mistake was a common one—namely, to trust the initial ad hoc observations made early in the winter of 1981–1982. Unique sign is left in fresh snow by a ferret when digging out a prairie dog. A survey protocol was developed that was never successful because snowfalls with no wind rarely occurred, and it was difficult to assemble the number of observers needed to quickly cover the large area potentially occupied. Valuable time passed, and in the spring, the scientists reported that there were at least nine ferrets in the population based on the ad hoc searches during the winter of 1981–1982.

Spotlighting for ferrets proved to be successful in August 1982, especially as young ferrets emerged from dens, and a minimum of 61 ferrets (adults and young) were reported to the committee in the fall. However, a standardized survey protocol had not been followed and neighboring prairie dog colonies had not been adequately sampled. In retrospect, most of the effort was probably spent in a "core" area where ferrets could readily be found. The scientists agreed to develop a standardized protocol for survey of available habitat.

Eighty-eight ferrets were reported to the committee during fall 1983, but there was still not a standardized protocol in use and some of the local prairie dog colonies were not being surveyed using a probabilistic sampling procedure. This time, there was some "serious discussion" concerning the need for standardized protocols and probabilistic sampling in areas at various distances from the core area.

In fall 1984, 129 ferrets were reported, enough to satisfy the committee and the State of Wyoming that some could be captured to start a captive breeding program without endangering the wild population. Maybe some should have been captured earlier—hindsight is always good. But disaster struck. An epidemic of canine distemper (*Morbilivirus*) killed most of the ferrets. A few remaining healthy ones were captured to start a captive breeding program. The rest is history.

My biggest "failure" as a statistician was not taking an active role with the scientists during the first year to help develop a standardized protocol for survey, including spreading the effort over prairie dog colonies at some distance from the initial discovery site. A stratified survey of prairie dog colonies based on spotlighting could have been in place in fall 1982. In hindsight, I am convinced that the ferret population would have been estimated to be more than 100, as in fall 1984. Perhaps a captive breeding program would have been started earlier with many more founding members. Maybe some of the pressure to study and protect the wild colony might have decreased, and distemper might not have gotten into the colony. Ferrets are being reintroduced into the wild, but the original colony at Meeteetse is gone.

Failure Due to Inadequate Surveys in Unexpected Habitats

Ray Czaplewski commented that in the beginning of the old-growth forest and northern spotted owl (*Strix occidentalis caurina*) controversy in the Pacific Northwest, scientists really only looked very carefully where they believed spotted owls should be and overlooked most everywhere else that did not fit with expectations. Czaplewski concluded that this was not how good science should work. A habitat model should be treated as a hypothesis that needs to be tested, not as a proven fact.

Failure Due to Lack of Proper Methodology

Doug Crowe indicated that his biggest failure was in an attempt to institute a census of aardvark (*Orycteropus afer*) in a region of Botswana, Africa. In his words, "It was a disaster. I'm not even sure they are rare. They feed on termites . . . and the sign of them tearing into termite mounds is everywhere. However, they are active only at night. At least that is what the books say. After a number of sleepless nights, we gave up the search. To date, I have yet to see a damned live aardvark!"

Steve Buckland has consulted with researchers on the problem of attempting to estimate the numbers of different species of reptiles on Round Island in Mauritius. These are small populations, and some are very cryptic, and they desperately need to be monitored. Short of sampling trees, then destructively sampling the tree to locate reptiles (the tree itself is also rare), Buckland concluded that he cannot come up with a methodology that would likely work.

Carl Schwarz observed that " . . . some disasters were caught early enough to be abandoned as infeasible without too much publicity. For example, we did a transect survey for abalone up in Strait of Georgia (Canada) but found no abalone. I'm afraid that we were able to hide the bodies without any documentation other than a one-page memo saying our preliminary survey showed serious shortcomings in the methodology." We should all appreciate Schwarz's honesty.

Failure Due to Sparse, Clustered, Mobile, or Extremely Seasonal Population

Charles Bomar indicated that his "worst" experience was in surveying grasshoppers in Arizona. He came across an area that had an extreme density of grasshoppers (above the levels for it to be considered a pest population). Upon a second visit a week later with representatives from the U.S. Department of Agriculture, no grasshoppers were found, it was 45.5°C outside, and the population was gone—either they moved, or died, or . . . ?

Jim Bodkin has been attempting to monitor densities of spiny lobsters (*Panuliris interruptus*) around San Nicolas Island, California, since 1986. Spiny lobsters are nocturnal, aggregating in crevices during the day, but are capable of long distance movements in response to unknown conditions. They are secretive and are commercially and recreationally harvested. Bodkin originally tried trapping lobsters using either a catch-per-unit (CPU) effort or mark-recapture method to estimate abundance. CPU estimates were highly variable, and only three individuals were recaptured out of about 1,200 marked. Bodkin eventually shifted to direct visual counting on reefs known to support lobsters. He concluded that the data were still messy and that the work should probably be considered a "failure" at sampling a population due to the high variance and very low power to detect change (see Chapter 16 in this volume for a further discussion of this issue).

Failure Due to Lack of Methodology and Funding

Chris Ribic (wildlife biologist, USGS Wisconsin Cooperative Wildlife Research Unit) described a project using a grid-based design to estimate at-sea entanglement of northern fur seals (*Callorhinus ursinus*) in net fragments drifting in the North Pacific. A survey with estimated acceptable precision turned out to be extremely expensive because net fragments were found at very low densities in the water and the problem was not of high enough priority to fund directly. Ribic indicated that her group tried to approach the problem using the High Seas Observer Program, in which sightings of marine mammals as well as debris are recorded in a standardized format, but the sparseness of the data made the attempt futile.

A respondent, who asked to remain anonymous, stated that his plans to sample rare populations have usually ended in holds on funding or denial of permits for development of new techniques in the study of rare populations. He indicated that past sampling efforts have produced little in way of understanding of either the habitat needs or population dynamics of the endangered and threatened species that he is concerned with. The uncertainty of the results and cost of new work on the rare species were not acceptable to the funding and/or permitting agencies. He concluded that he is left with a great deal of frustration with what he views as years of inadequate survey for some species listed as threatened or endangered under the Endangered Species Act.

Bits of Advice

Here I present a summary of survey responses identifying common aspects of studying rare populations, both problematic and otherwise. Survey respondents offer possible solutions to the problematic aspects of these studies.

Need for Coordinated Activity and Common Goals

Dan Goodman indicated that some of the more problematic studies he had been involved with had a common problem: insufficient coordination among multiple agencies or volunteer organizations performing the surveys, so that they focused only on sampling their particular territories. There was also a tendency to sample favored places where sighting rates were high, and protocols were used that generally did not allow for quantification of effort. Goodman concluded that, in fairness, these networks of researchers often formed before a specific statistical question was formulated, and attempts at statistical analysis had been afterthoughts that subsequently persuaded the participants to upgrade their methods and coordination.

One respondent who asked to remain anonymous indicated that "... probably the biggest problem in my mind about sampling rare populations is the politics that come with such populations." The species often are endangered or viewed as some kind of indicator of ecosystem health, and studies often are designed more to prove a point by one side or the other rather than to objectively look at whatever population parameters are important. These politics can ruin a study. His advice on designing studies of rare populations included:

- Make sure the study has a clear objective and all parties understand it and agree on it.
- Get a budget before the design stage, if at all possible.
- Try to study the entire range of possibilities.
- There are numerous study designs, data collection methods, and analysis techniques that can be combined to produce a scientifically acceptable outcome, and it is unlikely that any one person knows them all, so get help when possible.

Things Are Seldom Totally Black or White

Lowell Diller commented on a sampling procedure used to estimate the distribution of southern torrent salamanders (*Rhyacotriton variegatus*) and tailed frogs on forestlands owned by his employer. Both of the species have fairly narrow habitat requirements that create a highly clumped distribution in most streams. Diller's group used a stratified random sampling design to select stream segments across the ownership. Within each reach, they walked the entire length but searched only in suitable habitat. Because their objective was to determine occurrence (not abundance), the survey was terminated if animals were detected. Compared to randomly selected 20-m, fixed reaches that they had previously used, this technique was considered to be far more effective and efficient at determining presence of the species.

I would agree and disagree with Diller's design. The design does spread the effort out, and they undoubtedly determined presence with less effort on some units, but I wonder about their accuracy in determining absence on reaches that did not have much "suitable habitat." My advice would have been to put some effort into the "unsuitable" habitat of each reach and to consider use of methods in the recent literature on design of presence/absence studies (MacKenzie et al. 2002, 2003; Peterson et al. 2001; Chapters 8 and 9, this volume). However, based on his extensive experience, perhaps Diller could have talked me into using his design.

Densities of Rare Populations Often Are Higher Than Expected

Gary White wrote, " . . . something worth mentioning is the conservative effect of higher densities found than are usually predicted ahead of time." I really liked White's comment, because I also have found that densities often turn out to be higher than originally thought when dealing with rare populations. He recently set up a black-tailed prairie dog (*Cynomys ludovicianus*) survey for eastern Colorado, based on the line-intercept method and estimates of variance from the literature. The realized coefficient of variation was pretty close to the value from the literature, but the percentage of the area inhabited by prairie dogs was found to be two to three times larger than the values predicted. As a result, the estimates of density were higher and confidence intervals were much tighter than was predicted. White concluded, "I guess I should be disappointed that I could have designed a cheaper survey, but in reality, I'm happy that what I suggested is looking so good!"

Check with the Old-Timers

Jack Griffith advised anyone faced with designing a study to detect a rare population to first spend a lot of time seeking out "old-timers." His experience was that time and again they knew a lot about plants and animals that are now rare, and they provided valuable information that could be used in study design. Nick Markov's advice was to conduct interrogation surveys of experienced biologists before attempting long-term research on the distribution and abundance of carnivore or ungulate species in an area where they are rare. Bryan Manly's experience in surveying shellfish in New Zealand benefited from similar local information in design of the adaptive sampling procedure described above (see Chapter 6).

Presence/Absence Surveys

Presence/absence surveys are becoming popular for monitoring rare populations (e.g., MacKenzie et al. 2002, 2003; Peterson et al. 2001; Sargeant et al. 2002; Chapters 8 and 9, this volume). The methods are a combination of (1) design-based sampling, sometimes requiring multiple visits to sites; and (2) model-based estimation of site occupancy rates when detection probabilities are less than 1. The model-based component appears to provide a flexible framework that statisticians and biologists will find very useful.

Conclusion

From this statistician's point of view, rare biological populations are those with low probability of detection. The problems in surveying for detection, estimation of abundance, or study of rare populations are very similar whether the populations are small or large in abundance, sparse, elusive, or spatially clustered. However, I grant that successful studies of rare populations require more time, money, and patience.

Success in the study of rare populations requires two simply stated things: (1) a sampling procedure to spread the effort out over the entire area, perhaps with varying levels of effort in different strata; and (2) the development of appropriate field techniques to achieve a certain minimum probability of detection in sampled units. We might sample much less intensively in areas thought to have low densities, but we still should be sampling such areas. When the probability of detection is low, not only are the observed counts small and few individuals are available for study, but also the effect of imprecise estimates of the probability of detection introduces extreme variation in estimates of abundance and trend in population size.

Several respondents commented that they have purposely avoided rare populations, partly because of the difficulties in studying rare populations (in a biological and/or a statistical sense). National environmental/biological surveys in the United States that I am aware of—for example, the U.S. Forest Service's Forest Inventory and Analysis (FIA) and the USDA National Resources Inventory—do not focus on rare things. The U.S. EPA Environmental Monitoring and Assessment Program (EMAP) works with the states to design probability surveys for their state monitoring programs, however, to date, none of the surveys have focused on rare populations (Tony Olsen, environmental statistician, US Environmental Protection Agency). Olsen commented that EMAP has some plans in place to address the problems in their research program. Also, some regional monitoring programs have plans to monitor rare populations (e.g., Max et al. 1996). I encourage the scientific community to get involved in research to improve the situation.

I agree totally with Charles Bomar's comment that "the reality of rare is we don't know anything about the particular species." When a proper spatial and temporal sampling plan is developed and combined with field procedures that produce relatively high probability of detection, then the population is no longer "rare." That said, there are certainly many rare populations left to study: aardvarks, Pacific walrus, polar bear, abalone in the Strait of Georgia, Rio Grande silvery minnow, northern fur seals killed by net fragments in the North Pacific, etc. I encourage statisticians and biologists to continue to tackle the design of studies for these and other rare species.

References

Becker, E. F. 1991. A terrestrial furbearer estimator based on probability sampling. *Journal of Wildlife Management* 55:730–737.

Becker, E. F., M. A. Spindler, and T. O. Osborne. 1998. A population estimator based on network sampling of tracks in the snow. *Journal of Wildlife Management* 62:968–977.

Bodkin, J. L., K. A. Kloecker, G. G. Esslinger, D. H. Monson, and J. D. DeGroot. 2001. *Sea Otter Studies in Glacier Bay National Park and Preserve*. Annual Report 2000. USGS Biological Resources Division, Alaska Biological Science Center, Anchorage.

Bodkin, J. L., and M. S. Udevitz. 1999. An aerial survey method to estimate sea otter abundance. pp. 13–26 in G. W. Garner, S. C. Amstrup, J. L. Laake, B .F .J. Manly, L. L. McDonald, and D. G. Robertson, eds., *Marine Mammal Survey and Assessment Methods*. Balkema, Rotterdam, Netherlands.

Bonar, S. A., M. Divens, and B. Bolding. 1997. *Methods for Sampling the Distribution and Abundance of Bull Trout/Dolly Varden*. Research Report No. RAD97-05, Washington Department of Fish and Wildlife, Olympia.

Brown, J. A., and B. F. J. Manly. 1998. Restricted adaptive cluster sampling. *Environmental and Ecological Statistics* 5:49–63.

Buckland, S.T., D. R. Anderson, K. P. Burnham, J. L. Laake, D. L. Brochers, and L. Thomas. 2001. *Introduction to Distance Sampling: Estimating Abundance of Biological Populations*. Oxford University Press, Oxford.

Byers, G. W., and S. R. Shaw. 1999. A new species of Boreus (Mecoptera: Boreidae) from Wyoming. *Journal of the Kansas Entomological Society* 72(3):322–326.

Clark, T. W. 1997. *Averting Extinction: Reconstructing Endangered Species Recovery*. Yale University Press, New Haven.

Cochran, W. G. 1977. *Sampling Techniques*. 3rd ed. Wiley, New York.

Gaston, K. J. 1997. What is rarity? pp. 30–47 in W. E. Kunin and K. J. Gaston, eds., *The Biology of Rarity: Causes and Consequences of Rare-common Differences*. Chapman and Hall, London.

Gilbert, J. R. 1999. Review of previous Pacific walrus surveys to develop improved survey designs. pp. 75–86 in G. W. Garner, S. C. Amstrup, J. L. Laake, B. F. J. Manly, L. L. McDonald, and D. G. Robertson, eds., *Marine Mammal Survey and Assessment Methods*. Balkema, Rotterdam, Netherlands.

Green, R. H., and R. C. Young. 1993. Sampling to detect rare species. *Ecological Applications* 3:351–356.

Holland-Bartels, L. E., ed. 2001. *Mechanisms of Impact and Potential Recovery of Nearshore Vertebrate Predators Following the 1989 Exxon Valdez Oil Spill. Exxon Valdez* Oil Spill Trustee Council Restoration Project Final Report (Restoration Project 95025-99025), USGS Biological Resources Division, Alaska Biological Science Center, Anchorage.

Jaramillo-Legorreta, A. M., L. Rojas-Bracho, and T. Gerrodette. 1999. A new abundance estimate for vaquitas: First step for recovery. *Marine Mammal Science* 15:957–973.

Kalton, G., and G. W. Anderson. 1986. Sampling rare populations. *Journal Royal Statistical Society, Series A* 149:65–82.

Karanth, K. U., and J. D. Nichols. 1998. Estimating tiger densities in India from camera-trap data using photographic captures and recaptures. *Ecology* 79:2852–2862.

_____, eds. 2002. *Monitoring Tigers and Their Prey: A Manual for Wildlife*

Researchers, Conservationists, and Managers in Tropical Asia. Centre for Wildlife Studies, Bangalore, India.

Kern, J. W., K. J. Jenkins, and L. L. McDonald. 1994. *Sampling Designs for Estimating Sex and Age Composition of Dall Sheep in Wrangell-St. Elias National Park and Preserve: Progress Report and Preliminary Assessment*. WRST Research and Resource Management Report, No. 94-1, National Biological Survey, Wrangell-St. Elias Field Station, Glennallen, Alaska.

Knutson, M. G., W. B. Richardson, D. M. Reineke, B. R. Gray, J. R. Parmelee, and S. E. Weick. 2004. Agricultural ponds support amphibian populations. *Ecological Applications* 68:669–684.

Lancia, R. A., J. D. Nichols, and K. H. Pollock. 1994. Estimation of number of animals in wildlife populations. pp. 215–253 in T. Bookhout, ed., *Research and Management Techniques for Wildlife and Habitats*. The Wildlife Society, Bethesda, MD.

MacKenzie, D. I., J. D. Nichols, J. E. Hines, M. G. Knutson, and A. B. Franklin. 2003. Estimating site occupancy, colonization, and local extinction when a species is detected imperfectly. *Ecology* 84:2200–2207.

MacKenzie, D. I., J. D. Nichols, G. B. Lachman, S. Droege, J. A. Royle, and C. A. Langtimm. 2002. Estimating site occupancy rates when detection probabilities are less than one. *Ecology* 83:2248–2255.

Manly, B. F. J., J. M. Ackroyd, and K. A. R. Walshe. 2002a. Two-phase stratified random surveys on multiple populations at multiple locations. *New Zealand Journal of Marine and Freshwater Research* 36:581–591.

Manly, B. F. J., L. L. McDonald, D. L. Thomas, T. L. McDonald, and W. P. Erickson. 2002b. *Resource Selection by Animals: Statistical Design and Analysis for Field Studies*. 2nd ed. Kluwer Academic Publishers, Dordrecht, Netherlands.

Max, T. A., H. T. Schreuder, J. W. Hazard, D. D. Oswald, J. Teply, and J. Alegria. 1996. *The Pacific Northwest Region Vegetation and Inventory Monitoring System*. Research Paper PNW-RP-493, U.S. Dept. of Agriculture, Forest Service, Pacific Northwest Research Station, Portland, Oregon.

McDonald, L. L., G. W. Garner, and D. G. Robertson. 1999. Comparison of aerial survey procedures for estimating polar bear density: Results of pilot studies in Northern Alaska. pp. 37–52 in G. W. Garner, S. C. Amstrup, J. L. Laake, B. F. J. Manly, L. L. McDonald, and D. G. Robertson, eds., *Marine Mammal Survey and Assessment Methods*. Balkema, Rotterdam, Netherlands.

McDonald, L. L., H. B. Harvey, F. J. Mauer, and A. W. Brackney. 1990. Design of aerial surveys for Dall sheep in the Arctic National Wildlife Refuge, Alaska. pp. 176–193 in J. A. Bailey, ed., *Proceedings of the Seventh Biennial Northern Wild Sheep and Goat Symposium*. Northern Wild Sheep and Goat Council, Clarkston, Washington.

Morrison, M. L., W. M. Block, M. D. Strickland, and W. L. Kendall. 2001. *Wildlife Study Design*. Springer, New York.

Otis, D. L., K. P. Burnham, G. C. White, and D. R. Anderson. 1978. Statistical inference from capture data on closed animal populations. *Wildlife Monographs* 62:1–135.

Peterson, J., J. Dunham, P. Howell, R. Thurow, and S. Bomar. 2001. *Protocol for Determining Bull Trout Presence*. Western Division American Fisheries Society. Available from http://www.fisheries.org/wd/committee/bull_trout/bull_trout_committee.htm (accessed September 9, 2003).

Rocke, T. E., and M. D. Samuel. 1999. Water and sediment characteristics associated with avian botulism outbreaks in wetland. *Journal of Wildlife Management* 63:1249–1260.

Samuel, M. D., E. O. Garton, M. W. Schlegel, and R. G. Carson. 1987. Visibility bias during aerial surveys of elk in north central Idaho. *Journal of Wildlife Management* 51:622–630.

Sargeant, G. A., M. A. Sovada, and C. C. Slivinski, C.C. 2002. Monitoring changes in carnivore distributions: A case study of swift foxes in western Kansas. Abstract submitted for the 2002 Annual Conference of The Wildlife Society. USGS Northern Prairie Wildlife Research Center, Jamestown, North Dakota.

Seber, G. A. F. 1982. *The Estimation of Animal Abundance and Related Parameters.* 2nd ed. Griffin, London.

Smirnov, E. N., and D. G. Miquelle. 1999. Population dynamics of the Amur tiger in Sikhote-Alin State Biosphere Reserve. pp. 61–70 in J. Seidensticker, S. Christie, and P. Jackson, eds., *Riding the Tiger: Meeting the Needs of People and Wildlife in Asia.* Cambridge University Press, Cambridge.

Smith, D. R., R. F. Villella, and D. P. Lemarie. 2001. Survey protocol for assessment of endangered freshwater mussels in the Allegheny River. *Journal of North American Benthological Society* 20(1):118–132.

Starr, P. J., and C. Schwarz. 2000. *Feasibility of a Bottom Trawl Survey for Three Slope Groundfish Species in Canadian Waters.* DFO Canadian Stock Assessment Secretariat Research Document 2000/156, Fisheries and Oceans, Ontario, Canada. Available at: www.dfo-mpo.gc.ca/csas.

Strayer, D. L., and D. R. Smith. 2003. A guide to sampling freshwater mussel populations. *American Fisheries Society Monograph* 8:1–110.

Thompson, S. K. 1992. *Sampling.* Wiley, New York.

Thompson, S. K., and G. A. F. Seber. 1996. *Adaptive Sampling.* Wiley, New York.

Tolle, T., and R. Czaplewski. 1995. An interagency monitoring design for the Pacific Northwest Forest Ecosystem Plan. In J. E. Thompson, ed. Analysis in *Support of Ecosystem Management: Analysis Workshop III.* USDA Forest Service, Ecosystem Management Analysis Center, Washington, D.C.

White, G. C., and K. P. Burnham. 1999. Program MARK: Survival estimation from populations of marked animals. *Bird Study* 46:S120–S138.

Williams, B. K., J. D. Nichols, and M. J. Conroy. 2002. *Analysis and Management of Animal Populations.* Academic Press, San Diego.

3

Separating Components of Detection Probability in Abundance Estimation: An Overview with Diverse Examples

Kenneth H. Pollock, Helene Marsh, Larissa L. Bailey, George L. Farnsworth, Theodore R. Simons, and Mathew W. Alldredge

Estimation of population abundance of rare and elusive species critically depends on the estimation of detection probability under a particular sampling method. If we ignore the issue of animals not being available, then we obtain an estimate of the size of the available component of the population rather than the total population size. The available component may be only a small portion of the total population. In addition, this component may vary with time and with important auxiliary variables in ways that are so complex that it is unsatisfactory for monitoring the population (see Chapter 4).

Animals have to be "available" to a sampling method to be detected. In many animal populations not all animals are available to be captured using a particular sampling method. There may be many reasons for this. For example, in an aerial survey of dugongs (sea cows) off the coast of Australia, some dugongs may be underwater and invisible to the observers searching for them in the aircraft. Even if animals are available, they still have to be detected. This perception process is also uncertain, so, for example, if a dugong is on the surface of the water, observers in the aircraft may still miss it.

In this chapter we consider in detail a model for detection probability that accounts for the processes of availability and perception. Methodology for estimating these two components of detection probability is illustrated with three diverse examples involving aerial surveys of marine mammals (dugongs), point counts of terrestrial birds, and capture-recapture studies of terrestrial salamanders. The statistical methodology used in the three examples is very different.

We will use the dugong survey as a first example of a solution to a general problem of lack of availability (Marsh and Sinclair 1989; Pollock et al. in press). We then will show that very similar conceptual problems arise in many other settings and biometricians are now seeking solutions to them. Two other problems we consider are (1) estimation of density of birds based on point counts in which birds are detected by their calls but birds do not always call (i.e., they are unavailable for auditory detection; Farnsworth et al. 2002); and (2) population estimation of terrestrial salamanders, which presents a similar conceptual and practical problem because salamanders may be underground where they cannot be counted or captured (i.e., they are physically unavailable to capture because they are not present on the surface; Bailey et al. 2004a,b). Many formulations ignore the unavailable part of the population, but doing so may be unsatisfactory unless this component is a very small and constant part of the population.

Absolute Abundance Estimator

In studying wild animal populations to estimate population size and other parameters, typically not all animals are detected and/or not all of the area is sampled, so that a census or complete count of the population is impossible. Therefore, the population size has to be estimated using sampling methods. Skalski (1994) presented a framework for combining finite sampling theory and density estimation of individually sampled units. Here, we consider the sampled units collectively in estimating abundance. Two key elements are the proportion of the area actually sampled and the detection probability of animals in the study area that is actually sampled. The counts can be adjusted to obtain the population size estimate using the canonical equation (Lancia et al. 1994; Williams et al. 2002),

$$\hat{N} = \frac{C}{p_{area}\,\hat{p}_d}, \qquad (3.1)$$

where $\hat{N}$ is the estimator of population size, C is count of animals seen, p_{area} is proportion of area sampled, and $\hat{p}_d$ is the estimator of the fraction of animals detected (i.e., seen or caught).

Here we assume that the proportion of the area sampled is known and that the sampling design for the area uses probabilistic sampling such as simple random sampling (Thompson 2002a). Note, however, that the population size estimator in equation (3.1) will vary depending on the sampling approach. For example, if stratified random sampling were used, then equation (3.1) would be used separately in each stratum and then summed across strata. Although this is an important component of the problem with its own complexities, we focus in the remainder of this chapter on the detection component.

There are many ways to estimate the detection probability, $\hat{p}_d$, such as using capture-recapture sampling, distance sampling, or many others. We have presented the standard formulation for abundance estimation, but we sometimes need a more general detection model that takes account of animals being unavailable. Availability can be defined by the presence or absence of animals on the surveyed areas or may be defined by the survey technique, which may or may not have the ability to detect the animal of interest.

General Absolute Abundance Estimator

A more general equation involves first taking account of availability of animals for detection and then the detection of animals that are available,

$$\hat{N} = \frac{C}{p_{area}\,\hat{p}_a\,\hat{p}_{da}}, \tag{3.2}$$

where $\hat{N}$ is the population size, C is the count of animals, p_{area} is the proportion of the area sampled, $\hat{p}_a$ is the estimator for the probability of being available, and $\hat{p}_{da}$ is the estimator for the conditional probability of detection given availability. This equation is not really new, but it is often not emphasized. If we use equation (3.1) when equation (3.2) is appropriate, we just obtain an estimate of the population size of the available component. We will see in our later examples that this may be a small component of the total population and furthermore that this component may vary with time and with important auxiliary variables in such complex ways that it is unsatisfactory for monitoring the population. We now proceed to the dugong aerial survey example to illustrate one methodology for estimating the components of detection probability and then how to use them in sound abundance estimation.

Aerial Survey Methodology That Accounts for Nonavailability of Dugongs

The dugong (*Dugong dugon*) is a member of the order Sirenia. Dugongs are herbivorous marine mammals that occur in some 37 countries in the Indo-West Pacific. In Australia, dugongs occur in seagrass habitats along some 15,000 km of the northern coast. The water turbidity in these areas is variable, ranging from clear to very turbid over a very short distance. Dugongs have a patchy distribution and often occur at very low densities. Marsh and her colleagues at James Cook University in Townsville, Australia, have studied dugongs extensively (e.g., see Marsh and Sinclair 1989). Here we will focus on the aerial survey methodology used to estimate the density and population size of dugongs.

Marsh and her colleagues used a stratified random sampling design with a strip sampling method. They chose strip sampling instead of line transect methodology because dugongs are at the surface for only 1–2 seconds, which makes it impossible to reliably estimate their distance from the transect line. Flights are conducted at a standard height (137 m) and speed (185 km/h). Each flight crew consists of a pilot and survey leader plus tandem teams of two independent observers on each side of the aircraft (six total). The two observers on each side of the aircraft survey a strip of width 200 m (Marsh and Sinclair 1989).

Modeling the availability process involves estimation of the probability of an animal being available ($\hat{p}_a$). This is a very difficult parameter to estimate for the dugong surveys and must be done external to the aerial survey with additional data collected in a dedicated study. Artificial dugong models were fitted with timed depth recorders and taken by boat to the test area off the coast. They were lowered into the water next to the boat. Observers in a helicopter at the standard aerial survey height determined the depth at which the models become visible as they were raised in the water column. This enabled Marsh and colleagues to divide the water column into available and nonavailable zones of detection under a range of different conditions of water clarity, depth, and sea state. The dive profiles of 15 dugongs fitted with timed depth recorders were recorded in a separate study. The combination of these data sets enabled the research team to estimate the probability of dugongs being available for detection under various combinations of depth, sea state, and turbidity. The fraction of time that animals spent above the depth where they were visible (available) was an estimate of the probability of being available. Table 3.1 gives estimated availability probabilities with standard errors. Notice the variability in these probabilities depending on depth and other water conditions. The assumptions for this availability process estimation are:

- The depth at which dugong models become visible is measured without error.
- The depth at which dugong models become visible is the same as for real dugongs.
- Depth profiles of individually monitored dugongs are representative of the whole population of dugongs being studied in the aerial survey.

Table 3.1

Estimated availability probabilities (p_a) and their standard errors for various strata of survey depths and turbidities calculated based on data from the dugong models and the individual dive profiles of telemetered wild dugongs.

Sea condition for survey	Water quality	Depth range	Visibility of sea floor	Depth of zone of visibility (m) of models[a]	Depth of zone of visibility (m) used to calculate p_a	p_a (SE)
Optimal						
	Clear	Shallow	Clearly visible	Visible at bottom	All	1
	Variable	Variable	Visible but unclear	2.44	2.5	0.652 (0.0452)
	Clear	> 5 m	Not visible	4.32	4.0	0.462[b] (0.057)
	Turbid	Variable	Not visible	1.23	1.5[c]	0.474 (0.0525)
Marginal						
	Clear	Shallow	Clearly visible	Visible at bottom	Visible at bottom	1
	Variable	Variable	Visible but unclear	1.21	1.5[c]	0.474 (0.0525)
	Clear	> 5 m	Not visible	0.69	1.5[c]	0.296[b] (0.0724)
	Turbid	Variable	Not visible	1.43	1.5[c]	0.474 (0.0525)

[a]Averaged for models 2.0 m and 2.5 m long.
[b]Based on records from four dugongs with mean, median, and modal maximum dives of > 6 m and a corresponding subset of the data from one dugong that spent considerable time in water > 5 m deep (see text for explanation).
[c]Based on minimum dive depth detectable on 15 telemetered wild dugongs (see text for explanation).

- The flight speed is fast enough for the dugongs to be available only for an "instant."
- There is independent availability of group members in a detected group of dugongs.

Modeling the probability of an animal being detected given that the animal is available involves using two independent observers on each side of the aircraft. The probability of a dugong being detected by at least one observer is p_{da} and is estimated in the following way. If n_1 is the number detected by the mid observer, n_2 is the number detected by the rear observer, and m_2 is the number detected by both observers, then $\hat{p}_1 = m_2/n_2$, $\hat{p}_2 = m_2/n_1$, and $\hat{p}_{da} = 1 - (1 - \hat{p}_1)(1 - \hat{p}_2)$. This is an extension and adaptation of the Lincoln-Petersen method (Seber 1982; Pollock et al. 1990) used in two-sample, capture-recapture studies. The assumptions are that counts within the 200-m strip are measured accurately, that there are no matching errors between the two observers, and that detection probabilities for all groups for each observer are equal.

We used Program MARK (White and Burnham 1999) to generalize these models to allow for detection probability conditional on availability to vary by seat (mid or rear), side (port or starboard), and location of the survey. We then used AIC (Akaike's information criteria; Akaike 1973) to pick the simplest model that explained the data adequately. Alternatively, model-averaged estimates can be used. MARK also can be used to determine if the detection probability conditional on availability is dependent on individual group covariates, such as size of group, sea state, glare, distance class, etc.

We used the generalized Horvitz-Thompson estimator for population size based on the detection probability p_i of each individual group of size g_i, such that $\hat{p}_i = p_{area}\hat{p}_{ai}\hat{p}_{dai}$ and $\hat{N} = \sum_{i=1}^{v}(g_i/\hat{p}_i)$, where v is the number of distinct groups detected. We used a simulation or parametric bootstrap method that included all sources of variation in the estimates to develop an estimator of the standard error of population size and density.

Although results of these surveys were quite precise (Table 3.2), they were expensive, as were the extra data needed for estimation of probability of being available for detection. The estimation of probability of being available was based on the dive profiles of only 15 dugongs. Also, the artificial models may have been detectable at different depths from real dugongs. Ideally, the study using the artificial models should have been repeated to broaden the range of conditions. Marsh and colleagues also

Table 3.2.

Population size estimates and standard errors for a recent aerial survey of dugong for the Northern Great Barrier Reef in 2000.

Block	Population estimate	Standard error
1	73	36
2	1613	661
3	1726	615
4	1041	380
5	2843	958
6	473	255
8	631	248
11	347	318
13	471	102
14	586	536
Total	9804	1550

plan to expand the sample of animals used in the depth profiles. The detection given available estimation also can have small biases using the two-observer method, even though a proper protocol was used (Marsh and Sinclair 1989; Pollock et al. in press). With these provisos, we emphasize that this study is an enormous improvement over not including the unavailable portion of the population in the analysis, as is the case in many other aerial surveys, including surveys over areas where the availability bias varies enormously.

Point Count Methodology That Accounts for Nonavailability of Birds

Bird point count studies have been widely used to assist in the management of terrestrial birds. For example, the Breeding Bird Survey run by the U.S. Fish and Wildlife Service (U.S. FWS) has 3,000 roadside routes in the United States and Canada. Each route is 25 miles, with 50 points per route. Each count is a 3-minute, unlimited-radius point count. The survey was begun by Chandler Robbins of the U.S. FWS in 1966 and has become a very important source of information on population trends for many species (Robbins et al. 1986; Peterjohn and Sauer 1993).

Traditionally, the basic point-count survey counted all birds seen or heard during a fixed time interval. The survey contained no estimate of detection probability; analyses generally assumed detection probability was invariant. This approach is now considered inadequate and antiquated. Two recent overview papers by Thompson (2002b) and Rosenstock et al. (2002) stressed the importance of properly estimating detection probability for sound inference based on point counts.

There are now four methods of estimating detection probabilities for point count surveys: distance methods, multiple observer methods, time-of-detection methods, and combinations of the three methods. Two of these methods, distance sampling (Buckland et al. 2001) and multiple observer methods (Nichols et al. 2000; Pollock et al. 2003), assume that birds are available to auditory detection (e.g., sing) during the point count. A new method, the time-of-detection method (Farnsworth et al. 2002), allows for birds not singing, which is a type of nonavailability of birds because in many forest bird point counts a bird is effectively undetectable if it does not sing during the sampling interval.

Distance sampling (Buckland et al. 2001) is founded on the premise that detection is based on a sighting function, $g(r)$, that is a monotonic declining function with distance r and assumes $g(0)$ is 1. Distance methodology assumptions are: birds that are very close to the station will always be detected; there is no movement of birds (i.e., attraction or repulsion) during the survey; all distances are measured without error; and sightings of different birds are independent events. In our context, an additional assumption is that all the birds sing and are thus available for detection during the point count.

Multiple observer methods of estimating detection probability involve primary and secondary observers (Nichols et al. 2000) or two independent observers (Pollock et al. 2003). For simplicity, we will only discuss using the Lincoln-Petersen Method with two independent observers who each map their bird locations independently and match the birds heard—only by observer 1, only by observer 2, or by both observers—to estimate the birds missed. Note that this is basically the same approach as that used by Marsh in her aerial surveys of dugongs described above. The model assumptions are (1) there is no change in the population of birds within the detection radius during the point count (i.e., the population is closed and birds are not moving in or out); (2) there is no double-counting of individuals, and birds are matched between observers accurately; (3) all birds of a given species are equally detectable for a given observer (no heterogeneity); and (4) if counts with limited radius are used, observers accurately assign birds to within or beyond the radius used. In our context, an additional assumption is that all the birds sing at least once during the time of the point count.

Time-of-detection sampling methodology is new and was presented in an important paper by Farnsworth et al. (2002). They noticed that most detections are by sound in forested habitats and for cryptic species. Birds can be detected only if they sing, but now we can estimate how many are unavailable (i.e., do not sing). Effective detection probability is now the product of the probability that a bird sings during the count interval and the probability that the song is heard by an observer. Note that this is just a breakdown in detection probability developed in the introduction but now adapted for the situation where detection is only by sound. (We emphasize that the bird is present in the survey area but is unavailable because it does not sing.)

Now we consider the simple case in which the point-count is divided into two equal time intervals. We define n_1 as number of birds first counted in a fixed-radius plot in period 1 and n_2 as the number of birds first counted in period 2 only. Further, we assume detection probability is the same for both periods such that $E(n_1) = Np$ and $E(n_2) = N(1 - p)p$, where N is the population size in all the fixed-radius (r) plots and p is the detection probability, and E is the expected value (mean). This is of the same form as the removal method used for animal trapping studies in which the trapped animals are not returned to the population (Zippin 1958; Seber 1982).

If we use n_1 and n_2 as estimates for the expected means in the previous equations, we can solve these intuitive moment equations for N and p and obtain estimators $\hat{p} = (n_1 - n_2) / n_1$ and $\hat{N} = n_1/\hat{p}$. The assumptions of the time-of-detection method are (1) there is no change in the population of birds within the detection radius during the point count (that is there is no movement in or out), (2) the observer can accurately follow each individual bird so no bird is counted twice, (3) all birds have a constant probability of being detected during each interval, and (4) observers accurately assign birds to within or beyond the radius used.

The two key assumptions are (1) and (2). If birds are moving around a lot, it will be difficult for observers to follow them and the probability of double-counting birds is significant. Further, birds may move into the cir-

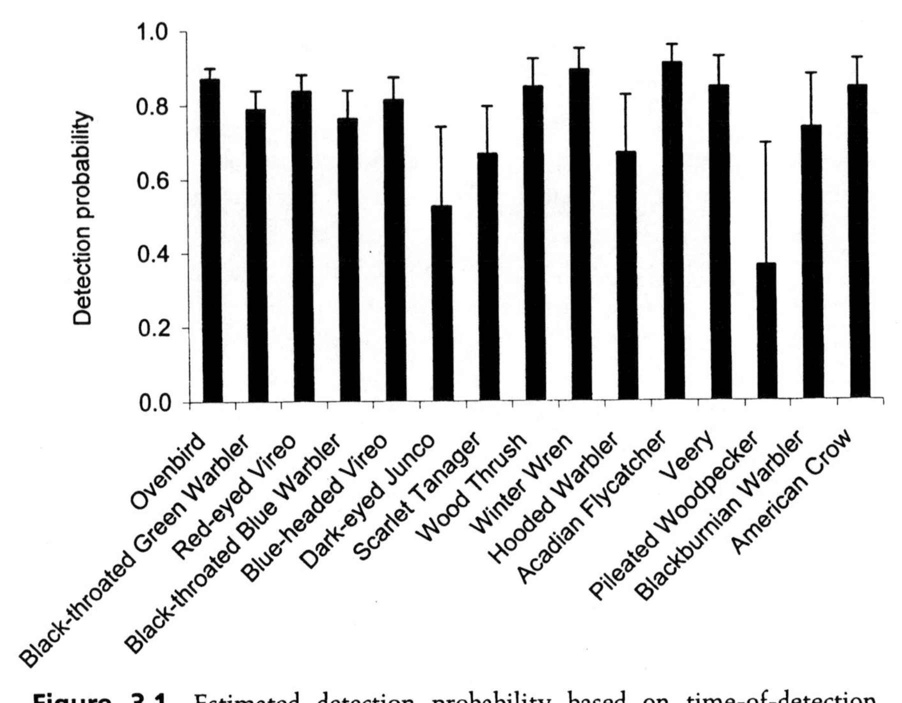

Figure 3.1. Estimated detection probability based on time-of-detection method during unlimited-radius counts for the 15 most frequently encountered species, ordered from most common (left) to least common (right) (from Farnsworth et al. 2002). Error bars represent one estimated standard error.

cle undetected and then sing and be detected. However, we believe there will be surveys where these assumptions are reasonable.

Many surveys have recorded birds counted into three time intervals (Ralph et al. 1995). The use of three intervals allows for heterogeneity of detection probability between birds. Farnsworth et al. (2002) discuss this and also extend the theory by allowing the multiple intervals to be of varying length. We illustrate with some results from Farnsworth and colleagues' point count surveys in the Great Smoky Mountains National Park (Figure 3.1).

We strongly recommend that combined methods be used in practice as much as possible. In particular we note that combined methods involving the time-of-detection method enable separation of probability of being *available* from probability of *detection, given available*. In our example, we were able to estimate only the overall probability of detection. There is also the need for careful evaluation of the effects of movement into the plot after the count has commenced.

Capture-Recapture Methodology That Accounts for Nonavailability of Salamanders

Bailey et al. (2004a,b) were concerned about population estimation for terrestrial salamander populations in the Great Smoky Mountains National Park in North Carolina and Tennessee. Salamanders are suspected to be reasonable indicator species of old growth forest health, and there is a need for more extensive monitoring programs of many amphibian species.

One objective of the study (Bailey et al. 2004a,b) was to evaluate the effectiveness and biases of salamander count indices that are very widely used because they are much less expensive than capture-recapture methods. Studies using count indices assume that detection probability is constant over time and space and further that all salamanders are available to be counted. Clearly, the first assumption is often violated. Further, there are special problems with sampling salamanders because they are very sensitive to desiccation and generally require cool, moist habitats. Thus, they often occur below the surface of the ground to avoid unfavorable moisture regimes on the surface and to escape predators. Salamanders below the surface are hence unavailable to be captured under standard sampling protocols that only sample the surface population.

Bailey et al. (2004a,b) used capture-recapture methods to study salamander abundance at different sites over 3 years. They established 15 sites during 1999 and 20 sites each year during 2000 and 2001 in the Mt. LeConte Quadrangle of Great Smoky Mountains National Park. An elastomer marking technique (Northwestern Marine Technology Inc., Shaw Island, Washington, USA) was used to mark more than 5,300 salamanders in 3 years. Every year, sites were sampled on 3–4 consecutive days (secondary samples) during each of four primary periods about 10–14 days apart. Individual capture histories were compiled for each site and each year (52 site-years in total were used for their analysis).

Bailey et al. (2004a,b) used primary and secondary sampling periods each year so that they could use the robust design (Pollock 1982; Pollock et al. 1990). This design allows combination of open and closed population models in one analysis. Pollock (1982) originally assumed it would be beneficial for allowing heterogeneity of capture probabilities between animals. Kendall and Nichols (1995) and Kendall et al. (1997) noted that this design could be very useful to estimate the temporary emigration probability (i.e., 1-the probability of an animal being available for capture). Their work has now given rise to extensive literature on temporary emigration models (see Williams et al. 2002).

For each site year, Bailey et al. (2004a) fitted 12 competing models with the robust design option in Program MARK (White and Burnham 1999). These models allowed temporary emigration or not; trap response present or not; and time variation in capture probabilities and population sizes between primary periods. They assumed that survival probability between the primary periods was 1 because it was so short (~10–14 days). They used the $QAIC_c$ criteria for model selection. This is an extension of the Akaike information criteria (Akaike 1973) to account for small samples and overdispersion. Bailey et al. found that temporary emigration models were selected 80.7% of the time and models that allowed a trap-shy response were selected 63.5% of the time. Temporary emigration matters a great deal; the average temporary emigration parameter estimate was 0.87 ($\widehat{SE} = 0.01$). The temporary emigration parameter estimates varied widely over years and sites, with a low of 0.62 and a high of 0.99. Dry, disturbed sites had higher values than wet, more pristine sites. On average only 13% of the animals were on the surface available for capture during each primary sampling period. Further, although the conditional probability of capture given available was 0.30 ± 0.01, the overall estimated probability of detection, which Bailey et al. (2004a) referred to as the effective capture probability, was only 0.04 ($\widehat{SE} = 0.002$; $0.13 \times 0.30 = 0.04$). Thus, using the robust, capture-recapture design, we can estimate overall detection probability and its different components for these salamanders.

Table 3.3 illustrates how including temporary emigration in the model affects estimates of both capture probability and surface (available) population size. In the absence of temporary emigration, estimates of population size are still unbiased if emigration is random, but refer to the total population, not just the available or surface population. Further, using unadjusted count indices to compare salamander populations over time and space would be very unwise because only a small and variable portion of the population is available to be counted at any time.

Discussion

We have shown with a few examples that nonavailability problems for population size estimation are extremely widespread. They come into play in studies such as aerial surveys of dugongs and other marine mammals (manatees, whales, dolphins), which spend a high proportion of their time

Table 3.3.

Time-specific estimates of conditional capture probability, $p_{i.}$, recapture probability, $c_{..}$, and population size, N_i, for salamanders on an undisturbed site (RO004, 2001). One hundred twenty-four individuals were caught; 22 individuals were captured on more than one sampling occasion. Note that n = number of animals captured in each primary period and m = number of n animals recaptured within each primary period. The "best" model (using $QAIC_c$ selection) included temporary emigration, time variation, and behavioral (trap-shy) effects. We also present models with and without temporary emigration and those that contain time variation in surface (available) population size. Standard error estimates are inflated by $\sqrt{\hat{c}} \bullet SE(\hat{\theta})$ with $\hat{c} = 1.54$.

		Models with temporary emigration					
		$\gamma, p_{i.}, c_{..}, N_.$[a]: $\hat{\gamma}_. = 0.85(0.06)$			$\gamma, p_{i.}, c_{..}, N_i$[b]: $\hat{\gamma} = 0.78(0.16)$		
Primary period	n	$\hat{p}_{i.}(SE)$	$\hat{c}_{..}(SE)$	$\hat{N}_.(SE)$	$\hat{p}_{i.}(SE)$	$\hat{c}_{..}(SE)$	$\hat{N}_.(SE)$
1	56	0.33 (0.10)	0.06(0.02)	79.98 (15.81)	0.34 (0.14)	0.06(0.02)	76.95 (20.09)
2	23	0.16 (0.07)		79.98 (15.81)	0.17 (0.12)		43.14 (24.31)
3	41	0.22 (0.09)		79.98 (15.81)	0.15 (0.19)		82.62 (75.83)
4	21	0.12 (0.07)		79.98 (15.81)	0.04 (0.04)		147.42 (163.46)

		Models without temporary emigration					
		$\gamma, p_{i.}, c_{..}, \hat{N}_.$[c]			$\gamma, p_{i.}, c_{..}, \hat{N}_i$[d]		
Primary period	m	$\hat{p}_{i.}(SE)$	$\hat{c}_{..}(SE)$	$\hat{N}_.(SE)$	$\hat{p}_{i.}(SE)$	$\hat{c}_{..}(SE)$	$\hat{N}_.(SE)$
1	3	0.05 (0.02)	0.06(0.02)	368.14 (108.75)	0.34 (0.14)	0.06(0.02)	76.95 (20.09)
2	2	0.02 (0.01)		368.14 (108.75)	0.04 (0.02)		162.36 (76.73)
3	3	0.03 (0.01)		368.14 (108.75)	0.02 (0.01)		460.23 (238.05)
4	2	0.01 (0.01)		368.14 (108.75)	0.01 (0.01)		730.46 (551.66)

[a] $\Delta AIC_c = 0.0$; AIC_c weight = 0.57.
[b] $\Delta AIC_c = 4.59$; AIC_c weight = 0.06.
[c] $\Delta AIC_c = 3.08$; AIC_c weight = 0.12.
[d] $\Delta AIC_c = 3.07$; AIC_c weight = 0.12.

underwater; bird point counts, in which birds may not sing and hence may be undetectable; and surveys for terrestrial salamanders that spend time underground and hence are undetectable. Some other situations in which nonavailability could be a problem include aerial surveys of terrestrial mammals (elk, deer, moose) in habitats with areas of very dense vegetation where some animals are so well hidden that they are not available for detection from the air; and surface counts of ant and other insect colonies, in which only the foraging component of the population is available for detection. Solutions to this problem are difficult and will usually be species- and sampling-method specific, as illustrated in this chapter. In some situations it may not be possible to estimate the probability of being available; however, we have shown that failing to include this component can cause severe problems of interpretation, especially when this probability is low and/or variable in space and time.

Common methods of estimating detection probability in animal counts are distance sampling (e.g., the point count method illustrated above) and multiple observer methods (e.g., dugong and bird surveys). These methods do not take into account animals being unavailable. Additional methods such as those illustrated here have to be used to correct for the unavailable portion of the population. This is particularly important if the portion unavailable fluctuates in space or time. Further, we have shown that for captured animals, standard capture-recapture methods cannot account for nonavailability unless the robust design is used. An exception is when special telemetry tags are used. In that case, for example, mark-resight aerial surveys using a marked subpopulation with radio tags could account for nonavailable animals that have moved outside the study area.

The time-of-detection method used to take account of non-singing birds can be employed routinely to estimate total detection probability, including the probability of being available. The biologist needs to record first-detection-time categories for all the birds; this is often quite feasible in practice. By contrast, in the dugong aerial survey, the information needed to correct for availability cannot be collected routinely due to expense and logistics. Similarly, in terrestrial salamander studies, it is not logistically feasible due to expense to conduct elaborate mark-recapture studies at all count locations. We believe, however, that in many examples, it is possible to design special studies into a larger count survey in a double-sampling framework (Thompson 2002a; Pollock et al. 2002). For example, a large sample of plots could have salamanders counted under a well-defined protocol with the detailed capture-recapture study restricted to a small random subsample of the plots in order to estimate the two detection probability components. These component estimates then could be used to adjust the counts appropriately.

When studies are planned to map the distribution of cryptic animals in space or to monitor the size of populations through time, we strongly recommend that methods be developed for separating and estimating the components of detection probability. Without this approach, it is unlikely that the results of the survey program will be reliable or useful.

Acknowledgments

We thank Dr. B. Brunhuber for editorial assistance.

References

Akaike, H. 1973. Information theory and an extension of the maximum likelihood principle. Pp. 267–281 in B. N. Petrov and F. Csaki, eds., *Second International Symposium on Information Theory*. Akademiai Kiado, Budapest, Hungary.

Bailey, L. L., T. R. Simons, and K. H. Pollock. 2004a. Estimating detectability parameters for plethodon salamanders using the robust capture-recapture design. *Journal of Wildlife Management*. 68:1–13.

_____. 2004b. Spatial and temporal variation in detection probability of plethodon salamanders using the robust capture-recapture design. *Journal of Wildlife Management*. 68:14–28.

Buckland, S.T., D. R. Anderson., K. P. Burnham, J. L. Laake, D. L. Borchers, and L. Thomas. 2001. *Introduction to Distance Sampling*. Oxford University Press, London.

Farnsworth, G., K. H. Pollock, J. D. Nichols, T. R. Simons, J. E. Hines, and J. R. Sauer. 2002. A removal model for estimating the detection probability during point counts divided into time intervals. *Auk* 119:414–425.

Kendall, W.L., and J. D. Nichols. 1995. On the use of secondary capture-recapture samples to estimate temporary emigration and breeding proportions. *Journal of Applied Statistics* 22:751–762.

Kendall, W.L., J. D. Nichols, and J. E. Hines. 1997. Estimating temporary emigration using capture-recapture data with Pollock's robust design. *Ecology* 78:563–578.

Lancia, R. A., J. D. Nichols, and K. H. Pollock. 1994. Estimating the number of animals in wildlife populations. pp. 215–253 in T. Bookhout, ed., *Research and Management Techniques for Wildlife and Habitats*. The Wildlife Society, Bethesda, Maryland.

Marsh, H., and D. F. Sinclair. 1989. Correcting for visibility bias in strip transect aerial surveys of aquatic fauna. *Journal of Wildlife Management* 53:1017–1024.

Nichols, J. D., J. E. Hines, J. R. Sauer, F. W. Fallon, J. E. Fallon, and P. J. Heglund. 2000. A double-observer approach for point counts. *Auk* 117:393–408.

Peterjohn, B. G., and J. R. Sauer. 1993. North American breeding bird survey annual summary 1990–1991. *Bird Populations* 1:1–15.

Pollock, K. H. 1982. A capture-recapture design robust to unequal probability of capture. *Journal of Wildlife Management* 46:752–757.

Pollock, K. H., M. Alldredge, J. D. Nichols, T. R. Simons, G. L. Farnsworth, and J. R. Sauer. 2003. Modeling availability and perception processes for detection in point counts of bird populations. Unpublished manuscript.

Pollock, K. H., H. Marsh, I. Lawler, and M. Alldredge. In press. Modeling availability and perception processes for strip and line transects: An application to dugong aerial surveys. *Journal of Wildlife Management*.

Pollock, K. H., J. D. Nichols, C. Brownie, and J. E. Hines. 1990. Statistical inference for capture-recapture experiments. *Wildlife Society Monographs* 107:1–97.

Pollock, K. H., J. D. Nichols, T. R. Simons, G. L., Farnsworth, L. L. Bailey, and J. R. Sauer. 2002. The design of large-scale wildlife monitoring studies. *Environmetrics* 13:105–119.

Ralph, C. J., S. Droege, and J. R. Sauer.1995. Managing and monitoring birds using point counts: Standards and applications. pp. 161–168 in C. J. Ralph, J. R. Sauer, and S. Droege, eds., *Monitoring Bird Populations by Point Counts*, USDA Forest Service General Technical Report PSW-GTR-149, USDA Forest Service, Albany, California.

Robbins, C. S., D. Bystrak, and P. H. Geissler. 1986. *The Breeding Bird Survey: Its First Fifteen Years, 1965–1979*. U.S. Fish & Wildlife Service Resource Publication 157, U.S. Fish & Wildlife Service, Washington, D.C.

Rosenstock, S. S., D. R. Anderson, K. M. Giesen, T. Leukering, and M. F. Carter. 2002. Landbird counting techniques: current practices and an alternative. *Auk* 119: 46–53.

Seber, G. A. F. 1982. *Estimation of Animal Abundance and Related Parameters*. 2nd ed. Macmillan, New York.

Skalski, J. R. 1994. Estimating wildlife populations based on incomplete area surveys. *Wildlife Society Bulletin* 22:192–203.

Thompson, S. K. 2002a. *Sampling*. 2nd ed. Wiley, New York.

Thompson, W. L. 2002b. Towards reliable bird surveys: Accounting for individuals present but not detected. *Auk* 119:18–25.

White, G.C., and K. P. Burnham. 1999. Program MARK: Survival estimation from populations of marked animals. *Bird Study* 46:S120–S139.

Williams, B. K., J. D. Nichols, and M. J. Conroy. 2002. *Analysis and Management of Animal Populations*. Academic Press, San Diego.

Zippin, C. 1958. The removal method of population estimation. *Journal of Wildlife Management* 22:82–90.

4

Indexes as Surrogates to Abundance for Low-Abundance Species

Paul B. Conn, Larissa L. Bailey, and John R. Sauer

Wildlife sampling is rarely based on censuses, that is, complete counts of animals. Instead, animals detected at a site during a sampling occasion are generally only a partial count of the actual population size. These counts (C) are related to the actual population size (N) at a sampling site by an unknown probability of detection (p) (see Chapter 2). Historically, many wildlife surveys have been based on counts, and several continent-scale surveys such as the North American Breeding Bird Survey (Robbins et al. 1986) still collect only count data at sample sites. However, it is well known that counts are biased estimators of N (i.e., p is generally less than 1) and that comparisons of abundances based on C are invalidated when probability of detection varies among sites (Barker and Sauer 1995; Sauer and Link 2004). To analyze count data, the investigator must rely on the index assumption that p does not vary in a systematic way.

To enhance the credibility of wildlife studies, a primary focus of ecological statistics has been the development of methods for estimation of p and hence N, using C and supplemental information. Biologists now have access to a variety of estimation procedures tailored for most commonly used ecological sampling techniques, such as mark-recapture and distance sampling (e.g., see Williams et al. 2002).

Despite these recent methodological developments, controversy continues among biologists regarding the value of count-based indexes as proper metrics for evaluating population change. The appeal of count index surveys is undeniable. Estimation of detection rates requires expertise and additional expense, and it is often cheaper and simpler to implement surveys in which only a count index is collected. Also, deficiencies of surveys based on indexes often are not obvious because the information needed to reasonably evaluate index assumptions is not collected. Although strong statements have been published regarding the need for caution in use of count-based measures such as minimum number known alive (M_{t+1}; Nichols and Pollock 1983), point counts (Barker and Sauer 1995), and other indexes of abundance (Anderson 2001), indexes continue to be used for estimation and modeling.

Recently, defenses of indexes have appeared in the literature, rationalizing the use of these measures for monitoring and comparative analyses of populations (e.g., Slade and Blair 2000; McKelvey and Pearson 2001; Hutto and Young 2002). Often, these defenses have focused on the logistical complications associated with implementing techniques that estimate detectability, especially when questions remain about the ability of these techniques to satisfy their key assumptions, and hence, represent unbiased and precise estimators. Occasionally, simulation is used to document robustness of count-based indexes to variation in detection rates over time (e.g., McKelvey and Pearson 2001).

In practice, the need for continued discussion about appropriate analysis of index counts is extremely relevant for conservation because population surveys still are being designed and implemented using indexes. Moreover, many analyses of these surveys continue to ignore the deficiencies of indexes and treat them as though they were censuses (e.g., Bart et al. 2003).

Discussion of appropriate use of indexes should be structured by the intuitive principle that any inference from survey data should be directed at N and not at C. Unfortunately, naive analyses of survey data using raw counts often make inferences about C. If additional information has been collected from the survey (e.g., through capture-recapture or distance methods), inference based on N is straightforward using standard analysis procedures such as those provided in programs MARK (White and Burnham 1999) or DISTANCE (Buckland et al. 2001). If estimation of detection rates suggests that index assumptions are valid (e.g., detection probabilities are constant), investigators then can choose to conduct analyses based on counts (Skalski and Robson 1992). In contrast, if no information exists on p, investigators should conduct a model-based analysis in which factors that influence p can be controlled, such as through sightability models (Samuel et al. 1987) or hierarchical models (Link and Sauer 2002). Often, however, investigators do not have sightability information or observable covariates for factors that influence detectability, which is problematic because detectability is frequently influenced by uncontrollable factors (Anderson 2001). In these cases, simulation provides the only means for evaluating how well C performs as an index of N. By hypothesizing the ways that p varies in time and space, simulations allow us to evaluate actual hypotheses tests based on N, in the context of the observed C. In effect, the simulation model provides an analysis of the sensitivity of any inference to potential variation in p and thus should be an essential component of any analysis of count indexes.

A simple formulation of treatment-specific detection probabilities sometimes permits one to assess importance of hypothesized variation in detection probabilities in conjunction with inference based on count indexes. Sauer and Link (2004) described such an approach in the context of comparing mean count indexes between treatments. However, computer simulations permit more detailed evaluation of the value of count indexes relative to a variety of alternative models with varying assumptions about temporal and spatial variation in detection probabilities. Simulations have become important as a tool for enhancing the credibility of count indexes or for pointing out the need for estimation of detection rates (Hilborn et al. 1976; Efford 1992; McKelvey and Pearson 2001).

In this chapter, we describe essential components for simulations that evaluate a capture index for a low-abundance, small-mammal population. Small-mammal studies provide a convenient framework for such a discussion because the question of whether to base population analysis on count indexes or capture-recapture population estimators that incorporate capture probabilities has received a great deal of attention (Hilborn et al. 1976; Jolly and Dickson 1983; Nichols and Pollock 1983; Nichols 1986; Seber 1986; Pollock et al. 1990; Hallett et al. 1991; Efford 1992; Skalski and Robson 1992; Rosenberg et al. 1995; Slade and Blair 2000; McKelvey and Pearson 2001). Most of the investigators who have addressed this question have concluded that model-based estimators are preferable for abundance estimation. Arguments for using counts have relied on comparisons between capture-recapture estimators and count-based statistics, either through simulation (Hilborn et al. 1976; McKelvey and Pearson 2001) or through correlative analyses using real data (Slade and Blair 2000). Hilborn et al. (1976), Slade and Blair (2000), and McKelvey and Pearson (2001) concluded that under some circumstances, count statistics perform similarly to capture-recapture estimators in diagnosing population changes. However, these conclusions were based on restrictive assumptions about the range of true capture probabilities, as well as the number of capture occasions per sampling period (Jolly and Dickson 1983), and in some cases the simulations were shown to be erroneous (Efford 1992).

We follow the approach to simulation described by McKelvey and Pearson (2001), but extend their simulations, focusing on a few essential components that we view as necessary to provide a comprehensive evaluation of how the capture index performs in the context of possible variation in p over treatments. In these comparisons, issues of model selection become important, and we describe several aspects of model selection that were not evaluated by McKelvey and Pearson (2001).

Count Indexes and Small-Mammal Populations

McKelvey and Pearson (2001) used simulation to compare performance of the count statistic M_{t+1} (number of unique individuals captured or minimum number alive [Otis et al. 1978]) relative to capture-recapture estimators for discerning changes in population size when sample sizes were small. They used program CAPTURE (Otis et al. 1978; White et al. 1982; Rexstad and Burnham 1991) to compute estimates of population size for three closed population, mark-recapture models: (1) the null model for constant capture probability (Otis et al. 1978); (2) the model for individual heterogeneity using the jackknife estimator of Burnham and Overton (1978, 1979); and (3) a time and heterogeneity model with estimation based on sample coverage (Chao et al. 1992; Lee and Chao 1994). They calculated population estimates, coefficients of variation (CV), and the percentage of time that each "default" estimator successfully discerned a 25% population decline between two time periods (from $N_1 = 50$ to $N_2 = 37$) for each of these models and then compared these statistics to the count index, M_{t+1}. Mark-recapture estimators clearly performed better when the same model was used to generate encounter histories as was used to estimate population size. However, the M_{t+1} statistic performed better when the number of simulated sources of variation in capture probabilities (time, individual heterogeneity, or behavioral response following first capture) was allowed to randomly vary between the two successive sample periods or when behavioral response to trapping was included in the population attributes of the generating model.

To provide a comprehensive simulation of the consequences of variation in p, we made several modifications in the simulation design used by McKelvey and Pearson (2001):

1) We allowed the expected value of capture probabilities in each period to differ. McKelvey and Pearson (2001) allowed underlying capture probabilities to vary between sampling periods but imposed constant expected value of capture probabilities for each period. Systematic variation in capture probability associated with treatments is perhaps the primary concern associated with population (or count) indexes and is a necessary component of any simulation study of indexes.
2) We generalized the analysis to permit model selection. McKelvey and Pearson (2001) relied primarily on comparisons between M_{t+1} and default capture-recapture estimators in program CAPTURE. Availability of a variety of new model selection procedures may obviate the need to use default capture-recapture estimators, leading to selection of models tailored to specific sampling scenarios.
3) We considered a wide range of simulated detection probabilities to ensure realistic simulations. In McKelvey and Pearson (2001), the relatively high values for expected capture probabilities (0.45) and number of trapping periods per session (4) led to a situation in which the total number of animals uniquely identified per session approached the true population size. Realistic simulations of the capture index must accommodate a range of possible detection rates to ensure that M_{t+1} is not a census. This consideration reflects the need for better information on detection probabilities to permit better simulation of count indexes.

Expected Probability of Capture

McKelvey and Pearson (2001) allowed capture probabilities to vary randomly between consecutive sampling periods when determining the proportion of simulations for which various estimators were able to detect a 25% population decline (ordinal ranking simulations). However, the expected probability of capture was the same during each period. It has long been known that if capture probabilities are the same between two consecutive periods, then the ratio of count statistics provides a good estimate of population increase or decrease with a greater degree of precision than would be possible by using mark-recapture estimators (Skalski and Robson 1992). Thus, simulations based on equal expected capture probabilities conform to the usual conditions for which the performance of count indexes is satisfactory.

Realistic simulations of variation in capture histories must contain variation in expected probability of capture. For instance, if capture probability covaried with population size, then the expected value of capture probabilities will differ between two consecutive sampling periods in which there was a population decline. This might be expected, for example, in cases of trap saturation at high population densities or if the attractiveness of baited traps to individual animals increases in times of limited resources. Capture probabilities could also be highly correlated with environmental and habitat conditions likely to affect population size.

In contrast to the design of McKelvey and Pearson (2001), we conducted ordinal ranking simulations with varying expected probabilities of capture. We simulated capture histories for population sizes of $N = 50$ and $N = 37$ to reflect a 25% population decline between two time periods and used a variety of possible population attributes (heterogeneity, time effects, behavioral response to trapping, and combinations thereof) that were assumed fixed for both populations. Levels of these effects are detailed in McKelvey and Pearson (2001). We allowed the expected probability of capture, $E(p)$, to increase from 0.45 in the $N = 50$ population to $0.45 + *$ in the $N = 37$ population, where $* = 0.05$, 0.10, or 0.20. We then calculated the percentage of 2,000 simulations for which the null, jackknife, or M_{t+1} estimators were smaller for the $N = 37$ population than for the $N = 50$ population. We did not examine the performance of Chao's sample coverage estimator (Chao et al. 1992; Lee and Chao 1994). In all but one case, the jackknife estimator predicted population declines of 25% equally or better than M_{t+1} (Table 4.1). The performance of M_{t+1} was particularly poor when true capture probabilities were heterogeneous.

Although we have followed McKelvey and Pearson (2001) with respect to the metric for comparing our results, it is important to recognize that ordinal ranking comparisons are a very simplistic way to evaluate the performance of population estimators. Count statistics may perform even more poorly when the quantity under investigation is a ratio of population sizes such as relative abundance or rate of change, a natural metric reflecting change (Skalski and Robson 1992). Note that in these simulations we used the same generating model for each sampling period. McKelvey and Pearson (2001) reported better relative performance of M_{t+1} when the generating model was allowed to vary randomly from year to year. Although the random addition of time-based variation between primary sampling

Table 4.1.

*Proportion of 2,000 ordinal-ranking simulations for which a population decline of 25% (from $N_1 = 50$ to $N_2 = 37$) was detected by each of three types of default estimators when the expected probability of capture changed from 0.45 to 0.45 + *. The simulated sampling model could include time effects (M_t), trap-shy behavior (M_b), individual heterogeneity (M_h), or any combination thereof. In addition, the null model (M_0) assumed constant capture probability within individual sampling periods. Levels of each of these effects are described in McKelvey and Pearson (2001).*

Simulated model	* = 0.05			* = 0.10			* = 0.20		
	Null	Jackknife	M_{t+1}	Null	Jackknife	M_{t+1}	Null	Jackknife	M_{t+1}
M_0	0.995	0.989	0.983	1.000	0.993	0.983	1.000	1.000	0.969
M_t	0.994	0.980	0.973	0.994	0.987	0.959	1.000	0.998	0.954
M_b	0.886	0.977	0.977	0.929	0.981	0.969	0.979	0.991	0.962
M_h	0.971	0.962	0.925	0.970	0.965	0.911	0.951	0.971	0.848
M_{bh}	0.890	0.957	0.932	0.906	0.943	0.917	0.957	0.964	0.863
M_{th}	0.970	0.953	0.915	0.970	0.971	0.909	0.961	0.969	0.875
M_{tb}	0.942	0.960	0.975	0.975	0.971	0.966	0.993	0.985	0.956
M_{tbh}	0.863	0.935	0.920	0.889	0.940	0.906	0.932	0.948	0.855

periods may have been reasonable, we doubted if random additions and deletions of extremely high levels of behavioral response and heterogeneity made as much biological sense.

Model Selection

It is well known that model misspecification in capture-recapture estimation can lead to biased estimators. Consequently, analyses that misspecify model structures produce erroneous comparisons of count indexes and capture-recapture estimators. This is evidenced in McKelvey and Pearson's (2001) simulations, in which all three "default" estimators they considered (null, jackknife, and C_{th}) were highly biased estimators when applied to simulated populations that exhibited behavioral responses to trapping (M_b, M_{tb}, M_{bh}, M_{tbh}). In contrast, the jackknife and null estimators exhibited less bias and had lower or similar CVs compared to the count index when the data-generating model lacked behavioral effects.

We contend that under the levels of behavioral response simulated by McKelvey and Pearson, their default estimators would rarely be suggested by model selection procedures in either program CAPTURE or program MARK (White and Burnham 1999). More appropriate models, recommended by model selection procedures, are likely to be less biased and equally or more precise estimators than the count index.

To illustrate this point, we used 10 randomly selected simulated data sets for each of the population scenarios containing a behavioral response (M_b, M_{tb}, M_{bh}, M_{tbh}). For each data set, we allowed model selection procedures in program CAPTURE to select the most appropriate model given the data. Table 4.2 presents the model-selection results and corresponding average population estimates based on model M_b in program CAPTURE (Zippin 1956). An estimator including behavioral effects was selected a high proportion of the time and produced reasonable estimates of population size.

Because model selection procedures in program CAPTURE often perform poorly with sparse data (Menkens and Anderson 1988), we analyzed the same data sets using the closed captures procedure in program MARK (White and Burnham 1999), where AIC_c values could be used to evaluate relative model performance using a subset of four closed-population models (M_b, M_{tb}, M_{bh}, M_{tbh}). We considered all models with $\Delta AIC_c < 2.0$ to be appropriate for a given data set (Burnham and Anderson 1998) (Table 4.3). An estimator containing a behavioral effect was selected in most scenarios where a behavioral effect was included in the data-generating model. Program MARK seemed to do a better job of identifying behavioral effects than program CAPTURE, although sample sizes were admittedly small.

Several developments in model selection have been made in recent years that deserve further consideration when investigators are interested in population estimation and population sizes are small. Stanley and Burnham (1998) reported that a model-averaging approach (Buckland et al. 1997; Burnham and Anderson 1998), combined with a new classifier for program CAPTURE, increased performance in the ability to select an appropriate closed, capture-recapture model. More recently, the development of heterogeneity models based on finite-mixture distributions (Norris and Pollock 1995, 1996; Pledger 1999, 2000) has allowed the fitting of all eight closed capture-recapture within a generalized likelihood framework (Pledger 1999, 2000). This advance has already been included in program MARK (White and Burnham 1999) and is readily available for use by biologists. Model selection is accomplished through likelihood ratio tests or AIC (see Pledger 1999, 2000), and model averaging within this framework also is possible. Some of the heterogeneity estimators often were not avail-

Table 4.2.

Average population estimates ($N = 50$) and selection proportions for default estimators when the generating model contained a behavioral effect (models M_b, M_{bh}, M_{tb}, M_{tbh}). For each generating model, the summaries shown are derived from 10 data sets and the proportion chosen represents the proportion of data sets for which the designated model was chosen "best" via model selection procedures in program CAPTURE. The numbers in parentheses are percent CV, which were derived from simulation results.

	Capture-recapture estimator								
	Null		Jackknife		C_{th}		Behavioral		M_{t+1}
Model	Propn. Chosen	$\hat{N}$(CV)	Propn. Chosen	$\hat{N}$(CV)	Propn. Chosen	$\hat{N}$(CV)	Propn. Chosen	$\hat{N}$(CV)	$\hat{N}$(CV)
M_b	0.0	83.10 (13.14)	0.0	84.20 (10.26)	0.0	106.00 (51.86)	0.4[a]	47.40 (5.09)	44.20 (8.32)
M_{bh}	0.0	84.10 (30.20)	0.0	84.30 (9.12)	0.0	96.30 (29.88)	0.8[a]	45.80 (5.99)	43.70 (7.86)
M_{tb}	0.4	63.40 (18.73)	0.0	72.60 (15.92)	0.0	85.30 (57.92)	0.2[b]	50.10 (12.25)	45.00 (8.51)
M_{tbh}	0.2	84.10 (29.64)	0.0	77.90 (12.70)	0.0	101.40 (66.39)	0.5[a]	45.80 (14.29)	40.80 (11.25)

[a] Remaining models chosen all include behavioral effects (M_{tb}, M_{bh}, M_{tbh}).
[b] Remaining models chosen include M_{tb} (0.2) and M_{th} (0.2).

Table 4.3.

Proportion of 10 data sets for which different closed-population models (M_o, M_b, M_t, M_{tb}) were considered appropriate by AIC_c values (calculated in program MARK) and the "best" model selected by program CAPTURE (i.e., the model with the value closest to 1.0). We did not consider models containing heterogeneity.

	MARK model selection				CAPTURE model selection			
Generating model	M_o	M_b	M_t	M_{tb}	M_o	M_b	M_t	M_{tb}
M_b	0.1	0.7	0.0	0.3	0.2	0.4	0.0	0.2
M_{bh}	0.0	0.8	0.1	0.5	0.0	0.8	0.0	0.2
M_{tb}	0.4	0.6	0.1	0.5	0.4	0.2	0.0	0.4
M_{tbh}	0.2	0.6	0.0	0.7	0.3	0.6	0.0	0.1

able in program MARK for data sets of McKelvey and Pearson (2001), but this may have been an artifact of the process in which encounter histories were created. In contrast, Pledger (2000) was able to fit most models to data sets of pocket mice (*Perognathus parvus*), cottontail rabbits (*Sylvilagus floridanus*), and deer mice (*Peromyscus maniculatus*) with similar population sizes (approximately N = 62, 135, and 46, respectively). The possibility of fitting all closed models in program MARK is therefore likely dependent on the quality of the data set but is not just limited to large data sets. We recommend that future work be directed at the investigation of questions of model selection and model averaging when not all models in the model set are estimable.

Initial Conditions and Sampling Design

In real world situations, the true range of capture probabilities is never known; instead, it must be estimated via statistical models. As a simplified example, we consider the case in which only one trapping session is conducted for each of two populations. In the absence of trapping response and heterogeneity, it may be reasonable to assume that the distribution of the two counts is C_1 ~ binomial (N_1, p_1) and C_2 ~ binomial (N_2, p_2), where C_1 and C_2 are the counts, N_1 and N_2 are the population sizes, and p_1 and p_2 are the underlying probabilities of detection for populations 1 and 2, respectively. In this context, we define population as any two groups of individ-

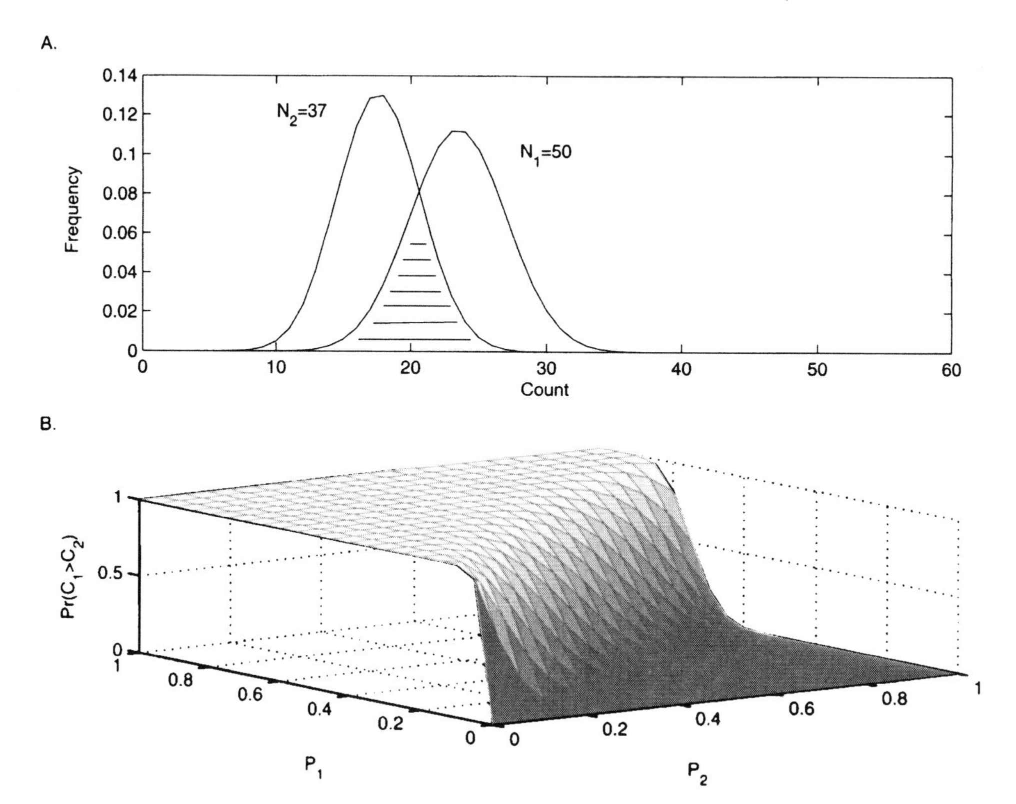

Figure 4.1. (A) Distribution of counts resulting from a binomial probability model with $p_1 = p_2 = 0.45$, $N_1 = 50$, and $N_2 = 37$. The barred area shows the probability that an incorrect inference about population decline will be made when using one count per sampling period. The probability of making an incorrect inference about these two populations is further developed in (B), which shows the probability of a correct inference about population decline for any two probabilities of detection when using one count per sampling period. The probability of a correct inference is defined as the probability that the count for population 1 (C_1) is larger than that for population 2 (C_2).

uals separated by time, space, or treatment. If we let $N_1 = 50$ and $N_2 = 37$ to reflect a 25% population decline, then the probability of diagnosing the population decline depends critically on the underlying capture probabilities (Figure 4.1). If more trapping occasions had been included, then the probability distribution for each of the counts would be a sum of conditional binomials, and the region of erroneous inference in Figure 4.1 would shrink. However, the ability of counts to diagnose population decline still would be a function of unknown capture probabilities.

We again note that the range of values for average capture probabilities coupled with the number of trapping occasions used in McKelvey and Pearson's (2001) simulations led to a condition in which the expected number of animals captured per primary sampling period approached the true population size. Bias from using count statistics as surrogates for population size decreases under these conditions (Hilborn et al. 1976; Jolly and Dickson 1983).

Further Considerations and Recommendations

Low-abundance species pose many difficulties for sampling. Issues associated with model selection, variation in detection probabilities, and difficulty in obtaining reasonable pilot data on detection probabilities have enticed many investigators to use count indexes such as M_{t+1} in lieu of more robust estimators that incorporate detection probabilities. Simulations that provide a superficial view of both the estimation procedures (through model selection) and the possibilities for differences in detection probabilities over time can further obfuscate the role of detectability in comparing count indexes over time. In our view, it is simply not defensible to use observed count statistics without first drawing inference about comparative capture (or detection) probabilities. Simulations can be used to define the magnitude of differences in detection probabilities among treatments that would compromise inference based on count indexes. Often, these simulations have heuristic value in establishing the credibility of inferences based on counts. However, in the absence of actual information on magnitude and variation in detection probabilities, it is difficult to assess whether count indexes provide misleading information about population trends.

Skalski et al. (1983) and Skalski and Robson (1992) provided a hypothesis-testing approach for evaluating the equal catchability of individuals in two populations when each population was sampled twice. Similarly, MacKenzie and Kendall (2002) developed methods under the same sampling situation but explored the use of hypothesis testing, equivalence testing, and model averaging for informing inference about equal detectability and estimation of relative abundances. When the number of trapping occasions exceeds two for each population or when there are more than two populations being compared over space or time, it is necessary to generalize the methods of Skalski et al. (1983), Skalski and Robson (1992), and MacKenzie and Kendall (2002). In these cases, one must show that the probability of being captured at least once during the primary sampling period is roughly the same for each population being compared.

In conclusion, we caution against the use of count statistics as estimators of population size unless the probability of being captured at least once in a primary sampling period is roughly the same for all populations being compared. Although simulations can provide insights into consequences of differences in detection probabilities, estimation of detection probabilities and an understanding of temporal and spatial variation in detection probabilities are generally needed to evaluate the simulation results. Such inferences about capture probabilities can be accomplished through fitting different statistical models to the data. Inferences based on raw counts are likely to be erroneous when there are trends in capture probability over time or when capture probability covaries with population size or some other characteristic of the populations being compared. Hence, these situations should be a component of any simulation that assesses the value of count indexes. For small-mammal population studies, using counts to compare across studies, sites, time, species, or trapping methods is particularly vulnerable to misleading results; thus, standardization of capture-recapture modeling as the de facto method of analysis for investigating questions about population size may improve the cohesion of the literature as a whole. We urge biologists to make use of the new model selection capabilities for closed models in program MARK (White and Burnham 1999) in addition to those included in program CAPTURE and, when possible, to make use of a priori knowledge about the species in question to guide selection of an appropriate statistical model.

Summary

Controversy still exists in the wildlife literature regarding use of count-based indexes to abundance for monitoring wildlife populations even though a variety of statistical methods have been developed for estimation and modeling of detection rates. For small populations and low-abundance species, it is often argued that estimation of detection probabilities is not feasible or adds little value to index data. Unfortunately, indexes often are not credible sources of information for management decisions because critics can easily construct scenarios in which the index assumption of consistency in detection probabilities is not valid.

Performance of count-based indexes for evaluating change in populations depends critically on underlying probabilities of detection, and any reasonable analyses of count-based indexes must demonstrate that variation in detection probabilities is not the source of variation in count indexes. Although such indexes may perform better than estimators that incorporate detection rates under some conditions, these conditions are not known a priori and therefore must be validated through procedures such as capture-recapture or distance sampling. If the researcher can demonstrate through these methods that capture probabilities are relatively constant over time and space or if the probability of detection of individuals within sampling periods can be shown to approach one, then count statistics may outperform capture-recapture estimators in correctly informing inference about population changes. In the absence of estimated detection probabilities, simulations must be used to document whether the index assumptions are likely to be valid.

Acknowledgments

We would like to thank Jim Nichols for inspiration and guidance in the formulation of this article. Bill Kendall, Michael Runge, and Marc Kéry also provided comments and discussion that greatly improved the quality of this paper.

References

Anderson, D. R. 2001. The need to get the basics right in wildlife field studies. *Wildlife Society Bulletin* 29:1294–1297.

Barker, R. J., and J. R. Sauer. 1995. Statistical aspects of point count sampling. pp. 125–130 in C. J. Ralph, J. R. Sauer, and S. Droege, eds., *Monitoring Bird Populations by Point Counts*. USDA Forest Service General Technical Report PSW-GTR-149, Pacific Southwest Research Station, Albany, California.

Bart, J., B. Collins, and R. I. G. Morrison. 2003. Estimating population trends with a linear model. *Condor* 105:367–372.

Buckland, S. T., D. R. Anderson, K. P. Burnham, J. L. Laake, D. L. Borchers, and L. Thomas. 2001. *Introduction to Distance Sampling*. Oxford University Press, Oxford.

Buckland, S. T., K. P. Burnham, and N. H. Augustin. 1997. Model selection: An integral part of inference. *Biometrics* 53:603–618.

Burnham, K. P., and D. R. Anderson. 1998. *Model Selection and Inference: A Practical Information-Theoretic Approach*. Springer-Verlag, New York.

Burnham, K. P., and W. S. Overton. 1978. Estimation of the size of a closed population when capture probabilities vary among animals. *Biometrika* 65:625–633.

______. 1979. Robust estimation of population size when capture probabilities vary among animals. *Ecology* 60:927–936.

Chao, A., S.-M. Lee, and S.-L. Jeng. 1992. Estimating population size for capture recapture data when capture probabilities vary by time and individual. *Biometrics* 48:201–216.

Efford, M. 1992. Comment—Revised estimates of the bias in the 'minimum number alive' estimator. *Canadian Journal of Zoology* 70:628–631.

Hallett, J. G., M. A. O'Connell, G. D. Sanders, and J. Seidensticker. 1991. Comparison of population estimators for medium-sized mammals. *Journal of Wildlife Management* 55:81–93.
Hilborn, R., J. A. Redfield, and C. J. Krebs. 1976. On the reliability of enumeration for mark and recapture census of voles. *Canadian Journal of Zoology* 54:1019–1024.
Hutto, R. L., and J. S. Young. 2002. Regional landbird monitoring: Perspectives from the northern Rocky Mountains. *Wildlife Society Bulletin* 30:738–750.
Jolly, G. M., and J. M. Dickson. 1983. The problem of unequal catchability in mark recapture estimation of small mammal populations. *Canadian Journal of Zoology* 61:922–927.
Lee, S.-M., and A. Chao. 1994. Estimating population size via sample coverage for closed capture-recapture models. *Biometrics* 50:88–97.
Link, W. A., and J. R. Sauer. 2002. A hierarchical analysis of population change with application to Cerulean Warblers. *Ecology* 83:2832-2840.
MacKenzie, D. I., and W. L. Kendall. 2002. How should detection probability be incorporated into estimates of relative abundance? *Ecology* 83:2387–2393.
McKelvey, K. S., and D. E. Pearson. 2001. Population estimation with sparse data: The role of estimators versus indices revisited. *Canadian Journal of Zoology* 79:1754–1765.
Menkens, G. E., and S. H. Anderson. 1988. Estimation of small-mammal population size. *Ecology* 69:1952–1959.
Nichols, J. D. 1986. On the use of enumeration estimators for inter-specific comparisons, with comments on a "trappability" estimator. *Journal of Mammalogy* 67:590–593.
Nichols, J. D., and K. H. Pollock. 1983. Estimation methodology in contemporary small mammal capture-recapture studies. *Journal of Mammalogy* 64:253–260.
Norris, J. L., and K. H. Pollock. 1995. A capture-recapture model with heterogeneity and behavioral response. *Journal of Environmental Ecology and Statistics* 2:305–313.
_____. 1996. Nonparametric MLE under two closed capture-recapture models with heterogeneity. *Biometrics* 52:639–649.
Otis, D. L, K. P. Burnham, G. C. White, and D. R. Anderson. 1978. Statistical inference from capture data on closed animal populations. *Wildlife Monographs* 62:1–135.
Pledger, S. A. 1999. *Finite Mixtures in Capture-recapture Models*. Ph.D. dissertation, Victoria University of Wellington, New Zealand.
_____. 2000. Unified maximum likelihood estimates for closed capture-recapture models using mixtures. *Biometrics* 56:434–442.
Pollock, K. H., J. D. Nichols, C. Brownie, and J. E. Hines. 1990. Statistical inference for capture recapture experiments. *Wildlife Monographs* 107:1–97.
Rexstad, E., and K. P. Burnham. 1991. *User's Guide for Interactive Program CAPTURE: Abundance Estimation of Closed Animal Populations*. Colorado Cooperative Fish and Wildlife Research Unit, Colorado State University, Fort Collins.
Robbins, C. S., D. Bystrak, and P. H. Geissler. 1986. *The Breeding Bird Survey: Its First Fifteen Years, 1965–1979*. U.S. Fish and Wildlife Service Resource Publication 157, U.S. Fish & Wildlife Service, Washington, D.C.
Rosenberg, D. K., W. S. Overton, and R. G. Anthony. 1995. Estimation of animal abundance when capture probabilities are low and heterogeneous. *Journal of Wildlife Management* 59:252–261.

Samuel, M. D., E. O. Garton, M. W. Schlegel, and R. G. Carson. 1987. Visibility bias in aerial surveys of elk in northwestern Idaho. *Journal of Wildlife Management* 51:622–630.

Sauer, J. R., and W. A. Link. 2004. Some consequences of using counts of birds banded as indices to populations. *Studies in Avian Biology*. In press.

Seber, G. A. 1986. A review of estimating animal abundance. *Biometrics* 42:267–292.

Skalski, J. R., and D. S. Robson. 1992. *Techniques for Wildlife Investigations*. Academic Press, San Diego.

Skalski, J. R., D. S. Robson, and M. A. Simmons. 1983. Comparative census procedures using single mark-recapture methods. *Ecology* 64:752–760.

Slade, N. A., and S. M. Blair. 2000. An empirical test of using counts of individuals captured as indices of population size. *Journal of Mammalogy* 81:1035–1045.

Stanley, T. R., and K. P. Burnham. 1998. Estimator selection for closed-population capture recapture. *Journal of Agricultural, Biological, and Environmental Statistics* 3:131–150.

White, G. C., D. R. Anderson, K. P. Burnham, and D. L. Otis. 1982. *Capture-recapture and Removal Methods for Sampling Closed Populations*. U.S. DOE Los Alamos National Laboratory Report Number LA-8787-NERP, Los Alamos, New Mexico.

White, G. C., and K. P. Burnham. 1999. Program MARK: Survival estimation from populations of marked animals. *Bird Study* 46:S120–S139.

Williams, B. K., J. D. Nichols, and M. J. Conroy. 2002. *Analysis and Management of Animal Populations*. Academic Press, New York.

Zippin, C. 1956. An evaluation of the removal method of estimating animal populations. *Biometrics* 12:163–189.

Part II

SAMPLING DESIGNS FOR RARE SPECIES AND POPULATIONS

Part II of this volume describes sampling designs for efficiently estimating abundance of rare species and populations, with a greater emphasis on the first stage of a two-stage design, that is, probability-based selection of sampling units. Consequently, Part II serves as a starting point when developing a design for sampling rare or elusive species.

Adaptive cluster sampling is a relatively recent design for sampling individuals that are spatially clustered within a relatively small portion of a study area. Although much recent work has been devoted to its theoretical development, Smith et al. (Chapter 5) venture beyond theory to provide an evaluation of adaptive cluster sampling when applied to real populations. They describe practical examples of this design and offer guidelines for its usage. In the next chapter, Manly describes an alternative adaptive approach to estimating abundance of rare species based on a two-phase, stratified sampling regime. Christman completes this section by reviewing the sequential sampling design, another form of adaptive sampling, and uses it to estimate abundance of a waterfowl population.

5

Application of Adaptive Sampling to Biological Populations

David R. Smith, Jennifer A. Brown, and Nancy C. H. Lo

Adaptive sampling is appealing because it mimics how biologists would like to collect data—at least more so than most statistical sampling techniques. When adaptively sampling, biologists search for a species of interest at predetermined locations, and if the species is found, searching continues nearby. This procedure usually produces a biased sample when applied to spatially clustered species because occupied habitat will be sampled disproportionately. Fortunately, Thompson (1990 and subsequent papers) showed how unbiased estimators of density and abundance could be obtained by following the adaptive cluster sampling procedure.

Additional appeal of adaptive cluster sampling can be attributed to its statistical properties. For species that tend to be rare and spatially clustered, adaptive cluster sampling has the potential to be efficient; that is, it can result in estimators of population density or abundance with smaller variance than conventional sampling methods for equal effort (Roesch 1993; Brown 1994; Smith et al. 1995; Christman 1997; Lo et al. 1997). Getting reliable information on rare and spatially clustered species can be challenging and costly, so any increase in reliability of information or reduction in survey cost is welcome and desirable.

It must be noted, however, that a theoretical increase in efficiency alone is not sufficient motivation to move a statistical method from theory to practice. There is little gain if a method, no matter how promising, is not put into practice. Paraphrasing John Tukey (1986:97), practical efficiency is equal to the statistical efficiency of a method times the probability that the method will be put into practice.

Despite its promise, there have been few real applications of adaptive cluster sampling in the published literature. By far, studies have focused on theoretical considerations and simulation studies (e.g., see Thompson 2003 and articles therein). The reason for this pattern, we believe, is two-fold. First, the methodology is still relatively young, and it takes time for new methods to work their way into common practice. Second, a set of challenges tempers the appeal of adaptive cluster sampling. In our view, challenges when applying adaptive cluster sampling include:

1) Increased efficiency is not guaranteed, and in fact efficiency depends critically on the spatial distribution of the target population.
2) The final sample size is random and, as such, not known prior to the survey.
3) Data collection can be complex under field situations.
4) Adaptive sampling may need to be modified for mobile animals or sensitive species and habitats.

Our objective in this paper is to present the challenges that a biologist will face when considering the application of adaptive sampling, to offer suggestions for overcoming those challenges, and to highlight areas where method development will improve the practical efficiency of adaptive sampling. We first introduce adaptive sampling and review its literature. We then outline the main challenges that a biologist faces when putting adaptive sampling into practice. Afterward, we present case studies to illustrate the application of adaptive sampling to biological populations. Finally, we summarize and discuss future directions.

Application of Adaptive Cluster Sampling

Since Thompson (1990) introduced the adaptive cluster sampling design, a substantial literature has developed. The literature can be classified into three categories: modifications of the basic design; simulation studies to provide guidelines on effective application; and most recently, applications to real biological populations from which practical issues have emerged.

Introduction and Review of Adaptive Cluster Sampling

Adaptive cluster sampling, which was created by Thompson (1990), has a large number of possible designs. With the most basic adaptive sampling

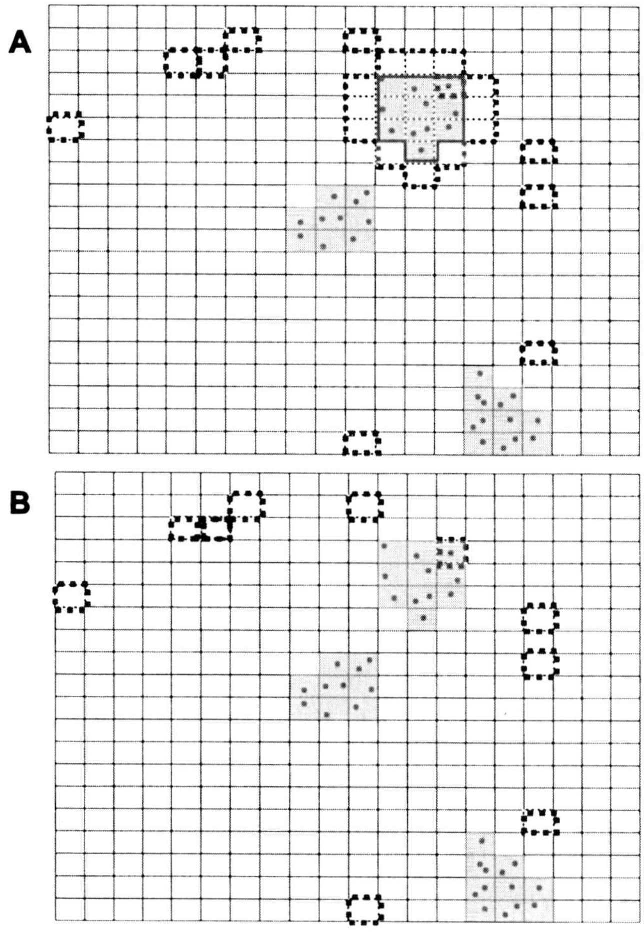

Figure 5.1. An example of adaptive cluster sampling. The study area consists of 400 units. The value within a unit is the count of black dots. There are three networks of black dots in the population. An initial simple random sample of 10 units is shown in (A). One of the initial units lands in a network of black dots, triggering adaptive sampling. The final sample of units is shown in (B).

design, initial sampling units are selected from a defined population according to a conventional probability-based sampling design. For example, the process could begin with simple random sampling without replacement as shown in Figure 5.1A. Observations in the initial sample units determine whether additional (adaptive) units are selected. If the observation in an initial unit meets some condition, adaptive units are selected in its neighborhood. The neighborhood of a unit, which includes itself and nearby units, can take a variety of shapes (Figure 5.2). However, a neighborhood must be symmetric in that if unit A is in the neighborhood of unit B then unit B is in the neighborhood of unit A.

For biological populations, the condition to adapt is typically based on the count of organisms exceeding some predefined level, such as a count > 0. In turn, if the count in an adaptive unit meets the condition then neighboring units are selected as long as the condition is met (Figure 5.1B). As a result of this adaptive process, clusters of sampling units are selected. Within each cluster there are units that meet the condition; this set of units is called a network. Because of neighborhood symmetry, if one unit in a network is selected, then all units in the network are selected. The number of units in a network is the network size. In addition, there are units in a cluster that do not meet the condition. These units are called edge units because in a cluster they define the network's edge. An initial unit that does not meet the condition is considered a network of size one.

There are several important choices to be made when implementing adaptive sampling. First, there is the initial sampling design, including sampling unit size and shape, selection scheme, and sample size. Second, there are choices that are particular to adaptive sampling, including the condition to determine whether adaptive units are selected and the configuration of the neighborhood. Typically, the condition to adapt is based on the observation of a unit meeting or exceeding a predetermined value. For example, a typical condition is to adapt if a species is present in a unit (i.e., count > 0). The neighborhood configuration is flexible (Figure 5.2), but a cross pattern (Figure 5.2B) is currently a common choice. Finally, there is a choice of estimators. Thompson (1990) derived two unbiased estimators, based on the Horvitz-Thompson and Hansen-Hurwitz estimators, for application to adaptive cluster sampling, and he showed how these estimators could be improved using the Rao-Blackwell method. In most compar-

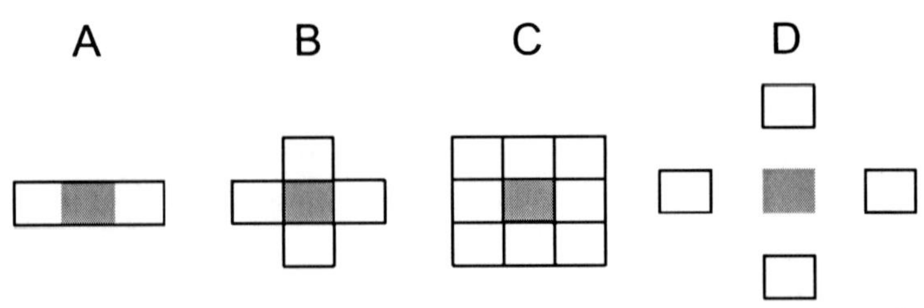

Figure 5.2. Possible neighborhood shapes for adaptive sampling. The gray-shaded unit represents a unit that meets the condition to adapt. The remaining units make up the gray unit's neighborhood.

isons, the Horvitz-Thompson estimator appears to be the superior choice (Salehi 1999, 2003).

The combination of these choices results in a wide variety of possible adaptive sampling designs. There has been considerable work on designs resulting from changes to the initial sampling design (see Thompson and Seber 1996 for an in-depth review of early work). Thompson developed adaptive sampling designs that use systematic and stratified sampling to select the initial sample (Thompson 1991a,b). Several authors have considered incorporating adaptive sampling in two-stage designs (Salehi and Seber 1997; Zhang et al. 2000; Muttlak and Kahn 2002; Christman 2003). Roesch (1993) and Pontius (1997) developed adaptive sampling designs that incorporate selection of the initial sample with probability proportional to size. Adaptive Latin square sampling was considered by Munholland and Borkowski (1996) and Borkowski (1999). Pollard and Buckland (1997) developed a strategy to combine adaptive sampling in a line transect survey. Palka and Pollard (1999) applied this strategy to survey harbor porpoise and found the strategy easy to implement and effective at reducing estimator variance.

Another area that has received considerable attention is the use of stopping rules to restrict adaptive sampling. Brown (1994) and Brown and Manly (1998) were the first to explore the performance of adaptive sampling with a stopping rule. The objective of Brown's stopping rule, which was triggered when a preset sample size was reached, was to limit the size of the final sample. Brown's stopping rule limited final sample size effectively but introduced some positive bias (in many cases) and the final sample size remained random (Salehi and Seber 2002). Lo et al. (1997) presented a restricted adaptive sampling strategy in which the stopping rule limited the amount of adaptive sampling per network. Lo and colleagues' strategy was applied to estimate Pacific hake larval (*Merluccius productus*) abundance, and they concluded that although the estimators were biased, restricted adaptive sampling resulted in a substantial variance reduction. Su and Quinn (2003) used simulation to evaluate a stopping rule that was similar to the one used by Lo et al. They added the stopping rule to an "adaptive sampling design with order statistics" and found that the magnitude of bias depended on the order statistic, stopping rule, and spatial distribution of the population.

Salehi and Seber (2002) presented unbiased estimators for designs similar to Brown's restricted adaptive cluster sampling design. They found in a simulated example that the unbiased estimators yielded smaller mean square errors (MSE) than biased estimators when sample size was small, but that the biased estimators had smaller MSE as sample size increased. Salehi and Seber (2002) developed stopping rules to reduce the final sample size. However, Christman and Lan (2001) looked at the problem from the perspective that conventional sampling might not generate a large enough sample or yield sufficient observations from a rare population. This prompted Christman and Lan (2001) to develop an inverse adaptive cluster sampling design that allows sampling until a predefined number of nonzero units are selected. Salehi and Seber (2004) presented an unbiased estimator for Christman and Lan's design and proposed a new design, the general inverse sampling design, that avoids selecting an unfeasibly large sample.

The issue of how to select an appropriate condition has received attention because the condition to adapt can have a profound effect on efficiency and expected final sample size (Smith et al. 1995; Brown 2003). If the condition is too liberal, adaptive sampling might be triggered too frequently and the expected final sample size might be excessively large—even too large to complete the planned survey. Conversely, if the condition is too restrictive, adaptive sampling might not be triggered at all and sampling might be insufficient to achieve desired precision. To help make an effective choice, Thompson (1996) suggested basing the condition on the order statistics from the initial sample. This approach, which Su and Quinn (2003) labeled *acsord* for "adaptive cluster sampling with order statistics," is feasible when the initial sample can be selected in its entirety prior to selecting the adaptive units. After the initial units are sampled, the observations are ordered, and the condition is set equal to the rth order statistic (Su and Quinn 2003). Of course, there remains the challenge of choosing the appropriate order statistic. Su and Quinn (2003) conducted extensive simulations of *acsord* sampling of five populations with various degrees of aggregation. They found that efficiency and relative bias were determined by the interaction between population characteristics, initial sample size, stopping rule, order statistic, and estimator used. Similarly, other simulation studies have shown complex effects of population characteristics and design factors on the efficiency of adaptive sampling (Smith et al. 1995; Christman 1997; Brown 2003).

Thompson (1994) identified analytically the factors that affect the efficiency of adaptive cluster sampling. There have been additional simulation studies to determine what affects efficiency (Brown 1994; Smith et al. 1995; Christman 1997; Brown 2003). In general, factors affecting efficiency fall into categories of population distribution, relative and absolute size of the expected final sample, unit size, or per unit sampling cost. For adaptive cluster sampling to even have a chance of being efficient the population must be geographically rare (meaning that organisms should occupy a small percentage of the units in the population) and the spatial distribution of the population should be highly aggregated or clustered. Although bias is not dependent on spatial distribution, efficiency is. Thus, prior knowledge of spatial distribution of the population is important when deciding whether to apply adaptive sampling. Brown (2003) showed how neighborhood definition and condition to adapt can affect the spatial characteristics of the population and improve the efficiency of adaptive sampling.

The closeness of the final and initial sample sizes is another important factor that determines efficiency. Analytically and empirically it has been shown that efficiency tends to be high when the expected final sample size is a small percentage increase of the initial sample size (Thompson 1994; Smith et al. 1995; Brown 2003). However, Christman (1997) noted that high efficiencies often are reached only when the expected final sample size is a large proportion of the population. This has implications on whether adaptive cluster sampling is practical in real applications. By necessity, sample sizes in real applications are a small proportion of the population, so achieving the high efficiencies observed in simulated studies might not be possible in real applications.

Thompson (1994) also noted that efficiency could be increased when per-unit sampling costs are taken into account. Adaptive sampling will tend to be efficient when travel cost to sampling units is high, so that cost of sampling neighboring units is less than sampling units at random, and when the cost of making observations on units not meeting the condition is less than on units that do meet the condition. This latter condition could be satisfied when the condition is based on an auxiliary variable that can be inexpensively measured—for example, by basing the condition on a rapid assessment or catch per unit effort (CPUE).

Additional work has focused on application of adaptive sampling when there are multiple variables of interest (Thompson 1993; di Battista 2002; Dryver 2003; Smith et al. 2003), when observations are incomplete and detectability is an issue (Thompson and Seber 1994), and when the objective of the survey is to describe spatial distribution or spatial prediction (Hanselman et al. 2001; Curriero et al. 2002; Chapter 14, this volume). Although di Battista (2002) found that adaptive sampling resulted in lower MSE for estimates of diversity for clustered populations compared to simple random sampling, he noted a bias and devised a jackknife procedure to reduce the bias. Dryver (2003) cautioned that the performance of adaptive cluster sampling depends on the covariance between species when the condition is based on one species but abundance of another species is being estimated. In multispecies assemblages of freshwater mussels, Smith et al. (2003) found that the probability of detecting rare species was greater in adaptive units. Thompson and Seber (1994) presented a general approach for incorporating detectability estimates. Pollard and Buckland (1997) combined adaptive sampling with distance sampling methods to account for imperfect detectability. The effect of preferential sampling on kriging methods was examined by Curriero et al. (2002), who used adaptive cluster sampling as an example of preferential sampling. Hanselman et al. (2001) used variograms to assess degree of spatial clustering and gauge the likely efficiency of adaptive cluster sampling of rockfish in the Gulf of Alaska.

Aside from adaptive cluster sampling, adaptive allocation sampling (Thompson et al. 1992) is an attractive alternative in that observations can be used to stratify the survey area even when prior knowledge of within-strata variances and means is unavailable. Various sampling designs are available under adaptive allocation (Thompson and Seber 1996). Some designs require two passes over the population area (cf. double sampling or two-phase sampling), whereas others require one pass where the level of sampling in a particular stratum depends on observations in the previous stratum. Estimators can be either design-based or model-based (Thompson et al. 1992).

The list of applications of adaptive sampling to real biological populations is small, but it is growing and covers a diverse set of taxa. It took several years from the introduction of adaptive cluster sampling (Thompson 1990) before the first real applications to biological populations were reported (Lo et al. 1997; Strayer et al. 1997; Woodby 1998). Up to that point applications to biological populations were based on simulations (Roesch 1993; Brown 1994; Smith et al. 1995; Christman 1997).

Lo et al. (1997) applied a restricted adaptive sampling design to estimate Pacific hake larval abundance. They concluded that their adaptive sampling scheme was easy to implement and resulted in a more precise estimator than a conventional alternative. Also, Lo et al. (1997) noted that adaptive sampling provided information on patch size, which is an interesting biological characteristic.

Strayer et al. (1997) applied adaptive cluster sampling to survey of freshwater mussels (Unionidae). Their design was essentially that presented by Thompson (1990) with simple random sampling of 0.25 m^2 quadrats at the initial sample and a cross-shaped neighborhood. Because freshwater mussel density varied among the multiple sites that were sampled, Strayer et al. (1997) used a different condition at each site; the condition was based on a rapid assessment of density that preceded adaptive sampling.

McDonald et al. (1999) applied an adaptive version of line transect sampling in an aerial survey of polar bears (*Ursus maritimus*). Their condition for adaptive sampling was the detection of polar bears or fresh seal kills along a 37-km transect line. The neighborhood was defined as parallel transect lines 9 km on each side of the initially sampled line. The condition was met on five transect lines, but neither polar bears nor fresh seal kills were found on any adaptively sampled lines.

Palka and Pollard (1999) combined adaptive and line transect sampling for a survey of harbor porpoises (*Phocoena phocoena*). Their design was based on a strategy presented by Pollard and Buckland (1997) (see also Pollard et al. 2002). Palka and Pollard (1999) concluded that the strategy was easy to implement in the field and resulted in more precise estimates of density compared to traditional line transect sampling.

Bradbury (2000 and pers. comm.) described an application of a modified adaptive sampling design to estimate density of red sea urchin (*Strongylocentrotus franciscanus*). His design was based on a systematic adaptive cluster sampling design modified for sampling in one dimension (Thompson 1991a). However, the final sample size was constrained by defining the neighborhood to include the units halfway between the initial systematically sampled units (Woodby 1998); in that way, the final sample size was constrained to be no more than twice the initial sample size. Bradbury (pers. comm.) concluded that the modified adaptive sampling design was easy to implement.

Systematic adaptive cluster sampling was also the basis for an application by Acharya et al. (2000) to assess rare tree species. The tree species under study were found in clusters, and Acharya et al. (2000) concluded that efficiency of adaptive sampling depended on cluster size, with greatest efficiency observed for the species that formed the largest clusters.

Conners and Schwager (2002) implemented adaptive cluster sampling in a hydroacoustic survey of rainbow smelt (*Osmerus morax*) in Lake Erie. Their field trial was limited to one initial transect with adaptive transect segments parallel to and 1.5 km from the initial transect. The length of the adaptive transect segments was equal to the distance over which the condition was met on the initial transect. Conners and Schwager (2002) concluded that application of adaptive cluster sampling is feasible but pointed out several potential problems, including the need for (1) real-time data processing to assess the condition; (2) accurate georeferencing to find the adaptive units; and (3) effective condition and neighborhood definition to control the expected final sample size.

Hanselman et al. (2003) applied adaptive cluster sampling to surveys of Gulf of Alaska rockfish (*Sebastes alutus, S. borealis*, and *S. aleutianus*) (see also Chapter 14, this volume). They based the condition on percentiles of past survey results, allowed a distance of 0.1 nautical miles (nm) between adaptive tows, and included a stopping rule to restrict the expected final sample size. Their evaluation focused on the effect of condition and species distribution on efficiency. Hanselman et al. (2003) compared adaptive cluster sampling to simple random sampling if sample size was equal to the final sample size minus the edge units and found adaptive cluster sampling to be efficient.

Smith et al. (2003) applied adaptive cluster sampling to surveys of freshwater mussels (Unionidae) at 24 independent sites. Their initial sample was selected systematically, the condition was species presence, and the neighborhood was the standard cross shape (Figure 5.2B). Smith et al. (2003) compared adaptive cluster sampling to simple random sampling if sample size was equal to the final sample size including edge units. They found that adaptive cluster sampling did not result in lower sampling error for fixed sample size. However, application of adaptive sampling substantially increased both the number of individuals sampled and the probability of detecting the presence of rare species.

Challenges When Implementing Adaptive Cluster Sampling

There are four important challenges that a biologist will face when contemplating the application of adaptive cluster sampling. Before applying adaptive sampling, biologists should ask themselves the following questions:

1) Should I apply adaptive cluster sampling to this population?
2) How large should I expect the final sample size to be?
3) How do I implement adaptive sampling under field conditions?

4) How can I modify adaptive sampling to account for species biology, behavior, and habitat?

Should I Apply Adaptive Cluster Sampling to This Population?

The basic challenge is to determine whether adaptive cluster sampling is appropriate for the population of interest—in other words, "When is it a good idea to apply adaptive sampling?" In general, adaptive sampling is a good idea when it is efficient *and* the uncertainty in final sample size is not too great. A sampling design is efficient when it leads to smaller variance for fixed cost compared to an alternative design, which is often simple random sampling. The remainder of this section addresses the issue of efficiency, and the next section discusses methods to reduce the uncertainty in the final sample size.

From a statistical point of view, conventional cluster sampling tends to be efficient when clusters comprise most of the variation in the variable of interest. The same rule applies to adaptive cluster sampling, which is efficient when the within-network variance is a high proportion of total variance. In addition, efficiency of adaptive sampling tends to increase when the final sample size is close to the initial sample size.

From a practical point of view, adaptive cluster sampling tends to be efficient when organisms are clustered and the clusters are geographically rare. Geographic rarity means that the sampling units that are occupied by organisms are a small proportion of all units in the study area. Rarity of clusters is the most important characteristic. Individuals could be rare, but not clustered. In that case, networks would be small—typically a solitary unit—and the within-network variance would be negligible. Conversely, individuals in a population could be clustered, but the clusters could be common and numerous. In that case, the expected final sample size would likely be much larger than the initial sample size. For adaptive cluster sampling to be efficient individuals in the population must be clustered and the clusters must be rare.

Many ecological populations are clustered. However, it does not follow that adaptive cluster sampling is appropriate for most clustered or ecological populations. For instance, Figure 5.3 shows three populations each with the same abundance, but with different spatial distributions. The population in Figure 5.3A has a few large clusters; its variance to mean ratio is 1.3, 6.75% of the units are occupied, and the within-network variance is 44% of total variance. The population in Figure 5.3B has many small clus-

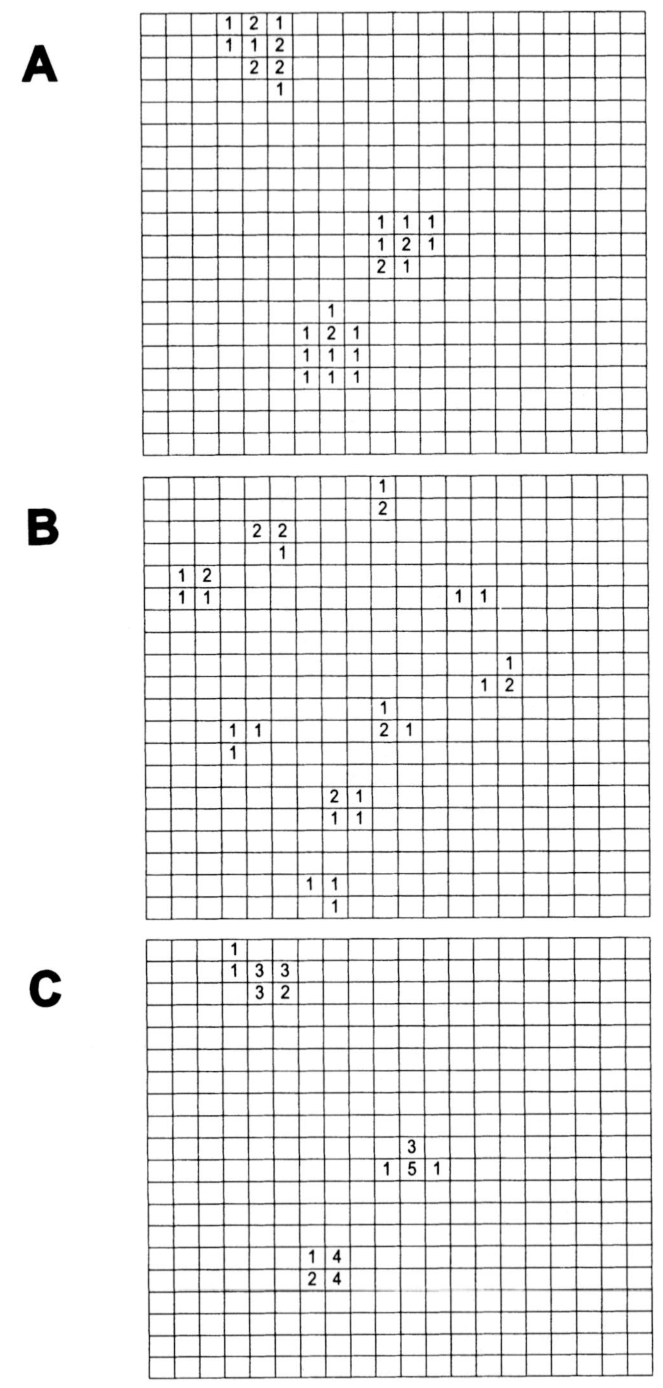

Figure 5.3. Three hypothetical populations showing various degrees of spatial clustering and rarity.

ters; its variance to mean ratio is 1.3, 6.75% of the units are occupied, and the within-network variance is 45% of total variance. The population in Figure 5.3C has a few small clusters; its variance to mean ratio is 3, 3.5% of the units are occupied, and the within-network variance is 48% of total variance. It is not obvious which populations would be appropriate for adaptive cluster sampling. Clustering appears in all populations. The degree of rarity, as measured by occupancy, is lowest in Figure 5.3C but does not vary between the other two populations. However, efficiency of adaptive cluster sampling differs among the three populations (Figure 5.4). Efficiency is defined here as the ratio of simple random sampling variance to adaptive sampling variance with sample size equal to the expected final sample size from adaptive sampling. For population 5.3A, efficiency of adaptive sampling depends on sample size with efficiency > 1 only for expected final sample size ≥ 90. A sample size of 90 translates to 23% of the study site being sampled, which seems high for ecological studies. For population 5.3B, adaptive cluster sampling is not efficient for a wide range of sample sizes. For population 5.3C, adaptive cluster sampling is efficient

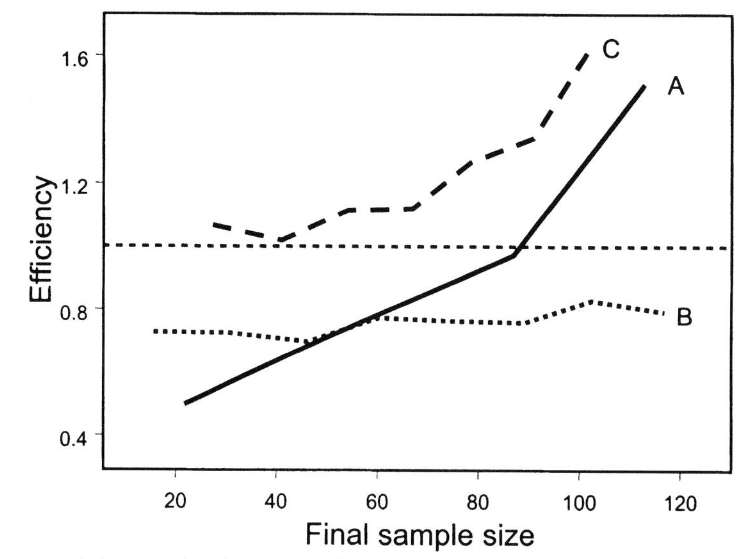

Figure 5.4. Results from simulated adaptive sampling of the populations shown in Figure 5.3. Efficiency is the ratio of simple random sampling variance to adaptive sampling variance with sample size equal to the expected final sample size from adaptive sampling. Efficiency greater than 1 indicates that adaptive sampling is the better design, and efficiency less than 1 indicates that simple random sampling is the better design.

and appropriate. So this example indicates that adaptive cluster sampling is efficient when clusters are small and geographically rare, that it is not efficient when clusters are small and numerous, and that efficiency depends on sample size when clusters are large and rare.

Other factors affect efficiency in addition to spatial distribution. For example, both the condition to adapt and the neighborhood definition affect efficiency through their effect on within-network variance and expected final sample size. Because the factors interact, simulation is invaluable for evaluating the efficiency of adaptive cluster sampling. Case study 1 illustrates the use of simulation to help design an adaptive survey.

How Large Should I Expect the Final Sample Size to Be?

Another challenge is to plan for uncertainty in the final sample size. Unlike conventional sampling designs in which sample size is fixed, the final sample size in adaptive sampling designs depends on what you find as you sample. The expected final sample size depends on the spatial distribution of the target population, the condition to adapt, the neighborhood definition, and whether stopping rules are employed in the sampling scheme. Because final sample size is random, techniques to predict final sample size are important for project planning. There are no hard and fast rules to predict final sample size, but there are some guidelines that are useful for anticipating final sample size in a qualitative sense.

Final sample size will tend to be highly variable in populations that contain only a few large clusters. If by chance the initial sample intersects a large cluster, many adaptive units will be sampled. If a large cluster is not intersected, the final sample size will be equal to the initial sample size. In populations that contain many small clusters, final sample size will tend to be much higher than the initial sample size.

The condition that triggers adaptive sampling will affect the size of networks in the population and, in turn, the final sample size. As the condition is made more restrictive by increasing the critical value, networks will, in effect, become smaller and adaptive sampling will be triggered less frequently, resulting in a smaller final sample size. Conversely, a liberal condition that triggers adaptive sampling much more often (a "hair-trigger" condition) will result in a final sample size considerably larger than the initial sample size.

Small neighborhoods (e.g., Figure 5.2A or 5.2B) will generate smaller final sample sizes than large neighborhoods (e.g., Figure 5.2C). Neighborhoods that contain discontinuities where adjacent units are "leap-frogged" (e.g., Figure 5.2D) will generate smaller final sample sizes than neighborhoods that include adjacent units (e.g., Figure 5.2B).

Stopping rules will reduce the maximum final sample size but will not eliminate variation in final sample size. Stopping rules also bias results, but the extent of the bias might not be large, depending on how it is implemented (Brown and Manly 1998). Several real-world applications have incorporated some sort of stopping rule (Lo et al. 1997; Woodby 1998; Hanselman et al. 2003) in which the potential for bias seems to have been outweighed by the need to control the final sample size. Case study 2 illustrates the use of a stopping rule.

How Do I Implement Adaptive Sampling in the Field?

Adaptive sampling can be a complex procedure to implement under field conditions. In addition to the usual challenge of conventional probability sampling, which is required to take the initial sample, biologists must navigate among adaptively sampled units within a cluster. The key to successful implementation of adaptive sampling is keeping careful records to track which adaptive units have been sampled and which units remain to be sampled. In addition, choice of neighborhood, stopping rules, and initial sample design taken can help ease implementation.

When sampling freshwater mussels in rivers, we find it helpful to start a map on the reverse side of our data sheet whenever adaptive sampling is triggered. For example, Figure 5.5A shows a grid that maps the units within a large adaptive cluster; the numbers refer to the count within a sampling unit. This grid map of a cluster came from one of our data sheets. Adaptive sampling was triggered when two mussels were observed in a unit (indicated by a circle in Figure 5.5). The remaining observations were mapped and recorded on the data sheet grid as sampling progressed. The map was indispensable for navigating within the cluster.

Implementation can be streamlined and simplified by choice of neighborhood. For example, the sampling units in case studies 2 and 3 are transects (or tows) and the neighborhoods are defined as transects that run parallel to, but some distance from, the condition-meeting transect. In this way sampling is simplified because you are adaptively sampling in only two directions rather than four (or more), similar to the neighborhood in Figure 5.2A. An extreme example of this was presented by Woodby (1998), in which the neighboring transects were placed halfway between the initial transects. In Woodby's design, adaptive sampling always stopped after one set of adaptive transects was sampled because sampling further would

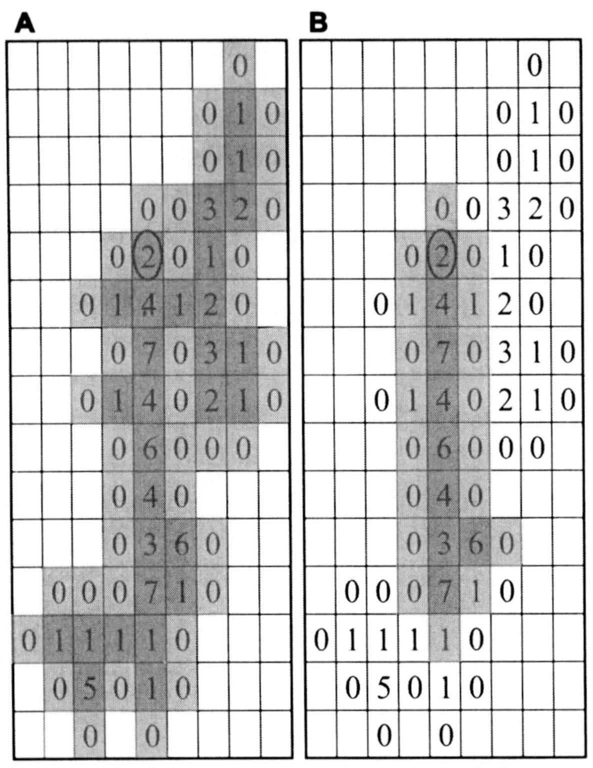

Figure 5.5. A cluster of units showing edge units in light gray and network units in dark gray. The initially sampled unit is circled and has a value of two. (A) shows the result when adaptive sampling is triggered whenever a selected unit's value is greater than zero. (B) shows the result when adaptive sampling is triggered whenever a selected unit's value is greater than one.

overlap other initial transects. Woodby's design essentially uses the neighborhood definition to create a stopping rule.

Stopping rules can ease implementation by reducing the spatial extent of adaptive sampling. Complexities associated with navigating to adaptive units and planning difficulties associated with an open-ended final sample size are diminished when adaptive sampling has a predetermined cutoff. Case study 2 provides a good practical example of the use of a stopping rule. Lo et al. (1997) incorporated a stopping rule in a survey of Pacific hake larvae. In this case, adaptive sampling would not have been possible without a stopping rule because of the large scale of the survey area.

Choice of design to take the initial sample will influence ease of implementation. Simple random sampling often is cumbersome under field conditions. Systematic sampling is an attractive alternative for taking the initial sample because it is relatively easy to implement (Thompson 1991a; Acharya et al. 2000; Hanselman et al. 2003; Smith et al. 2003). Systematic sampling with multiple random starts, which supports valid estimation of sampling variance and does not add much complexity to implementation, is preferred over single-start systematic sampling, which is commonly applied under field conditions. Systematic sampling is known to be an efficient sampling design for clustered populations, regardless of adaptive sampling (Christman 2000). Case studies 2 and 3 incorporate some form of systematic sampling for initial sample selection. Case study 1 does not include systematic sampling but uses simple random sampling instead because it is a computer simulation of freshwater mussel sampling. When we have actually applied adaptive sampling to freshwater mussel surveys, we have used systematic sampling to take the initial sample (Smith et al. 2003).

How Can I Modify Adaptive Sampling to Account for Species Biology, Behavior, and Habitat?

Special consideration is required when adaptive sampling is applied to species that are mobile, elusive, or sensitive to handling. The potential for double-counting is high when animals could be flushed into adjacent sampling units as a result of adaptive sampling. Imperfect detectability is an issue when sampling elusive animals whether sampling is adaptive or conventional (Thompson and Seber 1994). Because adaptive sampling tends to result in selection of occupied habitat, the potential for habitat disturbance is greater than in conventional sampling. If sampling is destructive or species or their habitats are sensitive to sampling, adaptive sampling might need to be modified to reduce disturbance.

A solution to problems caused by mobile species is to define neighborhoods that do not include adjacent units (e.g., Figure 5.2D). The separation between within-neighborhood units can be selected to exceed flushing distance. If aggregation size is smaller than the flushing distance, adaptive sampling methods would not be advantageous. There would be no value in adaptively sampling if the neighborhood jumps over the aggregation. Another approach for mobile species is to base the assessment on an index of species presence, as was done in case study 2.

Imperfect detectability is a pervasive issue in field studies of animal populations (Seber 1982). Numerous strategies have been developed to adjust for imperfect detectability, which would lead to underestimates of animal abundance if left unadjusted. In the context of finite population sampling, such as when sampling units are selected and then counts of animals are made on the selected units, imperfect detectability is dealt with by first adjusting for detectability on selected units, then expanding the adjusted per-unit counts to the entire sampling frame. This approach works for conventional sampling because selection of units does not depend on detectability, so adjustment for detectability can be handled prior to estimating totals or means. However, detectability can influence the selection of adaptively sampled units because the condition to adapt is typically based on a count of detected animals (Thompson and Seber 1994). For example, suppose that in truth an initial unit should trigger adaptive sampling because it is occupied by a species. In practice, neighboring units might not be selected because the animal in the initial unit might not be detected, in which case adaptive sampling would not be triggered. Thompson and Seber (1994) solve this dilemma by turning the problem around and first estimating the population of detectable animals and then adjusting that estimate by an independent estimate of detection.

Let us consider an example of how to design adaptive sampling to account for imperfect detectability. Detectability is an issue in surveys of freshwater mussels because they are benthic organisms that position themselves at various depths of the substrate. Some mussels are readily detectable at the substrate surface by visual or tactile observation, but some are buried below the surface and must be excavated for detection. To estimate total abundance or density, some amount of excavation is required and double sampling can be used to adjust for detectability in an optimal way that balances effort and precision (Smith et al. 2000). Double sampling in this case involves excavating a subsample of quadrats. The ratio of mussels detected on the substrate surface in excavated quadrats to total mussels in excavated quadrats is used to estimate detectability. Double sampling can be applied to an adaptive sampling design by excavating a subsample of the initial sample of quadrats. Adaptive sampling is used to estimate the detectable portion of the population; then the estimate of detectability from double sampling is used to adjust and estimate total abundance. Thompson and Seber (1994:219–220) provided a formula for incorporating an estimate of detectability into adaptive sampling estimates of abundance.

Some organisms are so sensitive that the mere act of sampling can be detrimental by interfering with survival or reproduction. In some situations, animals must be captured to be observed, habitat must be altered to collect animals, or plants must be removed to measure biomass. In such situations, sample size is of concern not only to control survey cost but also to reduce disturbance. The potential for disturbance is elevated for adaptive sampling because it tends to allocate effort into occupied habitat.

When sampling-related disturbance is a concern, a potential solution is to base the condition to adapt on a less invasive method of sampling. In that way, the impact from sampling the edge units, which form a large portion of the final sample, will be reduced or even eliminated. For example, American ginseng (*Panax quinquefolium*) is a rare, low-growing plant that is susceptible to trampling when being surveyed (John Young, U.S. Geological Survey, pers. comm.). One strategy to reduce disturbance during sampling would be to base the condition to adapt on a geographic information system (GIS)-based prediction of habitat (Boetsch et al. 2003). All sampling units (both initial and adaptive units) could be selected using GIS. The initial sample selection would be probabilistic and clusters of adaptive units could be selected based on predictions of habitat. In that way, edge units would never have to be visited in the field and travel to and among sampling units could be planned to minimize potential for disturbance and to reduce travel time.

Case Studies

We present three case studies to illustrate the design and application of adaptive sampling methods. These examples do not illustrate all potential challenges, but are based on our experiences and represent some of the practical challenges faced when implementing adaptive sampling.

Case study 1, from D.R.S., demonstrates the use of simulation to design an effective adaptive sampling survey. The objective was to estimate density of freshwater mussels, a class of often rare and endangered organisms. In this case, simulation helped to identify an adaptive sampling design that would be appropriate for populations similar to the study population and to plan for final sample size that would result from the design's implementation.

Case study 2, from J.A.B., discusses the application of adaptive sampling to a monitoring protocol for Australian brushtail possum (*Trichosurus vulpecula*), which is a nuisance species in New Zealand. In this case, the basic adaptive cluster sampling design had to be modified because the target organism could not be observed directly. Instead, an index of possum activity was observable, which created a time lag between selection of the sampling unit and the observed response.

Case study 3, from N.C.H.L., demonstrates the generality of adaptive sampling procedures. In an application to Pacific sardine (*Sardinops sagax*) assessment, adaptive sampling procedures are incorporated in a complex survey to allocate effort among strata and to reduce variance. This case study also illustrates application of adaptive sampling in a large-scale survey, where navigating among sampling units is a costly endeavor.

Case Study 1: Assessing the Effect of Condition to Adapt on Estimates of Freshwater Mussel Density

Freshwater mussels (Unionidae) represent a diverse assemblage including more than 300 species that are suffering an extinction rate higher than any other North American fauna (Ricciardi and Rasmussen 1999). Efficient sampling to assess and monitor mussel populations has become a critical need for malacologists and managers (Smith et al. 2001; Strayer and Smith 2003). The fauna tend to cluster and are often rare and at low density (Smith et al. 2003; Strayer and Smith 2003). Thus, freshwater mussels appear to be good candidates for adaptive cluster sampling.

Here we use freshwater mussels as an example to demonstrate how simulation can help determine how best to design an efficient adaptive cluster sampling survey. Application of adaptive cluster sampling is logistically feasible because freshwater mussels are not readily mobile and they can be observed directly (Smith et al. 2003; Strayer and Smith 2003). However, as in many real-world situations, selecting a proper value for the condition to adaptively sample presents a challenge. Simulations provide an excellent method to compare a range of alternatives prior to implementation. For this study we had access to a complete count (census) of freshwater mussels at a river site. Alternatively, a population can be generated based on a sample of data and assumptions about the spatial distribution.

Methods

We counted all mussels on a section of riffle habitat in the Cacapon River near Capon Bridge, West Virginia. The river section was approximately rectangular and measured approximately 40 m wide (bank to bank) by 90 m long. The substrate surface was thoroughly searched within a grid of 0.25 m^2 cells, and mussels were measured lengthwise and returned to the substrate. Figure 5.6A shows clustering in the population of freshwater mussels (*Elliptio complanata*) at the site (black squares in Figure 5.6A represent areas that were occupied by mussels).

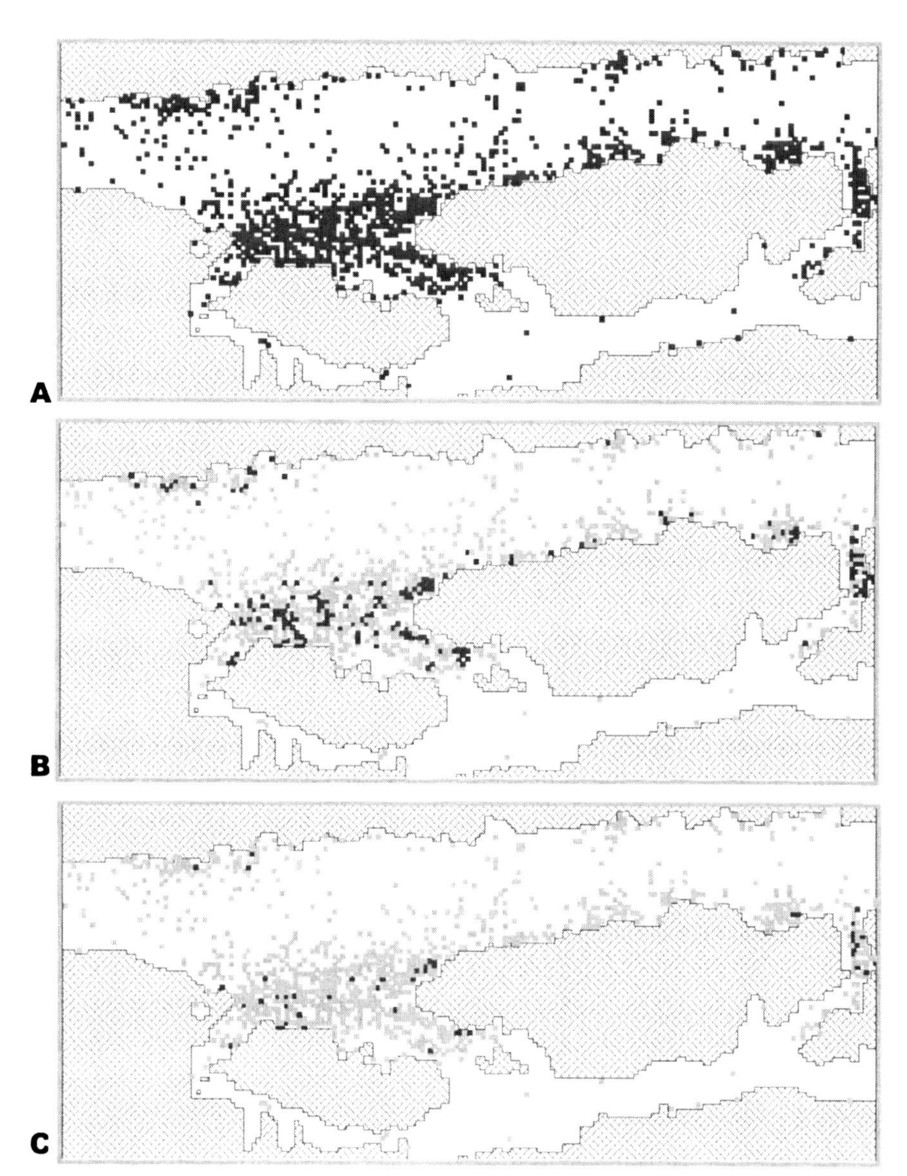

Figure 5.6. Population of freshwater mussels in a riffle in the Cacapon River at Capon Bridge, West Virginia. The cross-hatch indicates land. White indicates river substrate unoccupied by mussels. The remaining area is occupied by at least one mussel. Black squares show areas that meet the condition to adapt in a sample unit. Gray squares show areas that are occupied but do not meet the condition to adapt in a sample unit. The condition, which is based on the count per 0.25 m^2, changes among the panels. In (A), adaptive sampling would be triggered if any mussels were found. In (B), adaptive sampling would be triggered if at least three mussels were found. In (C), adaptive sampling would be triggered if at least five mussels were found.

We simulated the implementation of adaptive cluster sampling with 0.25 m^2 quadrats as the sampling unit, simple random sampling to take the initial sample, and a cross-shaped neighborhood (Figure 5.2B). Sampling was simulated using a range of conditions to adapt. The condition to adapt was of the form $y_i \geq c$, where y_i was the count of mussels in the ith quadrat and c was the critical or threshold value. We considered c from 1 to 5. For example, when any mussels are present in a quadrat and $c = 1$, then adaptive sampling is triggered. We compared within-network to population variance, initial to final sample size, and sampling efficiency that resulted from each condition. Efficiency was defined as the ratio of sampling variance from simple random sampling to adaptive cluster sampling given equal sample size, that is, sample size for simple random sampling was set to be the expected final sample size from adaptive cluster sampling. Simulations were replicated 1,000 times and results were averaged across the replications. Software used for this simulation can be found at http://www.lsc.usgs.gov/AEB/davids/acs/.

Results

The condition to adapt, in effect, partitions the study area into networks. In Figure 5.6, black squares indicate areas of the population that form the networks that meet the condition. Each panel in Figure 5.6 shows results from a different condition.

In this population, the condition to adapt had a strong effect on the within-network variance, ratio of final to initial sample size, and efficiency (Figure 5.7). For a condition of $y_i \geq 1$, the within-network variance was more than 40% of total variance, final sample size was over four times initial sample size, and efficiency was only 0.5. Within-network variance decreased, final sample size decreased, and efficiency increased as the condition became more stringent (Figure 5.7).

Discussion

As the condition becomes more restrictive the proportion of the population that meets the condition shrinks. This leads to two important results. First, the within-network variance decreases as the range of values within a network is truncated. For example, when the condition is $y_i \geq 1$ a network could contain values ranging from 1 to the maximum count; only units with counts equal to 0 would be excluded from networks that meet the condition. However, when the condition is $y_i \geq 5$ the range of a network's values would be limited from five to the maximum count. Second, as the

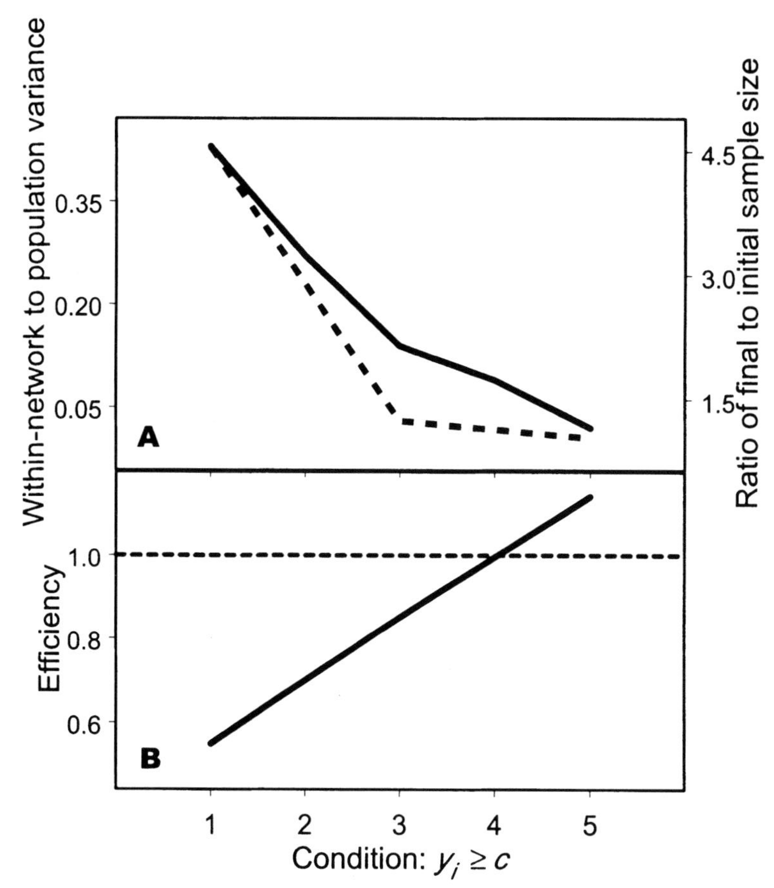

Figure 5.7. Results from simulated adaptive sampling of the population shown in Figure 5.6. Sampling was simulated for a range of conditions to adapt ($y_i \geq c$, where $c = 1, 2, 3, 4$, and 5). (A) shows ratio of within-network variance to population variance (solid line) and ratio of final to initial sample size (dashed line). (B) shows efficiency, which is the ratio of simple random sampling variance to adaptive cluster sampling variance with sample size fixed at the expected final sample size. Efficiency is a function of relationships shown in (A). Adaptive cluster sampling is more efficient when efficiency is greater than one, as indicated by the horizontal dashed line in (B).

condition becomes more restrictive, the expected final sample size approaches the initial sample size. This is caused by the reduction in network size—there are simply fewer sampling units in networks as the condition becomes more restrictive.

Interestingly, these two results act in opposite directions on the sampling efficiency. A reduction in within-network variance reduces efficiency, all else being equal. In contrast, efficiency increases as expected final sample size approaches initial sample size. The ultimate effect on efficiency depends on the net effect of the interaction between within-network variance and final sample size. In this case study, the reduction in final sample size had a greater effect on efficiency than reduction in within-network variance, but this will not always be true.

From the simulation we learned that adaptive cluster sampling would be a good idea only if we set the condition correctly. A restrictive condition was correct for this population. We also learned that it would have been a bad idea if the condition were set too liberally. In retrospect, it is apparent that the freshwater mussel population, although clustered, was not rare enough for adaptive cluster sampling to be efficient. By restricting the condition to adapt, we effectively made the population "rare," or at least the networks that met the condition became rare. If density was lower at the site, as Smith et al. (2003) found, a more liberal condition would have been feasible.

This result is specific to this population. In general, it is not necessary for efficiency to increase as the condition becomes more restrictive. This case study points out the utility of simulation before implementation. Simulation is a powerful method to evaluate efficiency across a wide range of alternative designs. In the absence of simulation we would not have been able to predict an efficient condition without a lengthy and expensive series of field trials.

Case Study 2: Monitoring Possum Abundance in New Zealand

The Australian brushtail (*Trichosurus vulpecula*) is a major environmental pest in New Zealand. These marsupials were first successfully introduced from Australia in 1858 to establish a fur industry. They rapidly spread so that today, about 70 million possums live throughout more than 90% of mainland New Zealand (Pracy 1974; Cowan 1990; Clout and Erickson 2000). Possums are considered a serious pest in New Zealand, primarily because they defoliate preferred plant species, predate bird eggs and chicks, and carry bovine tuberculosis, which poses a major threat to New Zealand's beef and venison industries (Green 1984; Cowan 1990; Brown et al. 1993; Coleman and Livingstone 2000).

Various government and private agencies in New Zealand expend considerable effort to control possums using either traps or poisons. Therefore, an accurate and efficient method for monitoring possums is necessary to assess whether control strategies are effective. Until recently, the primary index of possum density was based on lines of leghold traps. The major disadvantage of monitoring possums using traps is that it limits sample sizes to low levels because of the amount of labor required to transport and check traps (Brown and Thomas 2000; Thomas et al. 2003). A new monitoring method is being developed in New Zealand using a device called WaxTag, which contains a possum-specific attractant. The possum bites a wax block on the end of the tag, and the frequency of bite marks is used to calculate an index of possum density. Bite marks from other species can be distinguished from possum bite marks (Thomas et al. 1999).

Possum distribution is known to be clustered, and if residual hot spots can be detected, follow-up control can be targeted to specific locations. Adaptive sampling could be used to assess residual possum population size and to provide information on the spatial pattern of the remaining animals. The use of WaxTags creates the potential to use more informative survey designs, such as adaptive sampling, because large sample sizes are possible.

It is important to note that when using WaxTags for monitoring, the frequency of bite marks is considered an index of possum activity rather than an estimate of possum numbers because one possum can bite more than one WaxTag. An index based on possum activity is more biologically meaningful than an estimate for low possum densities because the environmental effect of one possum compared with the effect of multiple possums is of less concern than the environmental effect of some possums compared with no possums.

In this study, we assessed a practical application of using WaxTags and adaptive cluster sampling to monitor possums. The aim was to develop a survey protocol similar to the existing protocol for leghold traps but which also provided information on possum population size and spatial pattern. Thus, tag lines were set up using systematic sampling, and then additional lines were set up on either side of the initial lines during an adaptive phase. The new monitoring protocol needed to be modeled on the existing trap protocol because of the widespread use of the existing method. The possum control industry would be more likely to respond positively to incremental changes in the monitoring methodology than to a completely new model. For this reason, lines of WaxTags were used as the sample unit as prescribed for traps by the existing protocol.

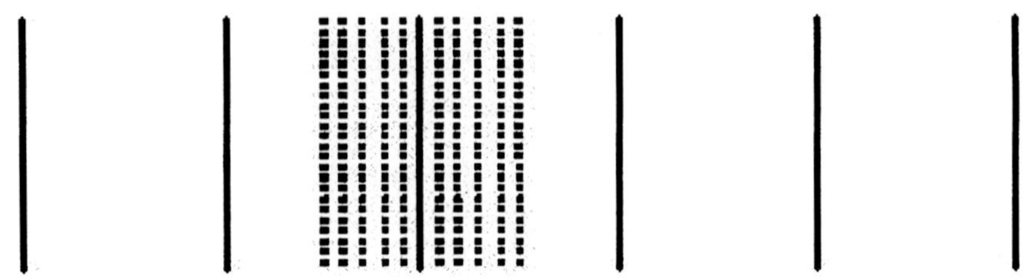

Figure 5.8. Layout of possum monitoring lines. There were fifteen lines at each site; six are shown here (solid lines). Lines were spaced 500 m apart, and along each line there were 10 WaxTag stations spaced 20 m apart. In the figure, the third solid line from the left "triggered" adaptive sampling and five parallel lines on both sides, spaced 50 m apart, were set (dashed lines). There were 10 stations spaced 20 m apart along each adaptive line.

Methods

Two sites were used to test adaptive sampling using WaxTags: Balmoral Forest and Eyrewell Forest in North Canterbury, New Zealand. At each site, an initial sample of 15 lines was placed at 500-m intervals (Figure 5.8). Each line consisted of 10 WaxTag stations spaced 20 m apart. At each station, two WaxTags were nailed to opposite sides of a tree. The WaxTags were left for three nights, and on the fourth day the number of stations with possum interference was recorded.

The adaptive phase of the study was designed according to the results from the initial lines. The three lines with the highest frequency of bite marks were chosen as the lines that "triggered" adaptive sampling, and five parallel lines spaced 50 m apart were set on both sides of each of these three lines. The WaxTags on these lines were left for a further three nights and the number of stations with possum interference was recorded.

The adaptive lines were all set at the same time rather than in a sequential pattern typical of most adaptive cluster sampling. This method was chosen because the sample units could not be assessed immediately. The lines were left out for three consecutive nights because possums are nocturnal and because it can be difficult to detect possums when populations are small.

Results

The possum population at Balmoral Forest was extremely small. After three nights, only three initial lines had any possum bite marks (Lines 6, 8, and 15; Table 5.1). On each of these lines, only one station had bite

Table 5.1.

Number of WaxTag stations with possum interference on 15 initial lines and 30 adaptive lines at Balmoral Forest, New Zealand. The initial lines were left out for three nights, and then five adaptive lines were set on each side of the three initial lines with the highest interference. Blank cells indicate no data because adaptive lines were not placed parallel to 12 of the initial lines.

	Initial line number														
Line	1	2	3	4	5	6	7	8	9	10	11	12	13	14	15
Adaptive line A						0		0							0
Adaptive line B						0		0							0
Adaptive line C						0		0							0
Adaptive line D						0		0							0
Adaptive line E						0		0							0
Initial line	0	0	0	0	0	1	0	1	0	0	0	0	0	0	1
Adaptive line F						0		0							1
Adaptive line G						0		0							0
Adaptive line H						0		0							0
Adaptive line I						0		0							0
Adaptive line J						0		0							0

marks and hence the condition to trigger adaptive selection was one (see Discussion section for elaboration on this point). The frequency of bite marks on the adaptive lines that were placed parallel to lines 6, 8, and 15 was also very low. In fact, after three nights, throughout the 30 adaptive lines, bite marks were recorded at only one station on the first adaptive line adjacent to line 15 (Table 5.1).

Higher rates of bite marks were observed at Eyrewell Forest than at Balmoral Forest; 13 lines detected possums (Table 5.2) at Eyrewell. Lines 6 and 11 clearly had the highest frequency of bite marks (eight and nine stations, respectively), but four lines had five stations with bite marks. However, the spatial pattern of possum activity differed among these lines. Only Line 1 had three adjacent stations with bite marks; the other three lines had only two adjacent stations with bite marks. Thus, the condition for adaptive selection was given an extra layer of complexity by counting first the number of stations with bite marks (five), and then second, the number of consecutive stations with bite marks (three). The spatial aggregation component was introduced into the condition because one goal of this study was to detect aggregates of possum activity.

Table 5.2.

Number of WaxTag stations with possum interference on 15 initial lines and 30 adaptive lines at Eyrewell Forest, New Zealand. The initial lines were left out for three nights, and then five adaptive lines were set on each side of the three initial lines with the highest interference. Blank cells indicate no data because adaptive lines were not placed parallel to 12 of the initial lines. All adaptive lines parallel to the three initial lines had possum interference, but only those shown in bold were above the adaptive selection condition.

	Initial line number														
Line	1	2	3	4	5	6	7	8	9	10	11	12	13	14	15
Adaptive line A	4					4					**8**				
Adaptive line B	3					**7**					**9**				
Adaptive line C	**7**					**6**					4				
Adaptive line D	**5**					4					**6**				
Adaptive line E	2					3					**8**				
Initial line	5	4	5	4	2	8	4	4	3	5	9	5	0	0	4
Adaptive line F	**9**					**6**					**10**				
Adaptive line G	7					5					**8**				
Adaptive line H	**6**					**8**					**5**				
Adaptive line I	**5**					**9**					**5**				
Adaptive line J	3					**9**					4				

Discussion

The sampling design used in this study is a modification of the usual adaptive cluster design. These modifications were made because of the specific challenges of monitoring low-density possum populations in New Zealand.

The first challenge was that the sample units in this case could not be assessed immediately. WaxTags, or any other device used to record possum activity, must be left overnight because the species is nocturnal. Furthermore, when animal numbers are low, the devices need to be left for more than one night (three is the standard practice) to ensure that resident possums have an adequate chance of encountering the device. As a result, it would have been too time-consuming to sequentially sample within a detected cluster. Instead, a maximum of five lines on either side of the initial triggered line was used, which limited the maximum size of a cluster but was the most practical option for a monitoring technique that will need to be both efficient and cost effective.

The practice of setting all five lines on either side of an initial line does raise some important analytical questions. First, the design restricts the maximum size of a network to five lines on either side of the initial line, which introduces bias because the network size and inclusion probability are incorrectly calculated for large networks. Second, given that this design yields data on the rate of bite marks from all 10 lines (five on either side), should all of this information be used to estimate the possum index, or should we use only the information that would have been gained if lines were sequentially set? For example, if we had set the adaptive lines sequentially next to Line 1 at Eyrewell Forest, the adjacent Adaptive Line 1E would have been set first. However, because it had a value of 2, which was less than the condition (5), the next four adaptive lines (1A–1D) strictly should not have been set. Similarly, for initial Line 6, an edge to the network would have been found in the first adaptive line on one side (Line 6E) and in the second adaptive line to the other side (Line 6G). We feel that the ideal analysis would use all of the data gathered from all of the lines because the additional information could add to a more robust analysis.

A more subtle question that arises is how to determine the value that triggered adaptive selection of adjacent lines. The initial line values given in Tables 5.1 and 5.2 represent the number of stations with bite marks recorded after the first three nights. However, at the end of the three nights when the adaptive lines were set, the initial lines were also reset so that any WaxTag with bite marks was replaced with a new WaxTag. A second value then was recorded for all of the lines after an additional three nights. In many cases, this second value did not correspond to the first value recorded on the initial lines. For example, Line 1 at Eyrewell Forest had an initial value of five stations with bite marks. However, when all of the bitten tags were replaced and the lines were set for an additional three nights, the Line 1 value increased to six stations with interference. Similarly, Line 6 had a first value of 8 but a second value of 6, and Line 11 had an initial value of 11 and a second value of 10. It seems sensible from a biological standpoint to use the initial values because those values triggered the adaptive selection. The second values, on the other hand, can be considered a measure of the temporal change in possum activity on that line. This temporal change could be a result of possums becoming conditioned to the presence of WaxTags or to any other source of temporal variation (e.g., changes in food supply or weather).

The second challenge in this application of adaptive cluster sampling is the practical limitations of setting and checking WaxTags. In particular, the survey effort must be divided into units of person-days, and the number of days required to complete the entire survey must be known in advance. This is because much of the possum monitoring in New Zealand is conducted by commercial businesses that need to run efficiently for the industry to be sustainable. Thus, it is not feasible to have surveys with unknown completion times, which is characteristic of traditional adaptive cluster sampling.

This challenge was met by using a constant number of lines in the initial sample that triggered adaptive selection as well as a predetermined number of adaptive lines. The sampling protocol was designed around the need to divide the work into units that one person could complete in a full day. Thus, the initial sample was set at 15 lines because one person can comfortably set this many lines in one day. Similarly, the 30 adaptive lines (i.e., 10 lines adjacent to three initial lines) could be completed in two person-days (i.e., two people in one day). In contrast, if we had chosen, for example, four initial lines and 10 adaptive lines, this would have equated to 2.67 person-days of work, which would make monitoring operations inefficient and costly.

The third challenge in this application is that the condition for adaptive selection was not known (nor could it be predicted) prior to the survey. Over time, it might eventually be possible to set the condition a priori, but in this trial we were able to overcome this problem by having a two-phase design. This design meant that lines had to be checked twice: once in Phase I to collect the initial sample values and to place out the adaptive lines and then again in Phase II to check all of the lines again. However, there are some advantages to this two-phase design. First, the survey effort required for an initial assessment of the possum activity level (the 15 initial lines) is clearly differentiated from the survey effort required for the second adaptive sampling phase. The first phase is analogous to current monitoring protocol, which uses traps, and it can be used to assess whether contractual target levels for control have been met. In contrast, the second phase can be used to concentrate follow-up control operations on hot spots or local areas of patchiness. There also are advantages to being able to separate these two phases for budgeting and financial accounting.

This example illustrates how the adaptive cluster sampling technique can be modified to help with a real biological problem with practical limitations. The challenges in using adaptive cluster sampling for monitoring possums are that the sample unit value can not be immediately gained in the field, that there are very practical limitations to sample sizes and expenditure of field effort, and that with limited knowledge the adaptive condition can not be set prior to sampling. Given these challenges, adaptive cluster sampling can be "adapted" and is a very useful tool for possum monitoring.

Case Study 3: Adaptive Allocation Sampling to Estimate Egg Production of Pacific Sardine

The Pacific sardine (*Sardinops sagax*) was once one of the more important fisheries off the west coast of the American continents. The estimated biomass peaked at 3.6 million metric tons (mt) in 1934 but fell to less than 100,000 mt in the late 1950s and mid-1960s as the fishery collapsed (Murphy 1966; MacCall 1979). In 1949, a survey was launched to help understand the decline of sardines and to monitor their population. The Southwest Fisheries Science Center of the National Marine Fisheries Service has been responsible for monitoring the spawning biomass of Pacific sardine by conducting a routine ichthyoplankton survey, commonly referred to as the California Cooperative Fisheries Investigations (CalCOFI).

The daily egg production method (Parker 1985; Hunter and Lo 1997) has been used to estimate spawning biomass of Pacific sardine (Wolf 1988a,b; Lo et al. 1996; Scannel et al. 1996; Barnes et al. 1997). The daily egg production method estimates spawning biomass by (1) calculating the daily egg production from ichthyoplankton survey data; (2) estimating the maturity and fecundity of females from adult fish samples; and (3) calculating the biomass of spawning adults. In this report, we concentrated on the ichthyoplankton survey.

Because sardine eggs are aggregated (Lo et al. 1996), efficient sampling methods have been sought. Before 1996, sardine egg production was estimated from plankton net sampling only, like CalVET (Smith et al. 1985). Since 1996, in addition to plankton nets, Bongo nets and the Continuous Underway Fish Egg Sampler (CUFES; Checkley et al. 1997) have been used to sample fish eggs (Hill et al. 1998, 1999).

Since 2001, we have used an adaptive allocation sampling design to estimate daily egg production. This design allocates additional net tows according to egg densities observed from the CUFES. Plankton net samples of eggs and yolk-sac larvae are allocated to the high-density area as determined by CUFES to estimate the daily egg production at age 0 (P_0), which then is incorporated in the daily egg production method to estimate spawning biomass.

METHODS

In 2002, we conducted a full-scale survey to estimate the spawning biomass of Pacific sardine (Lo et al. 2001). We sampled ichthyoplankton with plankton nets and CUFES aboard the R/V *McArthur* (March 21–April 19) and R/V *David Starr Jordan* (March 27–April 14). The Jordan segment of the survey was the routine CalCOFI April survey (http://swfsc.nmfs.noaa.gov/frd/CalCOFI/CurrentCruise.htm). In addition, we sampled adult sardine aboard R/V *David Starr Jordan* (April 14–25) after the routine CalCOFI cruise to estimate reproductive parameters.

We used egg counts from the CUFES from the 2002 survey to allocate placement of plankton net tows and to map the spatial distribution of the sardine spawning population. Following the adaptive sampling procedure, we towed plankton nets at 4-nautical mile (nm) intervals on each line after the egg density from each of two consecutive CUFES samples exceeded the critical value of 1 egg/min. Plankton net tows continued until the egg density from each of two consecutive CUFES samples was less than 1 egg/min.

We post-stratified the survey area into a high-density area (Region 1) and a low-density area (Region 2) according to the egg density from CUFES collections. We determined the stage of eggs from the plankton net tows and identified yolk-sac larvae from plankton and Bongo net tows in the high-density area. These responses were incorporated into a model of the embryonic mortality curve in the high-density area and later converted to the daily egg production, P_0, for the whole survey area. We employed this adaptive allocation sampling, which is similar to a 1997 survey of Pacific hake larvae (Lo et al. 2001), aboard the *McArthur* but not aboard the *Jordan* because the latter was conducting the routine CalCOFI survey.

We used eggs from plankton tows and yolk-sac larvae from both plankton and Bongo tows in Region 1 to compute egg production (P_0) assuming the embryonic mortality curve was exponential: $P_t = P_{0,1}\exp(zt)$, where P_t was daily egg or yolk-sac production/0.05 m^2 at age t days and z was the daily instantaneous mortality rate (Lo et al. 1996; Lo et al. 2001). We examined eggs for their developmental stages and converted them to age (Lo et al. 1996). Due to the small number of tows with eggs, we obtained egg production in Region 2 ($P_{0,2}$) by calibration: $P_{0,2} = P_{0,1} \times q$, where q was ratio of egg density in Region 2 to Region 1 from CUFES. The egg production for the entire survey area, P_0, was a weighted average of $P_{0,1}$ and $P_{0,2}$, where the weights were the area sizes.

We used the estimate of P_0 together with estimates of four adult parameters to compute the spawning biomass (B_s) according to

$$B_S = \frac{P_0 AC}{RSF/W_f}, \tag{5.1}$$

where A is the survey area in units of 0.05 m^2, C is the conversion factor from g to mt, $P_0 \times A$ is the total daily egg production in the survey area, and the denominator (RSF/W_f) is the daily specific fecundity (number of eggs/population weight (g)/day). The fecundity estimate was calculated from the daily spawning fraction or the number of spawning females per mature female per day (S), the average batch fecundity (F), the proportion of mature female fish by weight (sex ratio or R), and the average weight (in g) of mature females (W_f) (Parker 1985; Picquelle and Stauffer 1985; Lo et al. 1996; Lo and Macewicz 2002).

Regarding sampling gear, the diameter of the CalVET net frame was 25 cm, the tow was vertical to minimize the volume of water filtered per unit depth, the mesh size was 0.15 mm, and the tow depth was 70 m. The diameter of the Bongo net frame was 71 cm, the tow was oblique at a 45° wire angle, the mesh size was 0.505 mm, and the tow depth was 210 m when 300 m of wire was deployed. CUFES can be installed midship on a research vessel with the intake pipe over the side of the vessel or in the bowl. It extends 3 m below the water surface (see illustration in Checkley et al. 1997). Eggs were sieved from the water flow with the 0.5 mm nylon mesh of the CUFES concentrator.

Results

The survey area was post-stratified into a high-density area (Region 1) and a low-density area (Region 2, Figure 5.9). Region 1 encompassed the area where the egg density (eggs/min) in CUFES collections was at least 1/min. The rest of the survey area was Region 2 (Figure 5.9). One egg/min was equivalent to two to four eggs/plankton tow, depending on the degree of water mixing.

We collected 1,622 CUFES samples from *McArthur* (1,165) and *Jordan* (457) at intervals ranging from 1 to 47 min with a mean of 24.4 min and median of 30 min. In Table 5.3 we present gear-, region-, and vessel-specific incidence of eggs and yolk-sac larvae. Catches of eggs are shown in Figure 5.9 and catches of yolk-sac larvae are presented in Figure 5.10.

The daily egg production in Region 1 ($P_{0,1}$) was 2.33/0.05 m^2 (CV = 0.17, Lo and Macewicz 2002) and egg mortality was $z = 0.4$ (CV = 0.15) for an area of 88,403 km^2 (25,830 nm^2). The ratio (q) of egg density between Region 2 and Region 1 from CUFES samples was 0.056 (CV = 0.025). In

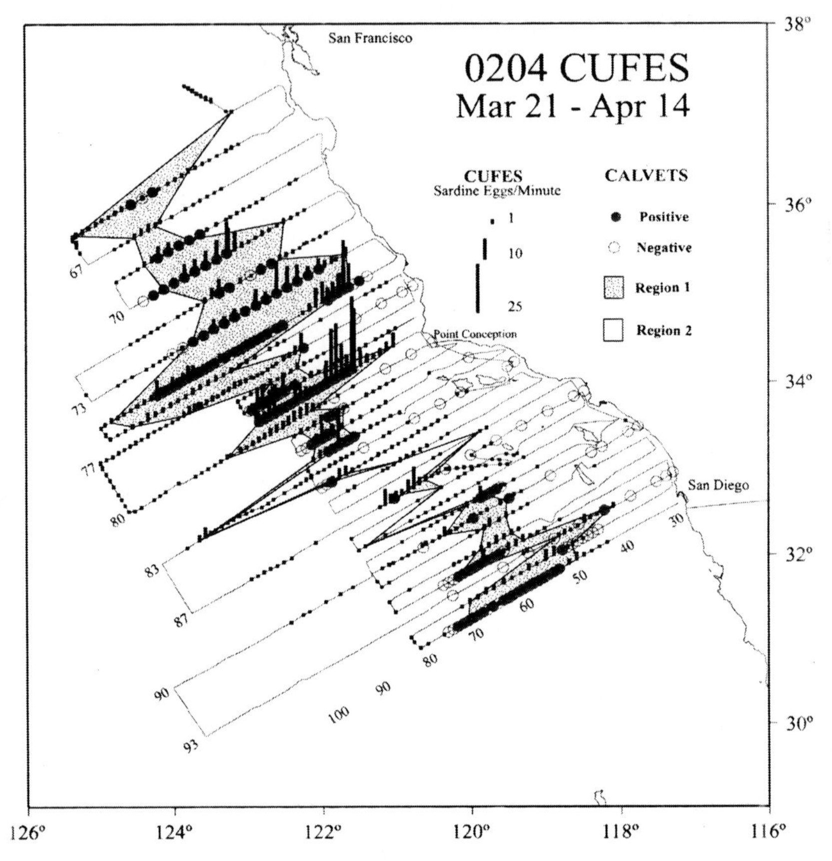

Figure 5.9. Sardine eggs from plankton net tows (solid circle denotes positive catch and open circle denotes zero catch) and from CUFES (stick denotes positive collection) in March–April 2002 survey. The numbers on line 93 are California Cooperative Fisheries Investigations (CalCOFI) station numbers. Region 1 is stippled area.

Region 2, egg production ($P_{0,2}$) was 0.13/0.05 m²/day (CV = 0.22) for an area of 236,679 km² (69,154 nm²). For the entire survey area of 325,082 km² (94,984 nm²), daily egg production was 0.728/0.05 m² (CV = 0.17) and egg mortality was 0.4 (CV = 0.15).

Discussion

We compared results from the 2002 survey to results from a conventional survey conducted in 1994 (Lo et al. 1996) to illustrate how changing to a CUFES-aided adaptive allocation design affected the Pacific sardine assessment (Table 5.4). We believe the comparison is instructive even though the two surveys differed somewhat in area and population size. The conven-

Table 5.3.

Number of positive tows of sardine eggs from plankton nets, yolk–sac larvae from plankton and Bongo nets, and eggs from CUFES in Region 1 (eggs/min ≥ 1) and Region 2 (eggs/min < 1) for both McArthur (Mc) and Jordan (Jord) cruises.

		Region								
		1			2					
Sampling Type	Outcome	Total	Mc	Jord	Total	Mc	Jord	Total	Mc	Jord
Plankton net eggs	positive	130	112	18	12	6	6	142	118	24
	Total	149	127	22	68	25	43	217	152	65
Plankton net yolk–sac	positive	83	76	7	29	13	16	112	89	23
	Total	149	127	22	68	25	43	217	152	65
Bongo net yolk–sac	positive	4	–	4	23	–	23	27	–	27
	Total	7	–	7	58	–	58	65	–	65
CUFES eggs	positive	453	389	64	372	252	120	825	641	184
	Total	495	428	67	1,127	737	390	1,622	1,165	457

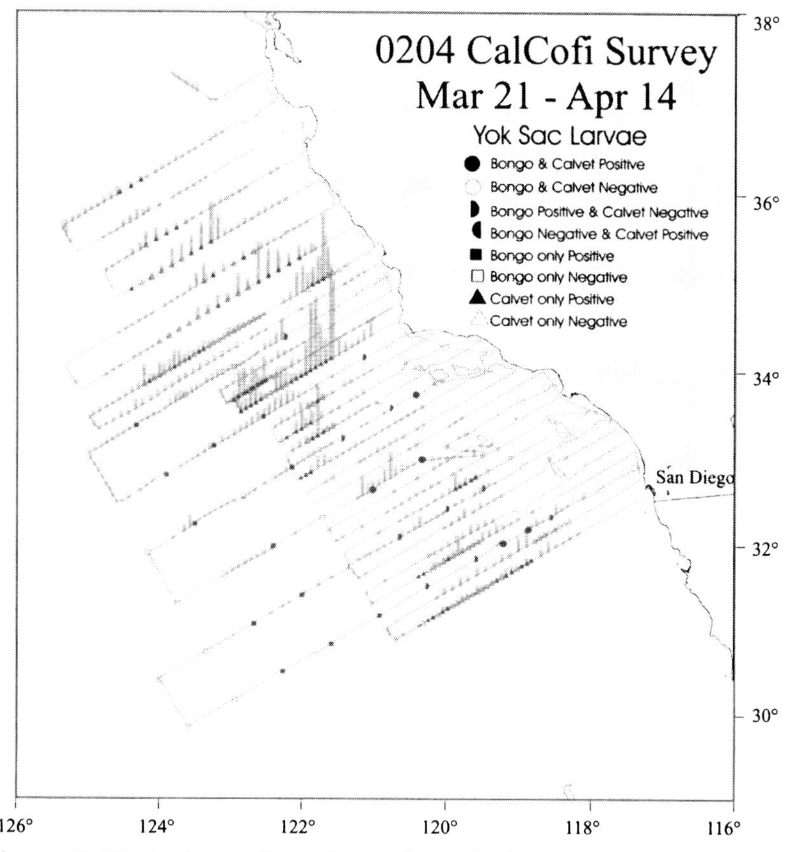

Figure 5.10. Sardine yolk-sac larvae from plankton net tows (circle and triangle) and from Bongo net tows (circle and square) in March–April 2002 survey. Solid symbols are positive and open symbols are zero catch.

Table 5.4.

Sardine daily egg production (P_0) from a conventional survey (1994) compared to an adaptive allocation survey (2002). We used the adaptive allocation survey observations from a CUFES to allocate plankton net tows.

	Survey	
Variable	Conventional	Adaptive allocation
Survey area (km^2)	380,175	325,082
Plankton tows		
Total	684	217
Positive for eggs	72	142
Percent positive	11	65
CUFES Samples		
Total	–	1,622
Positive for eggs	–	825
Percent positive	–	51
High-density stratum	–	91
Low-density stratum	–	33
Daily egg production		
P_0 (per 0.05 m^2)	0.19	0.73
CV	0.22	0.17
Daily specific fecundity (eggs/g)	11.39	22.94
Spawning biomass (mt)	127,102	206,033

tional 1994 survey covered a slightly larger area than the 2002 survey (380,175 km^2 vs. 325,082 km^2), and the total biomass of sardine was lower in 1994 than in 2002 (127,000 mt in 1994 vs. 206,000 mt in 2002).

An obvious difference in the results of the two surveys was that only 11% (74/684) of CalVET net tows were positive for sardine eggs in the 1994 conventional survey, whereas 65% (142/217) were positive in the 2002 survey (Table 5.4). This indicates that CUFES-aided adaptive allocation sampling was effective in allocating plankton net tows and thereby reducing ship time costs. The coefficients of variation for the estimates of P_0 were similar: 0.22 for the conventional survey compared to 0.17 for the CUFES-aided adaptive allocation survey. Thus, the variance penalty for using the ratio estimator q did not greatly diminish the benefit in using CUFES to post-stratify and allocate all plankton net tows to Region 1. This simple statistical comparison, however, does not reveal the greatest potential benefits in using CUFES-aided adaptive allocation sampling. Adaptive allocation would be most useful when the population is at a lower level, as it was in 1994, because at such levels one must survey a large area to ensure an unbiased estimator, but the population is probably concentrated in a very small fraction of the survey area. In addition, the high resolution maps of spatial distribution of eggs provided by CUFES have not as yet been incorporated into the survey design like Hanselman et al. (2001) did, but we plan to do so in the future. We also will develop new insights into the processes involved in selection of spawning habitats by the parents.

Any adaptive sampling requires a critical value to determine when to take additional observations. In our case, the critical value was an egg density from CUFES that triggered full water column sampling using the plankton net tows. We used a critical value of 1 egg/min, which was equivalent to 2–4 eggs/tow, depending on the degree of water mixing. In the past, the critical value was 2 eggs/min, which was equivalent to 4–8 eggs/tow. This range of critical values (2–8 eggs/tow) was similar to the value (5 eggs/tow) used in a stratified sampling design for an anchovy survey in Biscay Bay in Spain (Petitgas 1997).

An optimal critical value exists for each species and survey area. The critical value can be determined prior to the survey or during the survey using order statistics (Thompson and Seber 1996; Quinn et al. 1999). The extent to which the critical value can be fine-tuned to deliver an optimum balance between CUFES and plankton net tows for a particular region, species, and season is unknown. One factor is the large difference in catch ratios of eggs/plankton net tow to eggs/min from CUFES among years; this ratio ranged from 0.145 (2001) to 0.73 (1996), with most values around 0.25. This wide range does not support the idea of fine-tuning. These differences may overstate the expected variability for sardine because the areas were different; 1996 samples were taken over a very limited portion of the survey area, whereas in other years, the samples were collected from high-density spawning areas. Interestingly, our 2002 estimate (0.24) was similar to that we computed for sardines off the coast of South Africa (van der Lingen et al. 1998) and in previous years (Lo and Macewicz 2002).

In effect, the egg density from CUFES was used as an auxiliary variable to allocate plankton net tows. Fish eggs are constantly monitored while the CUFES is continuously pumping water. As a result, CUFES is a labor-intensive operation. To apply the adaptive allocation sampling using CUFES, fish eggs have to be easily identified by CUFES operators on the ship. Misidentification of eggs leads to large variance and possible bias. If that is not possible, other variables that are easier to measure and are somehow correlated with the variables of interest (sardine eggs in our case) can be used to base the adaptive allocation (Thompson and Seber 1996). Auxiliary variables include sea surface temperature, chlorophyll, plankton volumes for fish populations, and birds for some marine mammals.

Discussion and Future Directions

Ultimately, the practical efficiency (see Tukey 1986:97) of adaptive sampling will depend on the extent that the method is put into practice, which in turn depends on resolving (or alleviating) the challenges outlined in this chapter. It has been slightly more than a decade since Thompson (1990) introduced adaptive cluster sampling, and there is now a rich and growing base of literature focusing on adaptive sampling. However, practical application has lagged behind theoretical and methodological development. We note that 10 years after its introduction the Jolly-Seber method (Jolly 1965; Seber 1965) had not been practiced much, although it is now widely practiced in its many extensions. Hopefully, some of the material in this chapter will encourage statisticians to continue method development and stimulate biologists to experiment with adaptive sampling procedures. Although we outlined some challenges, we also offered possible solutions, and we firmly believe that more and better solutions will be discovered as biologists practice adaptive sampling on a variety of populations and under a variety of field conditions.

To help overcome the challenges that we outlined, we see a need for further method development on several fronts. First, guidelines need to be developed to help biologists identify populations that are candidates for adaptive procedures. Second, there is a need for continued work on strategies and alternative designs for restricting the final sample size. Third, because detectability is such a pervasive issue in animal ecology, methods of incorporating detectability into the finite population framework must be applied. Finally, user-friendly software would be helpful to simulate sampling before implementation and to analyze data from adaptive sampling designs.

We see at least three approaches to identifying candidate populations for adaptive sampling. Identification can be made for a specific population through a pilot survey (Salehi and Seber 1997), for a particular species (or taxonomic group) through experimental applications, ideally over multiple populations/sites (Lo et al. 1997; Hanselman et al. 2003; Smith et al. 2003), or for statistical populations that then could be compared to biological populations (Thompson 1994; Brown 2003). We expect that examples of the first and second approaches will multiply as biologists experiment with adaptive sampling procedures. Adaptive sampling might catch on within studies of certain taxa. For example, benthic organisms and pelagic fisheries are taxonomic groups that appear to be good candidates for adaptive sampling. We expect that statisticians will make substantial progress on establishing statistical criteria to guide application. Brown (2003) found that adaptive sampling generally performed well on statistical populations with small network sizes. The next step is to quantify the statistical criteria so that empirical measures can be used to guide application. For example, what are the network sizes or values for dispersion indices that correspond to appropriate application of adaptive sampling? Once we know the answers to that question, we can compare those values to population measurements taken in pilot surveys or to prior data to decide whether and how to apply adaptive sampling.

There will likely be substantial progress on methods to restrict the final sample size. In our experience, the open-endedness of the final sample size is a major deterrent to application of adaptive sampling. In applications to real populations, stopping rules have often been used because of the need to restrict final sample size (Lo et al. 1997; Hanselman et al. 2003; see also case study 2). Recent theoretical work by Salehi and Seber (2001) holds the promise that unbiased estimators will be derived for restricted designs. Other promising developments on the horizon include adaptive sampling designs that do not require a neighborhood. The absence of a neighborhood eliminates edge units and can remove much of the uncertainty in the final sample size. Neighborhood-free designs include adaptive allocation (Lo et al. 2001; see also case study 3) and sequential sampling designs (Christman 2003). A neighborhood-free adaptive sampling design called two-stage sequential sampling (TSS), recently developed by M. Salehi (Isfahan University of Technology, Isfahan, Iran), has been shown to perform well on a variety of populations compared to both conventional sampling designs and neighborhood-based adaptive sampling designs (Salehi and Smith, 2004). In the TSS design, an initial sample of secondary units (u_1) is selected within a sample of primary units. A condition is evaluated independently within each primary unit. If the condition is met, an additional sample of secondary units (u_2) is selected, but sampling stops there regardless of observations in the u_2 units. So under the TSS design the final sample size is restricted to be no more than $u_1 + u_2$ in each primary unit.

Because detectability is an issue for many species, biologists might be reluctant to even consider adaptive sampling until procedures that account for detectability have been well established and demonstrated on species at least closely related to the study species. Thompson and Seber (1994) presented methodology to account for detectability in adaptive sampling. Pollard and Buckland (1997) developed a technique to combine adaptive sampling with line transect sampling that adjusts for imperfect detectability. When applied in a survey of harbor porpoise, adaptive sampling reduced the variance in density estimates compared to traditional line transect sampling because increased observations resulted in improved estimates of detectability (Palka and Pollard 1999). Smith et al. (2000, 2001) demonstrated methods to incorporate detectability via double sampling in freshwater mussel surveys that use conventional finite sampling designs, and those methods can be extended to similar surveys that use adaptive sampling. Because mark-recapture sampling is an important technique used to account for imperfect detectability, combining mark-recapture and adaptive sampling would be productive.

Not many statistical techniques gain widespread acceptance without full-featured software that performs the necessary calculations. A software program called Visual Sampling Plan (VSP) produced by Battelle Memorial Institute (download at http://dqo.pnl.gov/vsp/vspsoft.htm) is a very powerful tool for sampling design, and it includes modern designs such as adaptive cluster sampling and ranked set sampling. VSP was developed to support contaminant monitoring and assessment, so the language used to describe sampling does not match how a biologist might discuss sampling. For example, sampling goals are framed in terms of comparisons to thresholds and reference values rather than estimating density, abundance, or biomass. Also, the range of adaptive sampling designs that are implemented in VSP is limited to selecting the initial sample by simple random sampling, two neighborhood shapes, and no option for stopping rules. We see a need for a similarly featured software package focusing on biological applications and including a much wider range of adaptive sampling designs.

Acknowledgments

We thank Steve Thompson for introducing us to adaptive cluster sampling and Bill Thompson for editing this volume. Mohammad Salehi M. and Frank van Manen provided comments that helped us improve the manuscript. D.R.S. would like to acknowledge Rita Villella and David Lemarié for cooperative work on developing efficient methods to sample freshwater mussels. J.A.B. thanks Laura Sessions and Malcolm Thomas for their helpful comments on the possum study and Fraser Maddigan for collecting the field data. N.C.H.L. thanks John Hunter for recommending use of CUFES to sample eggs of small pelagic fishes, such as sardine, anchovy, and jack mackerel. She also thanks Amy Hays, Valerie Growney, Beverly Macewicz, Ben Maurer, David Griffith, Ron Dotson, Richard Hasler, David Ambrose, Sharon Charter, William Watson and crew members of R/V *David Starr Jordan* and R/V *McArthur* for collecting all the egg, larval, and adult sardines. David Ambrose, Sharon Charter, William Watson, and Elaine Acuna staged sardine eggs. Thanks go to Richard Charter for making maps of sardine egg and larvae and trawl locations, William Watson for reading the report on sardine sampling, and Michelle DeLaFuente for organizing the report.

References

Acharya, B., G. Bhattarai, A. de Gier, and A. Stein. 2000. Systematic adaptive cluster sampling for the assessment of rare tree species in Nepal. *Forest Ecology and Management* 137:65–73.

Barnes, J. T., M. Yaremko, L. Jacobson, N. C. H. Lo, and J. Stehly. 1997. *Status of the Pacific Sardine (*Sardinops sagax*) Resource in 1996*. USDOC NOAA Technical Memo, NOAA-TM-NMFS-SWFSC-237, NTIS No. PB97-167761, National Marine Fisheries Service, La Jolla, California.

Boetsch, J. R., F. T. van Manen, and J. D. Clark. 2003. Predicting rare plant occurrence in Great Smoky Mountains National Park, USA. *Natural Areas Journal* 23:229–237.

Borkowski, J. J. 1999. Network inclusion probabilities and Horvitz-Thompson estimation for adaptive simple Latin square sampling. *Environmental and Ecological Statistics* 6:291–311.

Bradbury, A. 2000. Stock assessment and management of red sea urchins (*Strongylocentrotus franciscanus*) in Washington. *Journal of Shellfish Research* 19:618–619.

Brown, J. A. 1994. The application of adaptive cluster sampling to ecological studies. pp. 86–97 in D. J. Fletcher and B. F. J. Manly, eds., *Statistics in Ecology and Environmental Monitoring*. University of Otago Press, Dunedin, New Zealand.

_____. 2003. Designing an efficient adaptive cluster sample. *Environmental and Ecological Statistics* 10:95–105.

Brown, J. A., and B. F. J. Manly. 1998. Restricted adaptive cluster sampling. *Environmental and Ecological Statistics* 5:49–63.

Brown, J. A., and M. D. Thomas. 2000. *Residual Trap-Catch Methodology for Low-Density Possum Populations*. Contract Report UCDMS 2000/6, University of Canterbury, Christchurch, New Zealand.

Brown, K., J. Innes, and R. Shorten. 1993. Evidence that possums prey on and scavenge birds' eggs, birds, and mammals. *Notornis* 40:169–177.

Checkley, D. M. Jr., P. B. Ortner, L. R. Settle, and S.R. Cummings. 1997. A continuous, underway fish egg sampler. *Fisheries Oceanography* 6:58–73.

Christman, M. C. 1997. Efficiency of some sampling designs for spatially clustered populations. *Environmetrics* 8:145–166.

_____. 2000. A review of quadrat-based sampling of rare, geographically clustered populations. *Journal of Agricultural, Biological, and Environmental Statistics* 5:168–201.

_____. 2003. Adaptive two-stage one-per-stratum sampling. *Environmental and Ecological Statistics* 10:43–60.

Christman, M. C., and F. Lan. 2001. Inverse adaptive cluster sampling. *Biometrics* 57:1096–1105.

Clout, M., and K. Ericksen. 2000. Anatomy of a disastrous success: the brushtail possum as an invasive species. pp. 1–9 in T. L. Montague, ed., *The Brushtail Possum: The Biology, Impact and Management of an Introduced Marsupial.* Manaaki Whenua Press, Lincoln, New Zealand.

Coleman, J., and P. Livingstone. 2000. Fewer possums: Less bovine Tb. pp. 220–231 in T. L. Montague, ed., *The Brushtail Possum: The Biology, Impact and Management of an Introduced Marsupial.* Manaaki Whenua Press, Lincoln, New Zealand.

Conners, M. E., and S. J. Schwager. 2002. The use of adaptive cluster sampling for hydroacoustic surveys. *ICES Journal of Marine Science* 59:1314–1325.

Cowan, P.E. 1990. Brushtail possum. pp. 68–98 in C. M. King, ed., *The Handbook of New Zealand Mammals.* Oxford University Press, Auckland.

Curriero, F. C., M. E. Hohn, A. M. Liebhold, and S. R. Lee. 2002. A statistical evaluation of non-ergodic variogram estimators. *Environmental and Ecological Statistics* 9:89–110.

di Battista, T. 2002. Diversity index estimation by adaptive sampling. *Environmetrics* 13:209–214.

Dryver, A. L. 2003. Performance of adaptive cluster sampling estimators in a multivariate setting. *Environmental and Ecological Statistics* 10:107–113.

Green, W.Q. 1984. A review of ecological studies relevant to management of the common brushtail possum. pp. 483–499 in A. P. Smith and I. D. Hume, eds., *Possums and Gliders.* Australian Mammal Society, Chipping Norton, Australia.

Hanselman, D. H., T. J. Quinn II, J. Heifetz, D. Clausen, and C. Lunsford. 2001. Spatial inferences from adaptive cluster sampling of Gulf of Alaska rockfish. pp. 303–325 in G. H. Kruse, N. Bez, A. Booth, M. W. Dorn, S. Hills, R. N. Lipcius, D. Pelletier, C. Roy, S. J. Smith, and D. Witherells, eds., *Spatial Processes and Management of Marine Populations.* Lowell Wakefield Fisheries Symposium Series No. 17.

Hanselman, D. H., T. J. Quinn II, C. Lunsford, J. Heifetz, and D. Clausen. 2003. Applications in adaptive cluster sampling of Gulf of Alaska rockfish. *Fisheries Bulletin* 101:501–513.

Hill, K. T., L. D. Jacobson, N. C. H. Lo, M. Yaremko, and M. Dege. 1999. *Stock Assessment of Pacific Sardine for 1998 with Management Recommendations for 1999.* California Department of Fish and Game Marine Region Administrative Report 99-4, California Department of Fish and Game, Sacramento.

Hill, K. T., M. Yaremko, L. D., Jacobson, N. C. H. Lo, and D. A. Hanan. 1998. *Stock Assessment and Management Recommendations for Pacific Sardine in 1997.* California Department of Fish and Game Marine Region Administrative Report 98-5, California Department of Fish and Game, Sacramento.

Hunter, J. R., and N. C. H. Lo. 1997. The daily egg production method of biomass estimation: some problems and potential improvements. *Ozeanografika* 2:41–69.

Jolly, G. M. 1965. Explicit estimates from capture-recapture data with both death and immigration—Stochastic model. *Biometrika* 52:225–247.

Lo, N. C. H., Y. A. Green Ruiz, M. J. Cervantes, H. G. Moser, and R. J. Lynn. 1996. Egg production and spawning biomass of Pacific sardine (*Sardinops sagax*) in 1994, determined by the daily egg production method. *California Cooperative Oceanic Fisheries Investigations Report* 37:160–174.

Lo, N. C. H., D. Griffith, and J. R. Hunter. 1997. Using restricted adaptive cluster sampling to estimate Pacific hake larval abundance. *California Cooperative Oceanic Fisheries Investigations Report* 38:103–113.

Lo, N. C. H., J. R. Hunter, and R. Charter. 2001. Use of a continuous egg sampler for ichthyoplankton survey: application to the estimation of daily egg production of Pacific sardine (*Sardinops sagax*) off California. *Fisheries Bulletin* 99:554–571.

Lo, N. C. H., and B. Macewicz. 2002. *Daily Egg Production and Spawning Biomass of Pacific Sardine (Sardinops sagax) of California in 2002.* Southwest Fisheries Science Center Administrative Report LJ-02-40, National Marine Fisheries Service, La Jolla, California.

MacCall, A. D. 1979. Population estimates for the waning years of the Pacific sardine fishery. *California Cooperative Oceanic Fisheries Investigations Report* 20:72–82.

McDonald, L. L., G. W. Garner, and D. G. Robertson. 1999. Comparison of aerial survey procedures for estimating polar bear density: Results of pilot studies in northern Alaska. pp. 37–52 in G. W. Garner, S. C. Amstrup, J. L. Laake, B. F. J. Manly, L. L. McDonald, and D. G. Robertson, eds., *Marine Mammal Survey and Assessment Methods*. Balkema, Rotterdam, Netherlands.

Munholland, P. L., and J. J. Borkowski. 1996. Simple Latin square sampling +1: A spatial design using quadrats. *Biometrics* 52:125–136.

Murphy, G. I. 1966. Population biology of the Pacific sardine (*Sardinops caerulea*). *Proceedings of the California Academy of Sciences* 34:1–84.

Muttlak, H. A., and A. Khan. 2002. Adjusted two-stage adaptive cluster sampling. *Environmental and Ecological Statistics* 9:111–120.

Palka, D., and J. H. Pollard. 1999. Adaptive line transect survey for harbor porpoises. pp. 3–12 in G. W. Garner, S. C. Amstrup, J. L. Laake, B. F. J. Manly, L. L. McDonald, and D. G. Robertson, eds., *Marine Mammal Survey and Assessment Methods*. Balkema, Rotterdam, Netherlands.

Parker, K. 1985. Biomass model for egg production method. pp. 5–6 in R. Lasker, ed., *An Egg Production Method for Estimating Spawning Biomass of Pelagic Fish: Application to the Northern Anchovy* (Engraulis mordax). NOAA Technical Report No. 36, National Marine Fisheries Service, La Jolla, California.

Petitgas, P. 1997. Use of disjunctive Kriging to analyze an adaptive survey design for anchovy, *Engraulis encraicolus*, eggs in Biscay Bay. *Ozeanografika* 2:121–132.

Picquelle, S., and G. Stauffer. 1985. Parameter estimation for an egg production method of northern anchovy biomass assessment. pp. 7–16 in R. Lasker, ed., *An Egg Production Method for Estimating Spawning Biomass of Pelagic Fish: Application to the Northern Anchovy* (Engraulis mordax). NOAA Technical Report No. 36, National Marine Fisheries Service, La Jolla, California.

Pollard, J. H., and S. T. Buckland. 1997. A strategy for adaptive sampling in shipboard line transect surveys. *International Whaling Commission Report* 47:921–931.

Pollard, J. H., D. Palka, and S. Buckland. 2002. Adaptive line transect sampling. *Biometrics* 58:862–870.

Pontius, J. A. 1997. Strip adaptive cluster sampling: Probability proportional to size selection of primary units. *Biometrics* 53:1092–1096.

Pracy, L.T. 1974. *Introduction and Liberation of the Possum (*Trichosurus vulpecula*) into New Zealand*. Information Series No. 45, New Zealand Forest Service, Wellington.

Quinn, T. J., II, D. Hanselman, D. Clausen, C. Lunsford, and J. Heifetz. 1999. Adaptive cluster sampling of rockfish populations. *Proceedings of the American Statistical Association, Biometrics Section* 11–20. American Statistical Association, Alexandria, Virginia.

Ricciardi, A., and J. B. Rasmussen. 1999. Extinction rates of North American freshwater fauna. *Conservation Biology* 13:1220–1222.

Roesch, F. A., Jr. 1993. Adaptive cluster sampling for forest inventories. *Forest Science* 39:655–669.

Salehi M., M. 1999. Rao-Blackwell version of the Horvitz-Thompson and Hansen-Hurwitz in adaptive cluster sampling. *Environmental and Ecological Statistics* 6:183–195.

_____. 2003. Comparison between Hansen-Hurwitz and Horvitz-Thompson estimators for adaptive cluster sampling. *Environmental and Ecological Statistics* 10:115–127.

Salehi M., M., and G. A. F. Seber. 1997. Two-stage adaptive sampling. *Biometrics* 53:959–970.

_____. 2001. A new proof of Murthy's estimator which applies to sequential sampling. *Australian and New Zealand Journal of Statistics* 43:901–906.

_____. 2002. Unbiased estimators for restricted adaptive cluster sampling. *Australian and New Zealand Journal of Statistics* 44:63–74.

_____. 2004. A general inverse sampling scheme and its application to adaptive sampling. *Australian and New Zealand Journal of Statistics*. In press.

Salehi M., M., and D. R. Smith. 2004. Two-stage sequential sampling: A neighborhood-free adaptive sampling procedure. *Journal of Agricultural, Biological, and Environmental Statistics*. In press.

Scannel, C. L., T. Dickerson, P. Wolf, and K. Worcester. 1996. *Application of an Egg Production Method to Estimate the Spawning Biomass of Pacific Sardines off Southern California in 1986*. Southwest Fisheries Science Center Administrative Report LJ-96-01, National Marine Fisheries Service, La Jolla, California.

Seber, G. A. F. 1965. A note on the multiple recapture census. *Biometrika* 52:249–259.

_____. 1982. *The Estimation of Animal Abundance and Related Parameters*. 2nd ed. Griffin, London.

Smith, D. R., M. J. Conroy, and D. H. Brakhage. 1995. Efficiency of adaptive cluster sampling for estimating density of wintering waterfowl. *Biometrics* 51:777–788.

Smith, D. R., R. F. Villella, and D. P. Lemarié. 2001. Survey protocol for assessment of endangered freshwater mussels in the Allegheny River, Pennsylvania. *Journal of the North American Benthological Society* 20:118–132.

_____. 2003. Application of adaptive cluster sampling to low-density populations of freshwater mussels. *Environmental and Ecological Statistics* 10:7–15.

Smith, D. R., R. F. Villella, and D. P. Lemarié, and S. von Oettingen. 2000. How much excavation is needed to monitor freshwater mussels? pp. 203–218 in P. D. Johnson and R. S. Butler, eds., *Proceedings of the First Freshwater Mollusk Conservation Society Symposium*. Ohio Biological Survey, Columbus.

Smith, P. E., W. C. Flerx, and R. P. Hewitt. 1985. The CalCOFI vertical egg tow (CalVET) net. pp. 27–32 in R. Lasker, ed., *An Egg Production Method for*

Estimating Spawning Biomass of Pelagic Fish: Application to the Northern Anchovy (Engraulis mordax). NOAA Technical Report No. 36, National Marine Fisheries Service, La Jolla, California.
Strayer, D. L., S. Claypool, and S. Sprague. 1997. Assessing unionid populations with quadrats and timed searches. pp. 163–169 in K. S. Cummings, A. C. Buchanan, C. A. Mayer, and T. J. Naimo, eds., *Conservation and Management of Freshwater Mussels II: Initiatives for the Future*. Upper Mississippi River Conservation Committee, Rock Island, Illinois.
Strayer, D. L., and D. R. Smith. 2003. A guide to sampling freshwater mussel populations. *American Fisheries Society Monograph* 8:1–110.
Su, Z., and T. J. Quinn II. 2003. Estimator bias and efficiency for adaptive cluster sampling with order statistics and a stopping rule. *Environmental and Ecological Statistics* 10:17–41.
Thomas, M. D., J. A. Brown, and R. J. Henderson. 1999. Feasibility of using wax blocks to measure rat and possum abundance in native forest. *New Zealand Plant Protection* 52:125–129.
Thomas, M. D., J. A. Brown, F. W. Maddigan, and L. A. Sessions. 2003. Comparison of trap-catch and bait interference methods for estimating possum densities. *New Zealand Plant Protection* 56:81–85.
Thompson, S. K. 1990. Adaptive cluster sampling. *Journal of the American Statistical Association* 85:1050–1059.
_____. 1991a. Adaptive cluster sampling: Designs with primary and secondary units. *Biometrics* 47:1103–1115.
_____. 1991b. Stratified adaptive cluster sampling. *Biometrika* 78:389–397.
_____. 1993. Multivariate aspects of adaptive cluster sampling. pp. 561–572 in G. P. Patil and C. R. Rao, eds., *Multivariate Environmental Statistics*. North Holland/Elsevier Science Publishers, New York.
_____. 1994. *Factors Influencing the Efficiency of Adaptive Cluster Sampling*. Center for Statistical Ecology and Environmental Statistics Technical Report 94-0301, Pennsylvania State University, University Park.
_____. 1996. Adaptive cluster sampling based on order statistics. *Environmetrics* 7:123–133.
_____. 2003. Editorial: Special issue on adaptive sampling. *Environmental and Ecological Statistics* 10:5–7.
Thompson, S. K., F. L. Ramsey, and G. A. F. Seber. 1992. An adaptive procedure for sampling animal populations. *Biometrics* 48:1195–1199.
Thompson, S. K., and G. A. F. Seber. 1994. Detectability in conventional and adaptive sampling. *Biometrics* 50:712–724.
_____. 1996. *Adaptive Sampling*. Wiley, New York.
Tukey, J. 1986. *Collected Works of John Tukey*, vol. 3. CRC Press, New York.
van der Lingen, C. D., D. Checkley, Jr., M. Barange, L. Hutchings, and K. Osgood. 1998. Assessing the abundance and distribution of eggs of sardine, *Sardinops sagax*, and round herring, *Etrumenus whiteheadi*, on the western Agulhas Bank, South Africa, using a continuous, underway fish egg sampler. *Fisheries Oceanography* 7:135–147.
Wolf, P. 1988a. *Status of the Spawning Biomass of Pacific Sardine 1987–1988*. California Department of Fish and Game Marine Resources Division Report to the Legislature, California Department of Fish and Game, Sacramento.
_____. 1988b. *Status of the Spawning Biomass of Pacific Sardine 1988–1989*.

California Department of Fish and Game Marine Resources Division Report to the Legislature, California Department of Fish and Game, Sacramento.
Woodby, D. 1998. Adaptive cluster sampling: efficiency, fixed sample sizes, and an application to red sea urchins (*Strongylocentrotus franciscanus*) in southeast Alaska. pp. 15–20 in G. S. Jamieson and A. Campbell, eds., *Proceedings of the North Pacific Symposium on Invertebrate Stock Assessment and Management*. Canadian Special Publication of Fisheries and Aquatic Sciences No. 125, National Research Council of Canada, Ontario.
Zhang, N., Z. Zhu, and B. Hu. 2000. On two-stage adaptive cluster sampling to assess pest density. *Journal of Zhejiang University* 26:617–620.

6

Two-Phase Adaptive Stratified Sampling

Bryan F. J. Manly

Stratified random sampling is one of the more commonly used methods for estimating density and other characteristics of a biological population within a defined geographical area (Cochran 1977:Chapter 5; Thompson 1992:Chapter 11). With this method the total area is divided into a number of discrete subareas or strata such that within each of these strata the variable X of interest (e.g., the number of individuals per square meter) is expected to be relatively constant in comparison with the variation that exists over the entire area. A simple random sample then is taken from each stratum and the mean or total of X is estimated for that part of the population. An unbiased estimator of the overall population mean of X then is obtained by averaging the means for the individual strata, weighting them by the areas involved. An unbiased estimator of the overall population total is simply the sum of the estimates for the different strata.

The motivation for using stratified sampling is that if X has a small variance within each of the strata then the stratum means and totals will be estimated accurately even with small samples. Hence the population mean and total of X also will be estimated accurately. Stratified sampling is therefore a tool for eliminating effects of some of the population variation in X from the estimation of the population parameters.

If enough is known about the characteristics of a population, the sample allocation to strata can be optimized. Suppose that there are m strata, with U_i sample units in stratum i, and with the standard deviation of X equal to σ_i for these units. Also, let u_i denote the sample size in stratum i, and $u_T = \sum_{i=1}^{m} u_i$ denote the total sample size in all strata. Then it is well known (Cochran 1977:98) that the variance of the estimator of the population mean or total of X is minimized for a fixed total sample size when the sample size in stratum i is set at

$$u_i = \left(\frac{U_i \sigma_i}{\sum_{j=1}^{m} U_j \sigma_j} \right) u_T . \quad (6.1)$$

When estimating the density of an animal or plant population the stratification employed often will be based on habitat characteristics that are thought to be associated with the presence of the organism (e.g., riparian, conifer, aspen, etc.). At other times strata may simply be geographical areas because detailed habitat information is not available or the habitat associations of the organism are not well understood. In either case, the practical utility of equation (6.1) will be limited because only rough guesses will be available for the within-stratum standard deviations of X.

It is often the case that all that can be done is to assume that the standard deviation will be low in areas with a low density of animals and high in areas with a high density of animals. The sample then is allocated to strata based on guessed relative mean densities and the assumption that standard deviations are, for example, proportional to the mean densities. In practice, therefore, the sample allocation to strata may be far from optimal, even with a good stratification of the geographical area with respect to the variable X.

To overcome this problem in the context of fisheries stock assessment, Francis (1984) suggested a two-phase approach to sampling, with a similar scheme offered independently by Jolly and Hampton (1990). Francis proposed that a first-phase, stratified random sample should be taken using the best available information for the choice of strata and the allocation of the sample, based on equation (6.1). Then, using the results obtained from the first phase, a second sample should be taken to increase the number of sample units in those strata where this is expected to be most effective for reducing the variance of the estimator of the population mean or total. The idea, therefore, is to use the information obtained from the first-phase sample to compensate in the second phase for any shortcomings in the sample allocation.

Francis (1984) was concerned only with the estimation of one fish species in one fishery. However, his two-phase adaptive sampling method is easily generalized for use in situations where there are several populations to be estimated at several different locations (Manly et al. 2002). An example of such a situation that is described in more detail below involved the need to estimate abundance of three species of shellfish at 11 beaches around Auckland, New Zealand. In this case a limited amount of sampling effort was available, and there was a need, as far as possible, to get good estimates of shellfish numbers for all three species on all 11 beaches. To achieve this objective, the first phase of sampling consisted of taking 75 transect samples from each of the 11 beaches, with an appropriate stratification for each beach based on local knowledge. At the second phase another 145 transects were sampled. These were allocated out to the strata within beaches in such a way that the average of the coefficients of variation (CV) over 20% was reduced as much as possible for the estimates required from the study (the population size for each of the shellfish species for each of the beaches where the species was present).

In Section 2 of this chapter I provide some details for the Francis (1984) method for two-phase sampling of a single population and its generalization for several populations at several locations. Section 3 describes the results of some simulation studies of the original method and its generalization. Section 4 describes the shellfish example in more detail. I then summarize some recently obtained results concerning the use of bootstrapping for bias reduction and variance estimation in Section 5. Finally, I discuss the usefulness of the adaptive, two-phase stratified sampling design.

Details of the Two-Phase Sampling Design

As proposed by Francis (1984) for one population at one location, the first phase of the two-phase, adaptive stratified sampling design involves taking a conventional random stratified sample of size u_{T1} from the population, with the sample sizes in the different strata chosen to approximate the optimal allocation of equation (6.1), based on whatever prior knowledge is available. From the first phase data, the variances in the different strata can be estimated; these are assumed to be good approximations for the true variances. These variances are used to determine how best to allocate the second-phase sample of size u_{T2} to the strata.

In choosing the fraction of the total sample to allocate to the first phase of the survey, there must be a compromise between having u_{T1} large enough to give good estimates of variance and having u_{T2} large enough to make effective use of the information from the first phase. As a rule of thumb, allocating 75% of the units to the first phase and 25% to the second phase seems reasonable (Francis 1984).

At the second phase, the u_{T2} sample units are allocated out one by one to the strata. The first unit is allocated to the stratum where its use will give the largest reduction in the variance of the estimated population total or mean. The second unit is allocated to the stratum where its use will give the largest reduction in the variance of the estimated total or mean, given the allocation of the first extra unit, and so on. The process of adding units continues until all u_{T2} units have been allocated. No actual sampling is carried out during this second-phase allocation exercise, so the variances assumed for the strata remain equal to the estimates from the first-phase sample throughout the allocation process. Following the allocation, the extra second-phase sampling proceeds. The data then are analyzed as if they came from a conventional stratified random sample with a total size of $u_{T1} + u_{T2}$.

Treating the results of a two-phase stratified sample as a conventional stratified sample leads to a negative bias in the estimators of population mean and total. However, this generally seems quite small in comparison with the standard error (Francis 1984, 1991; Jolly and Hampton 1990, 1991). As demonstrated by Brown (1999), the two-phase design generally has better properties than the alternative adaptive cluster sampling method that was proposed by Thompson (1990), except for very highly clustered populations.

Francis (1984) considered two algorithms for two-phase sampling. The first involved estimating within-stratum variances with their sample values in the usual way, and the second involved assuming that the within-stratum variances are proportional to the squares of the means for these strata. The second algorithm seems to be appropriate for fisheries stock assessment, but here only the first algorithm is considered because of its more general applicability for populations in which the assumption that the variance is proportional to the mean squared may not be reasonable.

With Francis' original sampling design, there was only one observation on each sample unit, where this might typically be the population density per unit area for a single species. This can be generalized to the case in which each sample unit provides $S > 1$ observations, where S is number of species. For example, these observations might be the densities per unit area of the S species present in the region. The species then would represent several populations.

As a further generalization it can be assumed that the populations of interest occur at L different locations and that there is the need to estimate the mean or the total for each population at each location. For example, with the Auckland area shellfish study mentioned earlier, there were $L = 11$ beaches (the locations), at each of which there was one or more of the $S = 3$ species of shellfish (the populations). It was the individual beaches that were stratified, with the same strata applying for each of the shellfish species.

It is convenient to describe the generalized sampling design and analysis in terms of the problem of estimating the population totals for each of S species at each of several locations, assuming that the observations made on a sample unit consist of the densities per unit area for the S species. In fact, the allocation of sample units to strata within locations would be the same for estimating the mean densities rather than the population totals.

Suppose that an ordinary stratified random sample is taken at each of the locations. Let A_{ij} represent the area of stratum j at location i, u_{ij} be the number of units randomly sampled from stratum j at location i, $\bar{x}_{ijk}$ be the mean density per unit area for species k in stratum j at location i; and s_{ijk} be the sample standard deviation of density estimators for species k in stratum j at location i. Then the estimator of total population size for species k at location i is

$$\hat{N}_{ik} = \sum_{i} A_{ij}\bar{x}_{ijk}, \tag{6.2}$$

where the summation is over the strata at location i. The variance of this estimator (Cochran 1977:95) is

$$\hat{\text{Var}}(\hat{N}_{ik}) = \sum_{i} \frac{A_{ij}^2 s_{ijk}^2}{u_{ij}}. \tag{6.3}$$

Here, the sampling fraction is assumed to be small in all strata at all locations, so no finite population correction is required, although one could be added easily enough if necessary. The estimator of the CV in percentage terms is then

$$\hat{\text{CV}}_{ik} = 100 \times \left[\frac{\sqrt{\hat{\text{Var}}(\hat{N}_{ik})}}{\hat{N}_{ik}} \right]. \tag{6.4}$$

The proposed sampling design and analysis has the following steps:

1) At each of the L locations the study area is divided into strata using the best available information. The number of strata does not need to be the same at each location.
2) A first-phase sample with a total size of u_{T1} is allocated to the locations and strata. Possibly, but not necessarily, the same number of units is allocated to each location, and then equation (6.1) is used as far as possible when allocating units to strata within locations.
3) Means ($\bar{x}_{ijk}$) and standard deviations (s_{ijk}) are estimated using the first-phase sample data, for use in equations (6.2) to (6.4).
4) An optimization criterion is defined as

$$Z = P(\text{Mean CV over all populations and species}) + Q(\text{Maximum CV over all populations and species}) + R(\text{Mean of all CVs over } \alpha\%), \quad (6.5)$$

   where $P + Q + R = 1$, and $\alpha\%$ is some unsatisfactorily large value such as 20%.
5) A second-phase sample unit is allocated to the location and stratum where it gives the largest possible reduction in Z. This is determined by recalculating the CVs of equation (6.4) with the sample size increased by one, for each location and stratum, and finding the resulting change in Z. The location and stratum giving the maximum reduction then is chosen for use.
6) The sample size (u_{ij}) for the chosen location and stratum is incremented by one, whereas the sample mean ($\bar{x}_{ijk}$) and standard deviation (s_{ijk}) are left unchanged.
7) Steps 5 and 6 are repeated until u_{T2} second-phase units have been allocated. The second-phase data then are collected, and the data from both phases are combined for estimation using equations (6.2) to (6.4).

Different studies may require different values for the weights P, Q and R. For example, setting $P = 1$ and $Q = R = 0$ is appropriate where overall general accuracy of estimation is important, whereas setting $P = R = 0$ and $Q = 1$ places the emphasis on reducing the worse case CV.

A Simulation Study

Manly et al. (2002) carried out a simulation study to examine the properties of the generalized two-phase, adaptive stratified sampling design. This was based on real data from stratified surveys carried out in the past for the New Zealand Ministry of Fisheries. Ten model populations were set up using quadrat counts along transects from the past surveys, to represent 10 different locations. The quadrat counts were put in order based on their geographic locations within strata and then bootstrap resampled with replacement to produce new sets of data. There were two or three shellfish species recorded for each of the model populations, namely pipis (*Paphies australis*), cockles (*Austrovenus stutchburyi*), and wedge shells (*Macamona liliana*). The densities varied considerably among strata within locations and among locations. The authors concluded:

1) The estimated CV varied considerably for the estimation of different abundances, but the mean estimated CV was not very sensitive to the optimization criterion used.
2) The situation for the CVs actually obtained for estimation was similar to that for the estimated CVs in the sense that the different criteria produced about the same results, which suggested that the criterion used was not critical. Therefore, in general, it seemed reasonable to choose $P = Q = R = 1/3$ in equation (6.5) for use with real data.
3) Percentage biases were relatively small (less than ±5%) in general unless the corresponding CVs were large.
4) For CVs that were not too large (i.e., about 25% or less), the estimated CVs were nearly equal to the true CVs on average, but for situations where the true CV was large, there was a definite tendency for the estimated CV to be too low.
5) In comparison to stratified sampling with proportional allocation, the two-phase sampling process reduced the highest CVs and increased the lowest CVs (which is what was expected to happen).

Estimation of Actual Shellfish Populations

As mentioned before, the survey of shellfish on beaches around Auckland, New Zealand, involved 11 beaches and three shellfish species. The species were cockles, pipis, and tuatuas (*Paphies subtriangulata*), with tuatuas present on only one beach.

Information from local residents on likely areas of high density was used to define sampling strata. The first-phase sample was allocated out with 75 transects of four quadrats each randomly located on each beach, with the allocation to the strata within a beach being approximately proportional to twice the area if it was expected to have a high density for at least one species, and approximately proportional to the area otherwise.

The target estimated CV was 20% or less for each species on each beach. This was obtained from the first-phase samples for six of the beaches. For the other beaches, CVs ranged from 20% to 72% for pipis on one beach. However, the density of pipis on this beach was quite low, so it was not considered to be appropriate to give this too much weight in the optimization criterion. Therefore, the optimization criterion used was the mean CV for those over 20%.

Applying this criterion, an additional 145 transects were allocated out to the various strata on the beaches. Using all of the sample data the estimated CVs then were below 20% for all species and beaches, except for a few unimportant cases where the density of a species was very low. The overall result was considered to be quite satisfactory.

Bootstrapping for Bias Correction and Variance Estimation

Bootstrapping is a standard method for estimating the bias and variance of an estimator. This may or may not work well in practice (Manly 1997:Chapter 3), depending upon whether the data set being considered is large enough to represent the population of interest.

With the generalized two-phase sampling design, bootstrapping can easily be applied, as follows:

1) After all of the data have been collected, the densities of the different species in the sample units from the different strata at the different locations are used to represent a bootstrap population. By sampling these units with replacement, they can be made to represent the very large numbers of units in the real population.
2) The bootstrap population is sampled many times using the two-phase algorithm, with the total sample sizes at phases 1 and 2 fixed to be equal to those used with the real data. For each set of data obtained, the population sizes for each of the species at each of the locations are estimated. Each set of data then provides an estimated percentage relative bias, R = (Estimate – True Size)/(True Size), for each of the population size estimates. Here the "True Size" is the one that holds for the bootstrap population, which is the estimated size from the original set of data.

3) The mean values of R obtained from step 2, say R_m, are used as estimates of the true relative biases in estimated sizes. Corrected estimates then are calculated from R_m = (Initial Estimate – True Size)/(True Size), so that True Size = (Initial Estimate)/(1 + R_m). This latter equation suggests that the initial estimate of a population size can be corrected by dividing by (1 + R_m), which is how one obtains corrected values.
4) The variances of estimates, either corrected or not, are estimated by the observed variances of the bootstrap estimates.

Manly (2004) conducted a simulation study to examine whether bootstrapping was an effective tool with two-phase sampling. This was similar to the simulation study described above, but only involved model shellfish populations at five locations (beaches) using data obtained from past surveys.

The populations were sampled with a small total sample size (an average of 80 transects per location), which was expected to give very poor estimation of some population sizes with a moderate total sample size (an average of 160 transects per location), which was expected to give good estimates for most population size; and with a large total sample size (an average of 240 transects per location), which was expected to give good estimates for all of the population sizes. For the first-phase samples, 75% of the transects were allocated to the locations, with the same number of transects for each location, and proportional allocation to the strata within locations, based on area. Then at the second phase, Manly (2004) used the adaptive sampling algorithm to allocate the remaining 25% of the transects, one transect at a time.

Only 100 bootstrap samples were used for the simulation results because of the long simulation times that even this number required. With real data many thousands of bootstrap samples would be used, possibly leading to bootstrapping giving better results than those that were obtained in the simulation study. From the results of the study, Manly (2004) concluded:

1) Bootstrap bias correction can be expected to reduce but not necessarily eliminate biases.
2) Bias adjustment has very little effect on the variation in estimates.
3) Usually there is little advantage in estimating standard errors (and hence CVs) by bootstrapping instead of using equation (6.4) but occasionally bootstrapping will give variance estimates with lower sampling errors. If bootstrapping is used to remove bias, it might as well also be used to estimate standard errors.

4) The tendency to underestimate large CVs (> 20%) that was noted by Manly et al. (2002) persisted.

Discussion

Two-phase adaptive stratified random sampling is a useful design for situations in which the density of a plant or animal population is likely to vary with defined strata based on the habitat or general geographical areas but it is not possible to know in advance of sampling where the high densities are likely to occur. Except with very clustered populations, it is likely to give better estimation than the alternative adaptive sampling method of Thompson (1990) (see also Chapter 5, this volume), which is usually applied to rare populations. Two-phase adaptive stratified random sampling often will be easier to use than this alternative because there is no need to make the correct choices for sampling parameters like the trigger level for further sampling. The generalized version of two-phase adaptive sampling has proven to be particularly useful for surveys in which one needs to estimate population sizes for multiple species at multiple locations at the same time.

There is a price to pay in terms of bias in estimation because the usual stratified sampling estimators are used, although these do not account for the adaptive nature of the design. In particular, estimates of population size may be biased downward to some extent. However, the simulations that have been carried out all indicate that this bias will be small in percentage terms unless the CV is large, so that the estimation is poor, irrespective of any bias in the estimator. In addition, bootstrapping can be used to remove part of the bias. Some modified method of bootstrapping may possibly remove more of the bias, but this requires further investigation. Finally, two-phase sampling may be impractical with animals that are highly mobile. This is because their distribution in the study area may change completely between the time when phase 1 sampling is completed and phase 2 sampling begins.

References

Brown, J. A. 1999. A comparison of two adaptive sampling designs. *Australian and New Zealand Journal of Statistics* 41:395–403.

Cochran, W. G. 1977. *Sampling Techniques*. 3rd ed. Wiley, New York.

Francis, R. I. C. C. 1984. An adaptive strategy for stratified random trawl surveys. *New Zealand Journal of Marine and Freshwater Research* 18:59–71.

_____. 1991. Statistical properties of two-phase surveys: Comment. *Canadian Journal of Fisheries and Aquatic Science* 48:1128.

Jolly, G. M., and I. Hampton. 1990. A stratified random transect design for acoustic surveys of fish stocks. *Canadian Journal of Fisheries and Aquatic Science* 47:1282–1291.

_____. 1991. Reply to a comment by R.I.C.C. Francis. *Canadian Journal of Fisheries and Aquatic Science* 48:1128–1132.

Manly, B.F.J. 1997. *Randomization, Bootstrap and Monte Carlo Methods in Biology.* 2nd ed. Chapman and Hall, London.

_____. 2004. Using the bootstrap with two-phase adaptive stratified samples from multiple populations at multiple locations. *Environmental and Ecological Statistics* 11: In press.

Manly, B. F. J., J. M. Akroyd, and K. A. R. Walshe. 2002. Two-phase stratified random surveys on multiple populations at multiple locations. *New Zealand Journal of Marine and Freshwater Research* 36:581–591.

Thompson, S. K. 1990. Adaptive cluster sampling. *Journal of the American Statistical Association* 85:1050–1059.

_____. 1992. *Sampling.* Wiley, New York.

7

Sequential Sampling for Rare or Geographically Clustered Populations

Mary C. Christman

One of the more difficult tasks for the wildlife or conservation biologist is the estimation of the population size of a rare or elusive species. The costs of sampling a sufficiently large number of sites and the high probability of not observing the species of interest are the main considerations for sampling rare species. As a result, sampling often is done haphazardly with the intention of sampling only until some members of the population are located. The researcher who performs this type of sampling then often fails to incorporate the unusual sampling design into estimation.

I present examples of sampling designs that explicitly include sampling until sufficient numbers of individuals of a species are located. In addition, I give some recommendations for strategies for performing this type of adaptive sampling protocol.

Adaptive Sampling Designs

Sequential sampling is a type of adaptive sampling in which, at each observation in the sampling protocol, the decision to continue depends on the data recorded to that point. Hence, the sampling effort adapts to the data obtained during sampling. This differs from typical nonadaptive sampling designs where all sample sites are selected prior to data collection so that the sample size is fixed and known. As a result, adaptive sampling designs usually have random sample sizes because the researcher cannot determine a priori what will be observed during sampling.

There are many examples of adaptive sampling strategies. They include adaptive cluster sampling, adaptive allocation of sample sizes in stratified random sampling, adaptive allocation to different treatments in clinical trials, two-phase sampling, and sequential sampling. In adaptive cluster sampling, members of rare species are assumed to be clustered in space and hence sampling near sites known to have members of the species should increase the amount of information obtained. If members of a rare species are encountered, adjacent sampling sites are added to the sample according to the adaptive sampling protocol rules (see Chapter 5).

Adaptive allocation of sample sizes in stratified random sampling is performed when the stratum variances are unknown. Here, an initial sample is taken in each stratum and the sample variance of the estimator of interest is calculated. Then additional samples are allocated to the strata based on these initial estimates of strata variances. Adaptive allocation to treatments in clinical trials is another example of the use of data to allocate future samples. Here, subjects are initially assigned to treatments randomly. As the trial progresses the results of the treatments are evaluated. Future assignments of subjects to treatments are based on these intermediate results by adjusting the probability of a subject being assigned to a treatment according to the efficacy of the treatment (Hu and Rosenberger 2004). Two-phase sampling is similar to adaptive allocation in stratified random sampling except that the strata boundaries are determined based on the initial (first phase) sample (Thompson 1992; see Chapter 6, this volume).

Sequential Sampling

In sequential sampling surveys, data collection continues according to the initial design until some stopping rule is satisfied. Like the other examples, it also has a random sample size. There are many ways sequential sampling can be conducted, such as within strata or as part of multistage cluster sampling. Often it is based on a simple random sampling (SRS) design within strata or within clusters. All sequential sampling designs have predefined stopping rules that stipulate the requirements for sample termination (e.g., Christman 1993; Christman and Lan 2001).

The basic design for sequential sampling uses a probability-based sampling strategy to select units from a population (see below) until some condition regarding the observations is satisfied. For example, suppose one wishes to estimate the total abundance of a rare insect, a wolf spider (*Hogna* sp.), say, in a salt marsh. One approach would be to randomly select locations within the marsh and sample until spiders have been detected in at least 10 locations. Here, "population" is used in the statistical sense as the complete set of units (locations) with nonzero probability of being sampled. Hence, it is the set of all possible sites within the marsh that could have been selected for sampling of spiders. This is sometimes referred to as the sampling frame when the statistical population can be enumerated.

Inverse or sequential sampling stops when the stopping criterion has been met. At each stage in the data collection effort, the data accumulated are evaluated against this stopping criterion. If it is met, sampling stops. If not, another unit is observed and the criterion is again evaluated. Sampling continues until the criterion is met. When sampling is done using a probability-based design, the typical stopping rule is to sample until at least m of the units contain the rare species of interest.

Modifications of this rule exist as well. For example, one might sample until at least m of the units contain k or more members of the species of interest. Another modification is to start by randomly sampling u_0 units, where u_0 has been determined prior to any sampling. A stopping rule that could be applied is the following: if at least m units contain the rare species, then stop; otherwise, continue sampling until m units with rare species are observed. Here the minimum sample size is u_0, whereas in the first stopping rule example, the minimum sample size is $m \leq u_0$.

Sequential or inverse sampling can be independently conducted within strata and the usual stratified estimators can be used (Cochran 1977; Christman 2000). Alternatively, sequential sampling can be done within clusters at the second stage of two-stage cluster sampling. It also can be used at the first stage of cluster sampling to choose the number of clusters to be sampled.

Estimating the Mean Per Unit or the Population Total (*N*)

To estimate the mean per unit (e.g., number per meter^2) or the population total (N), we need an estimate of the proportion (π) of the population that meets the criterion used in the stopping rule. Because the sample size is random, the classical estimator of π based on a fixed sample size, $\hat{\pi} = m/u$

where m is the number of sampling units that met the criterion and u is the total number of units observed during sampling, is biased and should not be used. An unbiased estimator is given by $\hat{\pi}^* = (m - 1)/(u - 1)$ (Lehmann 1986). The variance of $\hat{\pi}^*$ is difficult to determine but has been shown to be bounded (Mikulski and Smith 1975):

$$\frac{1-\pi}{m} \leq \mathrm{var}(\hat{\pi}^*) \leq \frac{1-\pi}{m+\pi-2}. \qquad (7.1)$$

Conservative estimates of the variances of the estimators of the mean or total can be obtained using these bounds and some arguments concerning the likely upper bounds on π (Lan 1999; Christman and Lan 2001).

Sampling Until at Least m Units Are Observed that Satisfy the Stopping Criteria

It is relatively straightforward to estimate the mean per quadrat or population total (N) when the stopping rule is to sample until at least m of the units contain the rare species of interest. The equations are based on the specific stopping rule. Two criteria are commonly used. The typical stopping criterion is that at least m samples have one or more members of the rare species. The alternative is that at least m samples have k or more members of the rare species. The equations for each of these rules are shown in Box 7.1.

Example: Estimating the Abundance of Green-Winged Teal

Figure 7.1 shows a sampling frame that might be typical for studying the spatial distribution of a rare waterfowl species (modified from Smith et al. 1995). A map of a pond in southern Florida containing green-winged teal (*Anas crecca*) has been divided into 200 (U) equally sized quadrats (25 km^2 each). Random sampling will be done until at least two sampled quadrats contain teal. In Figure 7.1, a sample of $u = 20$ units contains $m = 2$ units with teal. An estimate of the proportion of quadrats that contain at least one duck is $\hat{\pi}^* = (m - 1)/(u - 1) = 0.0526$. The mean number of ducks per quadrat for the units containing ducks is $\bar{y}_1 = m^{-1}\sum_{i=1}^{m} y_i = 2^{-1}(55 + 24) = 39.5$. Combining the two estimates, the estimated mean per quadrat is $\hat{\pi}^*\bar{y}_1 = 0.0526(39.5) = 2.0777$ birds. The estimated total abundance of teal in the pond is $U\hat{\pi}^*\bar{y}_1 = 200(2.0777) = 415.54$ birds.

Box 7.1. Estimators for the average number of a rare species per sampling unit or for the population total when stopping the sampling effort is based on observing at least *m* units that satisfy the stopping criterion.

Criterion for stopping rule	*Popoulation mean*	*Population total*
m units must have 1 or more members of the rare species observed	$\hat{\pi}^*\bar{y}_1$	$U\hat{\pi}^*\bar{y}_1$
m units must have k or more members of the rare species observed	$\hat{\pi}^*\bar{y}_k + (1-\hat{\pi}^*)\bar{y}_{nk}$	$U\hat{\pi}^*\bar{y}_k + U(1-\hat{\pi}^*)\bar{y}_{nk}$

Here, $\hat{\pi}^* = (m-1)/(u-1)$ is the estimator of the proportion of the population that satisfies the criterion (see text); $\bar{y}_1 = m^{-1}\sum_{i=1}^{m} y_i$ is the sample mean number of members of the rare species per quadrat for the m sampling units in which they were observed; U is the total number of sampling units available for potential sampling; $\bar{y}_1 = m^{-1}\sum_{i=1}^{m} y_i$ is the sample mean number of members of the rare species per quadrat for the m sampling units in which there were k or more members of the species; and $\bar{y}_{nk} = (u-m)^{-1}\sum_{i=1}^{u-m} y_i$ is the sample mean number of members of the rare species per quadrat for the $u-m$ sampling units in which there were less than k members of the species.

0	0	0	0	0	0	0	0	0	0	0	0	0	0	0	0	0	0	0	0
0	0	0	0	0	0	0	0	0	0	0	0	0	0	0	0	0	0	0	0
0	0	0	0	0	0	0	0	0	0	0	0	0	0	35	0	0	0	0	0
0	0	0	0	0	0	0	0	0	0	0	0	0	0	0	0	75	0	0	0
0	0	0	0	0	0	0	0	0	0	0	0	0	0	0	0	0	55	13	0
0	0	0	0	0	0	0	0	0	0	0	0	0	0	0	0	0	0	0	0
0	0	0	0	0	0	0	0	0	0	0	0	0	0	0	0	0	0	0	0
0	0	0	0	0	0	0	0	0	0	0	0	0	0	0	0	0	0	0	0
0	0	0	0	0	0	0	0	0	0	0	0	0	0	0	0	0	0	0	24
0	0	0	0	0	0	0	0	0	0	0	0	0	0	0	0	0	0	0	0

Figure 7.1. Counts of green-winged teal in quadrats covering an area of southern Florida. Each quadrat covers 25 km^2. The sampling frame consists of 200 quadrats that contain a total of 202 birds (τ) in five of the quadrats ($\pi = 5/200 = 0.025$). The shaded quadrats are a sample of quadrats that might be randomly selected until the criterion "stop when two or more quadrats contain teal" is met.

"MIXED" STOPPING RULE

With the "mixed" stopping rule, sampling begins with a simple random sample of u_0 units. At the completion of sampling, if less than m of those units satisfies the stopping criterion, we add units sequentially until at least m units with the rare species are observed. As indicated in the last section, the stopping rule could be based on those units that contain at least one member or contain at least k (> 1) members of the species of interest. The estimators are given in Box 7.2.

Box 7.2. Estimators for the average number of a rare species per sampling unit or for the population total for the mixed stopping rule.

Let $u = u_0 + m$ be the final sample size. Here I show the estimators for the stopping rule in which the m units must contain $k \geq 1$ members of the rare species.

Population parameter	*Estimator*	
Mean	$\bar{y}$	if $u = u_0$
	$\hat{\pi}^*\bar{y}_k + (1 - \hat{\pi}^*)\bar{y}_{nk}$	if $u > u_0$
Total Abundance	$U\bar{y}$	if $u = u_0$
	$U\hat{\pi}^*\bar{y}_k + U(1 - \hat{\pi}^*)\bar{y}_{nk}$	if $u > u_0$

In this sampling design it is possible to observe more than m sampling units containing rare species in the first u_0 samples. Let m^* ($\geq m$) be the actual number observed. Then, $\hat{\pi}^* = (m^* - 1) / (u - 1)$ is the estimator of the proportion of the population that satisfies the criterion (see text), $\bar{y} = u_0^{-1}\sum_{i=1}^{u_0} y_i$ is the sample mean for the simple random sample, U is the total number of sampling units available for potential sampling, $\bar{y}_k = m^{-1}\sum_{i=1}^{m} y_i$ is the sample mean number of members of the rare species per quadrat for the m sampling units in which there were k or more members of the species, and $\bar{y}_{nk} = (u - m)^{-1}\sum_{i=1}^{u-m} y_i$ is the sample mean number of members of the rare species per quadrat for the $u - m$ sampling units in which there were less than k members of the species.

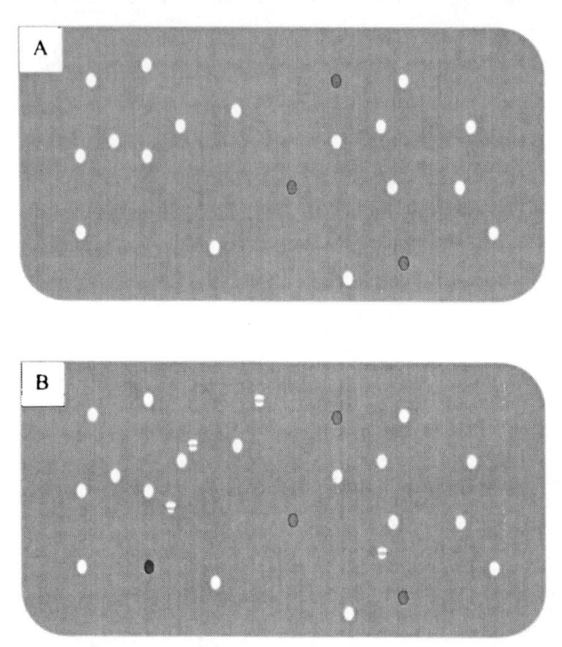

Figure 7.2. (A) The marsh with a sample of 20 randomly sited circular quadrats. Those that are dark gray contain > 25 spiders; the remainder (white) have < 25 spiders. (B) The marsh with a sample of 20 randomly sited circular quadrats plus five additional quadrats. The additional quadrats with less than 25 spiders are marked with lines and the one additional quadrat with more than 25 spiders is shaded black.

Example: Estimating the Abundance of a Spatially Rare Spider

The wolf spider, *Hogna* sp., is found in marshes along the east coast of the United States. It is generally in very low abundance (on the order of 0.5 spiders/m^2) but can be locally abundant (> 50 spiders/m^2) in some habitats during some seasons. Suppose we wish to estimate the abundance of *Hogna* sp. within a marsh located adjacent to a coastal bay. One approach would be to randomly select u_0 locations within the marsh with the intention of observing a 1-m radius circle around each chosen location. Each circle covers an area of π m^2. Our criterion for determining when to cease sampling is that we see at least four locations with more than 25 spiders per circle. Hence, if the initial sample contains four or more locations with more than 25 spiders, we are finished sampling. If not, then sampling continues sequentially until we do have four circles with more than 25 spiders. An example is shown in Figure 7.2. The initial sample of $u_0 = 20$ quadrats yielded {2, 1, 1, 27, 1, 2, 1, 0, 0, 0, 1, 41, 38, 0, 1, 1, 1, 0, 2, 0}, which contained only three sampled units with more than 25 spiders. The additional sequentially sampled units yielded {2, 0, 0, 1, 29}. So our final sample size is $u = 25$. We calculate the estimated proportion of marsh area that has at least 25 spiders. From Box 7.2 we get $\hat{\pi}^* = (m^* - 1)/(u - 1) = 0.125$. In this example, we do not have a sampling frame of U units. Instead we will estimate the mean number of spiders per unit area and multiply that by the total area of the marsh. To estimate the mean number of spiders per unit area, we calculate the mean for those quadrats with > 25 spiders: $\bar{y}_k = m^{-1}\sum_{i=1}^{m} y_i = 4^{-1}(135) = 33.75$ spiders/quadrat. Similarly, the mean for quadrats with ≤ 25 spiders is $\bar{y}_{nk} = (u - m)^{-1}\sum_{i=1}^{u-m} y_i = 21^{-1}(17) = 0.81$. Combining these two estimates, we obtain the estimated overall number of spiders per quadrat: $\hat{\pi}^*\bar{y}_k + (1 - \hat{\pi}^*)\bar{y}_{nk} = 0.125 \times 33.75 + (1 - 0.125) \times 0.81 = 4.93$ spiders/(πm^2) or 1.56 spiders/m^2. The total area of the marsh is 1 km^2 = 10^6 m^2, so our estimated total abundance of *Hogna* sp. in the marsh is 1.56×10^6 spiders.

Estimating the Variance of the Estimated Mean or Total

The above equations yield unbiased estimators of the mean or total (N) based on sequential sampling. Unfortunately, the variance of these estimators is not easily calculated using standard statistical methods (Christman and Lan 2001). Lower and upper bounds to the variance can be obtained using equation (7.1), but they are conservative and hence can lead to large confidence intervals. A reasonable alternative is to use percentile confidence intervals with bootstrapping (Efron and Tibshirani 1993). These have the added advantage that they do not require that the estimator be approximately normally distributed; it is not likely to be when estimating parameters for species with large variability in the abundance/sample unit.

For with-replacement random sampling, which is a reasonable assumption when the sample size (u) is less than 5% of the size of the sampling frame (U), one uses with-replacement bootstrapping methods (Efron and Tibshirani 1993). The same sampling strategy that was applied during actual sampling should also be used in the bootstrapping. An example is shown in Box 7.3.

If sampling is done without replacement then the bootstrapping approach needs to be modified. There are several alternative approaches for performing the bootstrapping (see Christman and Pontius 2000). An example of one is given in Box 7.4.

Box 7.3. With-replacement bootstrapping to obtain a 95% confidence interval for the mean number of *Hogna* sp. in a marsh.

The algorithm for bootstrapping is:

1. Denote the original sample as the pseudopopulation P.
2. Take a sample with replacement from P according to the sampling design used to originally collect the data; call this S_i.
3. Calculate the estimated mean or total for the sample S_i using the appropriate formula; call this E_i.
4. Repeat Steps 2 and 3 B times so that you have B estimated values: $E_1, E_2, \ldots, E_B$.
5. To obtain a 95% confidence interval of the population mean or total,
   a. Order the $\{E_1, E_2, \ldots, E_B\}$ from low to high.
   b. The lower bound is the 2.5th percentile value.
   c. The upper bound is the 97.5th percentile value.

When this is done using $B = 1{,}000$ and data from the wolf spider example in this chapter, we obtain the bootstrapped frequency distribution shown below. The 95th percentile interval for the mean number of spiders per m^2 is (0.888, 3.363).

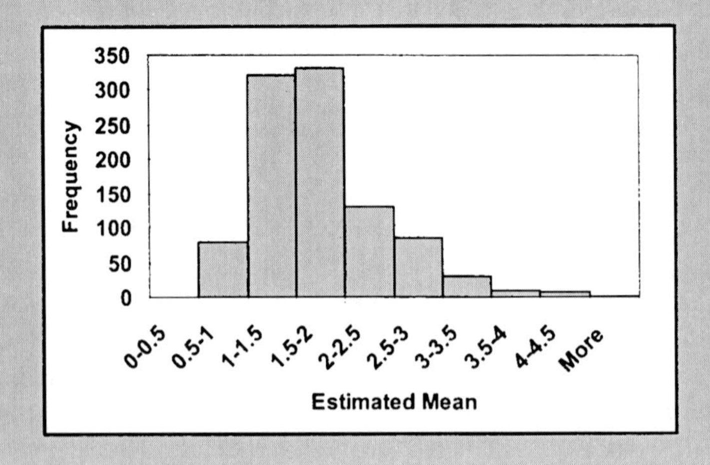

Box 7.4. Without-replacement bootstrapping to obtain a 95% confidence interval for the total number of green-winged teal in a pond in southern Florida.

The algorithm for bootstrapping is similar to with-replacement bootstrapping except that the pseudopopulation is defined differently. Here we require construction of a pseudopopulation whose size is the same as the original population from which we sampled. One approach is to scale up the sample in such a way as to reflect the estimated mean and proportion of quadrats with the rare species (see Booth et al. 1994). To do this we divide the sample into two parts: (1) the subsample of m quadrats that meets the criterion; and (2) the remaining $(u - m)$ observations that do not meet the criterion. Thus, we replicate subsample (1) $\lfloor(\hat{\pi}^* \times N)/m\rfloor$ times, where $\lfloor\ \rfloor$ refers to truncation. Then we replicate subsample (2) $\lfloor\{(1 - \hat{\pi}^*) \times N\}/(u - m)\rfloor$ times. When all of the replications are combined into one list, it should contain N measurements. If there are fewer measurements, we randomly select the necessary number of observations from the original sample to construct a pseudopopulation of size N. If there are too many, we randomly remove observations from the pseudopopulation until it is of size N.

Once the pseudopopulation has been constructed, bootstrap from the pseudopopulation where each bootstrap sample is taken without replacement and according to the original sampling method used. When this is done using $B = 1{,}000$, and data from the green-winged teal example in this chapter, we obtain the bootstrapped frequency distribution shown below. The 95th percentile interval for the mean number of birds per quadrat is (0.647, 18.333).

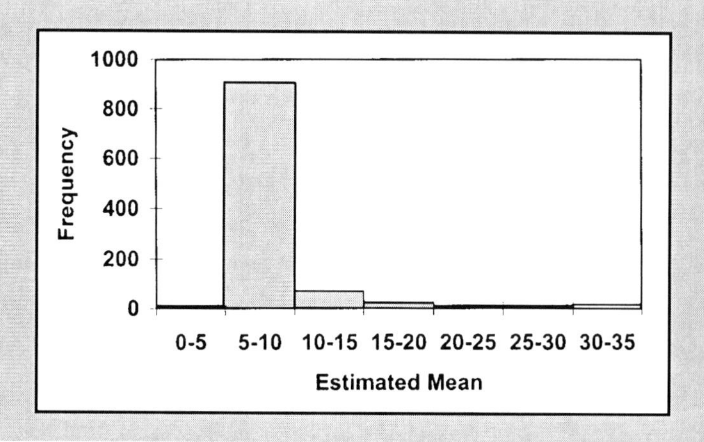

Discussion

The sequential sampling described here almost always has a higher variance associated with the design than the variance obtained under SRS with the same final sample size. Hence, it does not guarantee a more precise estimator. Its advantage is that the researcher is assured of a sample that contains at least m sampling units with the rare species of interest. This cannot be said for SRS. For example, for the green-winged teal study, the probability of observing an SRS of size $u = 20$ with no teal is 58.7%. A sample taken without replacement of size $u = 75$ is required to obtain a probability less than 10% (actually 9.75%). For the simple stopping rule of sampling until two quadrats with teal are observed, the average sample size also is around 75. However, the median is much less (approximately 55–60) and the probability of observing all zeroes is zero. Hence, a large sample size is required, but that is unfortunately the nature of the problem of accurately estimating abundances of rare species.

One alternative sampling strategy that would decrease the sampling variability as well as provide some control over the final sample size would be to stratify prior to sampling. A reasonable approach would be to stratify the study area into two strata: suitable habitat and unlikely habitat. Sequential sampling then could be performed in the stratum with likely habitat and an SRS could be taken in the other stratum. In this way, the sample size is somewhat controlled, but the ability to observe samples with the rare species still is guaranteed (Lan 1999).

References

Booth, J. G., R. W. Butler, and P. Hall. 1994. Bootstrap methods for finite populations. *Journal of the American Statistical Association* 89:1282–1289.

Christman, M. C. 1993. *On Estimation of the Number of Classes in a Population.* Ph.D. Thesis, George Washington University, Washington, D.C.

______. 2000. A review of quadrat-based sampling of rare, geographically clustered populations. *Journal of Agricultural, Biological and Environmental Statistics* 5:168–201.

Christman, M. C., and F. Lan. 2001. Inverse adaptive cluster sampling. *Biometrics* 57:1096–1105.

Christman, M. C., and J. R. Pontius. 2000. Bootstrap confidence intervals for adaptive cluster sampling. *Biometrics* 56:503–510.

Cochran, W. G. 1977. *Sampling Techniques.* 3rd ed. Wiley, New York.

Efron, B., and R. J. Tibshirani. 1993. *An Introduction to the Bootstrap.* Chapman and Hall, New York.

Hu, F., and W. F. Rosenberger. 2004. Variability, optimality, power: Evaluating response-adaptive randomization procedures for treatment comparisons. *Journal of the American Statistical Association.* In press.

Lan, F. 1999. *Sequential Adaptive Designs to Estimate Abundance in Rare Populations*. Ph.D. Dissertation, American University, Washington, D.C.
Lehmann, E. H. 1986. *Testing Statistical Hypotheses*. 2nd ed. Wiley, New York.
Mikulski, P. W., and P. J. Smith. 1975. A variance bound for unbiased estimation in inverse sampling. *Biometrika* 63:216–217.
Smith, D. R., M. J. Conroy, and D. H. Brakhage. 1995. Efficiency of adaptive cluster sampling for estimating density of wintering waterfowl. *Biometrics* 51:777–788.
Thompson, S. K. 1992. *Sampling*. Wiley, New York.

Part III

ESTIMATING OCCUPANCY

The objective of a monitoring program may be to track changes in presence or occupancy of a species within the area of interest. This type of information may be attractive to practitioners because it tends to be less costly to collect than other demographic data. Nonetheless, occupancy estimators, just like other population estimators, are susceptible to bias from incomplete and variable detectability. Unfortunately, this point is too often overlooked, and observed absence is treated as true absence, while ignoring the possibility that a species is present but undetected.

Part III describes two alternative approaches to estimating occupancy through correction of incomplete detection probability. In Chapter 8, MacKenzie et al. provide a review of occupancy estimation, present their latest work on developing a reliable estimator of occupancy, and apply their estimator to sampling rare populations. In the next chapter, Peterson and Bayley offer a Bayesian alternative for estimating probability of presence for rare or difficult-to-detect species. Poon and Margules complete this section by describing a practical sampling method for locating populations of rare plants in remote areas, which is essentially a form of species-presence survey.

8

Occupancy Estimation and Modeling for Rare and Elusive Populations

Darryl I. MacKenzie, J. Andrew Royle,
Jennifer A. Brown, and James D. Nichols

Animal sampling programs directed at populations and communities typically focus on one or more measurements of the population that characterize the system of interest (referred to here as state variables). These state variables are estimated at different places or times to allow inference about variation over time or space and to study the influence of environment and management. For a variety of logistical and other reasons, we believe that the proportion of area or sample units occupied by a species will frequently be a reasonable state variable for use in studying rare and elusive species. As for most estimation problems in animal ecology, it is important to recognize that detection probabilities are not 1 and that detectability must be incorporated into the estimation procedure (Lancia et al. 1994; Thompson et al. 1998; Yoccoz et al. 2001; MacKenzie and Kendall 2002; Williams et al. 2002). We present a brief history of occupancy estimation and then introduce recently developed approaches that we believe to be useful. These approaches include efforts to incorporate heterogeneity in site-specific detection probabilities associated with variation in abundance. We also describe a multiseason model permitting direct estimation and modeling of rate of change in occupancy and of the vital rates responsible for such change. Finally, we present an example analysis of avian point count data and an example study design for monitoring a rare native insect in New Zealand.

Occupancy as a State Variable

In this chapter we propose the proportion of an area, or proportion of suitable habitat units in an area, that is occupied (we will refer to this loosely as occupancy) as a possible state variable in investigations and monitoring programs for rare and elusive species. The fundamental reason for this proposal simply involves expense and efficiency. Although many methods have been developed for properly dealing with detectability and estimating animal abundance (Seber 1982; Lancia et al. 1994; Buckland et al. 2001; Williams et al. 2002), virtually all of these approaches perform best when sample sizes are relatively large. Adequate sample sizes (e.g., number of individuals observed or caught) are difficult to achieve for rare and elusive species.

Occupancy surveys involve searches of sample units for evidence of species presence. Frequently, such searches will be less expensive than collection of the ancillary data needed to estimate abundance. It is possible to use animal sign (e.g., tracks or feces) as an indicator of presence, and detection of sign will frequently be easier than observation or capture of animals. For example, indicators used to detect the presence of possums in New Zealand from bite marks in wax blocks are 33% faster to set and check compared to leghold traps. In addition, traps are more than 20 times heavier than the wax detection devices, and consequently, fewer traps can be set in the field (Thomas et al. 2003).

Inferences based on occupancy are not the same as those based on abundance, so the investigator must decide whether occupancy is an adequate state variable for management or scientific questions being posed. In many cases, abundance and occupancy are closely related. In the case of territorial species with size of sample units approximately the same as territory size, occupancy may be directly related to abundance by a constant (e.g., the estimated number of occupied cells might equal the estimated number of nesting pairs). In pest eradication programs, where abundance has been greatly reduced following a control program, the distinction between species presence and absence is much more important than that between small numbers of animals.

For some kinds of questions, occupancy is the state variable of primary interest. For example, the number or proportion of patches occupied is the state variable of interest in a large class of metapopulation models (Levins 1969, 1970; Hanski 1994, 1997, 1999), in models used to investigate extinction thresholds in territorial species (Lande 1987, 1988), and in models used to investigate potential changes in range and extinction probabilities associated with climate change (Travis 2003).

Occupancy data are commonly used to describe geographic ranges and to test hypotheses about relationships affecting range size and shape (e.g., Gaston and Blackburn 1996; Gaston et al. 1996). Occupancy data are also used to estimate relationships between occupancy and patch characteristics such as size and habitat (e.g., Robbins et al. 1989; Scott et al. 2002). A presence-absence matrix of species occurrence (occupancy) in spatial units has been termed "the fundamental unit of analysis in community ecology and biogeography" (Gotelli 2000; also see McCoy and Heck 1987) and is widely used for drawing inferences about species coexistence (e.g., Cole 1949; Pielou 1977). "Incidence functions" (Diamond 1975), the proportions of islands in a particular class (defined by such characteristics as species number and island area) that are occupied by species of interest, have seen extensive use in community ecology and metapopulation work (Hanski 1992, 1999).

Occupancy Estimation

Occupancy surveys involve searches of sample units, which may be either arbitrary spatial units (grid cells of specified size) or natural sampling units (e.g., patches of preferred habitat or stream habitat units). Searches are repeated during a relatively short time interval, during which the true occupancy status of each sample unit is not expected to change. For example, a natural wetland might be visited on five nights during a 2-week period, with investigators listening for anuran vocalizations and looking for anurans of a particular species of interest. Data resulting from such sampling can be summarized as a detection history, a sequence of 1's and 0's denoting detection and nondetection, respectively, of the target species. For example, a history of 01011 would denote a wetland in which the species was detected at sampling occasions 2, 4, and 5.

Virtually all of the applications of occupancy data noted thus far estimated occupancy as the proportion of sample units at which a species was detected. However, this estimation approach assumes that if the species was not detected it was not present, that is, that detection probability (probability of detecting species presence given that at least one individual of the species is located in the sampled area) is 1. This assumption will certainly not be met most of the time, and its failure can lead to biased estimates and misleading inferences (e.g., Moilanen 2002).

To our knowledge, Geissler and Fuller (1987) were the first to use repeated visits to estimate detection probability in a manner analogous to the approach used in closed-population capture-recapture models (Otis et al. 1978). They then used this information to estimate the number and proportion of sample units occupied. This basic approach has been used to estimate occupancy of spotted owl habitat units in both published (Azuma et al. 1990) and unpublished (Nichols, unpublished data) studies and has been recommended for other species more recently (Kéry 2002; Nichols and Karanth 2002; Stauffer et al. 2002). Some of these approaches used a two-step approach to estimate occupancy, first estimating the number of occupied units and then using this to estimate the proportion of units occupied. Others focused on the number of visits required to detect a species with specified probability, given presence.

Single-Season Model

MacKenzie et al. (2002) considered models in which occupancy probability was incorporated directly into the model likelihood, permitting direct estimation and modeling of this parameter. This model provides the basic structure used throughout the remainder of this chapter.

Likelihood

The probability of observing a particular detection history for the target species at a site can be simply determined by considering the processes that resulted in the observed data. For example, consider the detection history 1001. To observe this history, the species must be present (as it was detected at least once); it was detected in the first and fourth searches of the site, but not in searches 2 and 3. The probability of observing this detection history at site i ($\mathbf{H}_i$) could be expressed as $\Pr(\mathbf{H}_i = 1001) = \psi p_1(1 - p_2)(1 - p_3)p_4$, where ψ is the probability a site is occupied by the species, and p_t is the probability of detecting the species (given presence) during the tth independent search of a site. A similar expression is obtained for all sites at which the species is detected at least once. However, for sites where the species was never detected (e.g., $\mathbf{H}_i = 0000$), there are two possibilities. Either the species was present but went undetected in the searches (i.e., a "false absence"), or the species was genuinely absent from the site. Using this concept, the probability of never detecting the species at a site can be simply expressed as $\Pr(\mathbf{H}_i = 0000) = \psi\prod_{t=1}^{4}(1 - p_t) + (1 - \psi)$.

Assuming that the detection histories for all sites are independent, the model likelihood function is:

$$L(\psi,\mathbf{p}|\mathbf{H}_1,\ldots,\mathbf{H}_U)=\prod_{i=1}^{U}\Pr(\mathbf{H}_i). \quad (8.1)$$

Note that this requires both the probability of occupancy and the detection probability to be constant across all U sites (unless modeled as a function of covariates; see below). This can be simplified to:

$$L(\psi,\mathbf{p}|u_1,u_2,\ldots,u_T,u.)=\left[\psi^{u.}\prod_{t=1}^{T}p_t^{u_t}(1-p_t)^{u.-u_t}\right]\left[\psi\prod_{t=1}^{T}(1-p_t)+(1-\psi)\right]^{U-u.}, \quad (8.2)$$

where T is the number of searches of each location, $u.$ is the number of locations where the species was detected at least once, and u_t is the number of sites where the species was detected during the tth search.

Estimation

The likelihood in equations (8.1) and (8.2) can be maximized to obtain maximum likelihood estimates (MLEs) of the model parameters, which will generally require a numerical approach or specialist software such as Program PRESENCE (MacKenzie et al. 2003) or Program MARK (White and Burnham 1999). An alternative approach to using maximum likelihood would be a more computer-intensive approach such as Markov chain Monte Carlo (MCMC). This approach may be especially useful for investigating models that are difficult to fit using maximum likelihood techniques, such as a model with random effects. We foresee MCMC playing an increasingly important role for making inference in ecological statistics as appropriate computer software becomes available, although here we tend to focus on maximum likelihood methods.

Missing Observations

A common feature of studies of this type is that not all of the sites will be searched simultaneously (i.e., within a small time frame) or with an equal number of visits. Furthermore, in some instances a planned search of a site may not occur due to logistical constraints (such as limits on available resources) or unforeseen circumstances, such as a change in weather or a vehicle breakdown. In the interests of simplicity, whenever a site is not searched for any reason, we shall refer to it as a missing observation.

The above likelihood-based method can easily accommodate such missing values by reconsidering the probability of observing a given detection history. For example, suppose that a site was supposed to be searched on four occasions, but was not searched on day 3 (due to the surveyor being ill), and that the resulting detection history for this site was 10-1, where "–" represents the missing observation. The probability of this detection history may be simply expressed by completely ignoring the third search occasion so that neither p_3 nor $(1 - p_3)$ appear in the probability statement; that is, $\Pr(\mathbf{H}_i = 10-1) = \psi p_1 (1 - p_2) p_4$. As before, the model likelihood can be calculated by equation (8.1) and can be used to obtain MLEs or within an MCMC-type analysis.

One important point to note is that the search-specific detection probabilities relate to the searches conducted simultaneously across all sites and not to the first, second, third, etc. search of a specific site. For example, suppose that over a 5-day period a number of sites are to be searched multiple times for the target species. Consider the data below for two of the sites; the first site was searched on days 1, 2, 3, and 5, whereas the second site was searched on days 2, 4, and 5.

Site	Day 1	Day 2	Day 3	Day 4	Day 5
1	1	0	1	–	0
2	–	0	–	1	1

For the first site we would treat day 4 as a missing observation, and days 1 and 3 as missing observations for the second site, even though there was no intention of searching these sites on these days. The probability of observing the respective detection histories would be,
$\Pr(\mathbf{H}_1 = 101{-}0) = \psi p_1 (1 - p_2) p_3 (1 - p_5)$ and $\Pr(\mathbf{H}_2 = {-}0 - 11) = \psi(1 - p_2) p_4 p_5$.

Covariate Modeling

Often there may be covariate information available on certain variables that have been measured at the sites of interest. Two general categories of covariate could be measured: (1) those that are constant during a season (such as the local habitat type); and (2) those that may vary during the season (such as weather conditions). Detection probabilities may be a function of either type of covariate; for example, our ability to detect the species could vary in different habitats or change with temperature as the species becomes more or less active. However, because we assume sites are closed to changes with respect to occupancy, it does not seem reasonable to consider that occupancy may vary in relation to a covariate that changes during a season, but occupancy may vary in accordance with a constant variable reflecting, for example, habitat preferences by the species.

By including covariate information, it is possible to account for some forms of heterogeneity where probabilities vary between sites. This can be simply accomplished by using the logistic model, where occupancy and/or detection probabilities are expressed as functions of the covariates and logistic coefficients $\theta_i = \frac{\exp(X_i\boldsymbol{\beta})}{1 + \exp[X_i\boldsymbol{\beta}]}$, where θ_i is the probability of interest for site i, X_i is the row vector of covariate information for site i, and $\boldsymbol{\beta}$ is the column vector of coefficients to be estimated.

Model Assumptions and Consequences of Violations

Three main assumptions relate to the above model: (1) sites are closed with respect to changes in occupancy within the season; (2) model parameters are constant across sites (i.e., no heterogeneity not accommodated by covariates); and (3) species are detected both within and between sites independently.

The closure assumption is required so that if the species is detected at a site at least once during the season, then in searches of the site when the species was not detected, we know the species must have been present but undetected. Without the closure assumption, when a species is not detected during a search of a site we are unable to distinguish between a genuine and a false absence. If the closure assumption is violated, estimators of occupancy and detection probabilities may be biased. Although this has not specifically been assessed within a site occupancy context, based upon the findings of Kendall (1999), we would suggest that only if a species occupies and vacates sites in a completely random manner (i.e., the probability that the species was present at the site during survey j does not depend on whether the species was present at survey $j - 1$) would estimators remain unbiased, although the interpretation of the probabilities would change. What we have previously termed the proportion of sites occupied should be interpreted as the proportion of sites used by the species. The estimated probability of detection given species presence now consists of two confounded components: (1) the probability that the species is present at the sites during the particular search; and (2) the probability of detection given presence. We suggest that other types of violations of the closure assumption (emigration or immigration only, or Markovian processes where the probability of occupancy depends upon the occupancy state at the time of the previous survey) will lead to biased estimators.

Heterogeneity in detection probabilities will generally result in occupancy being underestimated. An important point is that some forms of heterogeneity may be modeled using covariate information such as habitat information or other local environmental conditions. By modeling the heterogeneity, its effect may be greatly reduced. Attempts to deal with unmodeled sources of heterogeneity adopt a view that each location, i, has a potentially distinct detection probability, p_i, arising from some latent distribution of detection probabilities. Many approaches have been developed within the capture-recapture context for modeling and estimation in the face of such variation (see closed capture-recapture model M_h, Otis et al. 1978; Williams et al. 2002). These approaches include jackknife (Burnham and Overton 1978; Otis et al. 1978), bootstrap (Smith and van Belle 1984), and sample coverage estimators (Chao and Lee 1992; Chao et al. 1992), finite mixture (or latent class) estimators (Norris and Pollock 1996; Pledger 2000), and models based on various continuous mixtures, including the logistic-normal (Coull and Agresti 1999; Fienberg et al. 1999) and beta mixtures (Burnham and Overton 1978; Dorazio and Royle 2003). One of the more likely sources of heterogeneity is variation among sites in the number of individuals of the species—that is, the probability of detecting the species will likely be greater at sites with higher abundances. We deal with this important source of heterogeneity separately (see Models Allowing for Variation in Abundance among Sites).

The third assumption that may be violated is independence of the surveys both within and between sites. Lack of independence of surveys within sites may occur, for example, by knowing where to look for particular members of the species at a site following detection of the species in an earlier survey. Although this could be allowed for as part of the analysis (by defining an indicator variable that equals zero for all surveys up to and including the first detection at the site and that equals 1 for all subsequent surveys), it highlights that some thought must be given during the design of a study to how observers are allocated to each survey of a site (e.g., by rotating observers amongst sites within a season). Movement of animals among multiple sites can produce dependence of surveys between sites and argues for careful consideration of plot size and proximity relative to animal movement capabilities. In instances of nonindependence, the effective sample size is smaller than the number of sites used in the study, suggesting that estimates are actually known less precisely than thought, that is, estimated standard errors are too small. To account for this, a variance inflation factor is required. One technique for obtaining a variance inflation factor is given below.

Assessing Model Fit

As in all modeling exercises, an assessment should be made of how well a model describes the data. The use of model-selection statistics (e.g., Akaike's Information Criteria, AIC; Burnham and Anderson 1998) is no substitute for assessing model fit, because such procedures are used to find the "best" approximation within a set of candidate models. There is no guarantee that a "best" model will also be a "good" model.

Recently, MacKenzie and Bailey (2004) have suggested a goodness-of-fit test for the model described above. They suggested that a simple Pearson chi-square statistic can be calculated, based upon the observed and expected number of sites with each possible detection history, h (O_h and E_h respectively; equation (8.3)), where the expected number of sites is estimated from the model, which may include individual covariates (equation (8.4)).

$$X^2 = \sum_h \frac{\left(O_h - E_h\right)^2}{E_h} \tag{8.3}$$

$$E_h = \sum_{i=1}^{U} \Pr\left(\mathbf{H}_i = h\right) \tag{8.4}$$

Rather than rely upon asymptotic results, MacKenzie and Bailey (2004) recommended using parametric bootstrapping to determine whether the observed chi-square statistic appears unusually large. They followed White and colleagues' (2001) suggestion that an overdispersion factor, $\hat{c}$, may be estimated as $\hat{c} = X^2_{Obs}/\overline{X}^2_B$, where $\overline{X}^2_B$ is the average of the test statistics obtained from the parametric bootstrap. The estimate of c can be used to adjust model selection procedures (see Burnham and Anderson 1998:53), and standard errors may be inflated by a factor of $\sqrt{\hat{c}}$.

MacKenzie and Bailey (2004) showed that this approach for assessing model fit provides reasonable estimates of $\hat{c}$ given various assumption violations (due to the model having an incorrect structure or lack of independence among sites) and that when the correct model is fit to the data, on average, $\hat{c} \approx 1$.

Models Allowing for Variation in Abundance among Sites

As noted above, an especially important assumption underlying the above modeling is that detection probability is constant across sites. Departures from this assumption can lead to biased estimates of occupancy rate and to incorrect inferences about factors that influence occupancy rate. We discussed above the use of covariate modeling, as well as approaches to dealing with heterogeneity not associated with covariates.

An important source of heterogeneity in detection is that induced by variation in abundance, N, among sites (Royle and Nichols 2003). To be specific, suppose that N_i animals exist at site i, and let r denote the detection probability for an individual animal. Then, the probability of detecting at least one animal during a search of site i is $p_i = 1 - (1 - r)^{N_i}$. Evidently, if N_i is large, the site-level detection probability will be large and vice versa. For example, suppose individual-level detection probability is 0.2. Then p_i = 0.200, 0.360, 0.488, 0.590, 0.672 for sites containing N_i = 1, 2, 3, 4, 5 individuals. Thus, if one expects occupied sites to contain between one and five individuals, there is considerable heterogeneity in p_i induced by this variation in abundance. Note that p_i is constant only if N_i is also constant across sites, a scenario that is unlikely to occur in many situations. However, we note that in some situations, for example, especially for larger N_i, heterogeneity in detection probability resulting from variation in abundances may be small and well approximated by models that do not explicitly incorporate such variation.

Royle and Nichols (2003) developed a model that accounts for heterogeneity in p that is a result of variation in abundance. Their model exploits this relationship between detection probability at the individual level (r) and detection probability at the site level (p). Importantly, because of the formal linkage between heterogeneity in p and variation in N, their model also yields information about abundance at the site level. In essence, information about abundance may be deduced from heterogeneity in p. Thus, simple detection/nondetection data may in fact yield information beyond the proportion of sites occupied.

Royle and Nichols' model expands on the basic occupancy model by considering additional model structure describing variation in abundance. In essence, the model allows for varying states of "occupancy." A site may be occupied by a single individual, two individuals, etc., which is the mechanism that induces variation in detection probability at the site level. One reasonable model for variation in abundance, based on a homogeneous Poisson point process to describe the distribution of animals in space, yields the result that N_i has a Poisson distribution with mean μ, where μ is the average number of individuals per site. Then, attention may be focused on the integrated likelihood, that is, the binomial sampling distribution for the number of detections at a site (y_i) integrated over possible states of N_i. Specifically, the likelihood is

$$L(r,\mu) = \prod_{i=1}^{U} \left\{ \sum_{N_i=0}^{\infty} Bin(y_i \mid T, p_i) Pois(N_i \mid \mu) \right\}, \qquad (8.5)$$

where $p_i = 1 - (1 - r)^{N_i}$ and U is the total number of sites. In practice, a reasonably large upper bound is used on the summation in equation (8.5). Equation (8.5) can be maximized numerically using conventional techniques to yield estimates of r and μ. Note that the occupancy rate under this model is a derived parameter, that is, $\psi = 1 - e^{-\mu}$. However, a more meaningful and general characterization of abundance might be based on the proportion of sites in each abundance class, that is, on the estimated Poisson distribution function.

Although maximization of equation (8.5) yields estimates of the mean abundance, it is possible to estimate N_i for each site using the estimated posterior mean of N_i, given the data for site i, and the estimated model parameters (see Royle and Nichols 2003). This is the classical (estimated) best unbiased predictor (BUP) of an unobservable random effect.

Other distributions for site-level abundance that allow for over- or underdispersion with respect to the Poisson model can be considered. For example, the negative-binomial is commonly used to model overdispersion that might arise due to clustering and the binomial to represent underdispersion (due to "inhibition"). Royle and Nichols (2003) had little success fitting the negative-binomial to detection/nondetection data and suggested that the overdispersion parameter may be weakly identifiable given data only on detection/nondetection. Given the availability of count data or index data (Royle 2004a), more general models are easily fitted. A preferable method of accommodating over- or underdispersion with respect to the Poisson model is to use covariates that are thought to influence abundance. Where available, a log-linear model relating μ_i (expected abundance at site i) to some covariate, say x_i, can be employed. For example, using repeated presence/absence of anurans on wetlands, the size of the wetland, denoted as x_i, is likely to influence abundance. In that case, a reasonable model might be $\log(\mu_i) = \beta_0 + \beta_1 x_i$. Under this model, the occupancy rate varies among sites in accordance with site-level variation in abundance, that is, $\psi_i = 1 - e^{-\mu_i}$. A less direct approach to dealing with such variation in abundance is to use the approach of MacKenzie et al. (2002), modeling detection probability as a function of a site-specific covariate believed to reflect abundance (e.g., pond size).

Extension to Count Data

Surveys of many organisms yield counts of individuals observed at each sampling occasion (e.g., birds). Although it may be advantageous to reduce such data to detection/nondetection for some purposes, there is considerable information about abundance in count data, and hence about occurrence and detectability. Royle (2004b) extended the basic idea of Royle and Nichols (2003) to accommodate counts.

The likelihood under this more general model is

$$L(r, \mu) = \prod_{i=1}^{U} \sum_{N_i = \max_i(y_{it})}^{\infty} \left\{\prod_{t=1}^{T} Bin(y_{it} \mid N_i, r)\right\} Pois(N_i \mid \mu),$$

where y_{it} is the count made at site i during visit t. Consideration of other mixing distributions for N_i is possible (see Royle 2004b), as are models that allow for the possibility of covariates that influence abundance and detection. As with detection/nondetection data, this model yields abundance information, and the relationship between occupancy and abundance is implicit.

Pseudo-Replicate Counts

Use of spatial subsamples as replicates is widespread in studies of animal communities (e.g., Williams et al. 2002). It is possible to exploit the model for counts described above in these problems in order to more properly recognize the relationships among detection, occurrence, and abundance. However, N_i in this situation is something of an abstraction (the number of animals in the general area from which the spatial replicates are obtained), and interpretation is not as clear as in the case where the counts are temporal (i.e., "true") replicates.

Let y_{ij} be the count made at spatial subsample j on primary unit i. We assume that abundance is reasonably homogeneous among subsamples within primary unit i. Then, the view that y_{ij} is binomial with index N_i and detection probability r might be reasonable. This model is strictly true only if abundance is precisely constant across all subsamples. In particular, it admits the possibility that individuals may be represented in multiple subsample counts (i.e., are counted more than one time). However, when this is not the case, the interpretation of N_i as a detection-bias-adjusted index of average abundance might be appealing anyway.

ABUNDANCE INDEX DATA

A further extension considers the situation in which y is not a count, but rather an ordinal *index* of abundance (Royle 2004a). For example, in calling surveys of anurans, an index that represents calling intensity is used. A common index takes on values 0 (no frogs heard) to 3 (a full chorus). As with previous types of problems, heterogeneity in p as a result of variation in abundance may be a concern. These index data yield information about heterogeneity in abundance under certain model assumptions.

Specifically, we suppose that each sample site has a population that is capable of generating a maximum index value of C_i. We view C_i as the true index state for site i, but note that index values less than C_i may be observed in the sample due to imperfect detection (e.g., a breeding pond may have sufficient numbers of a frog species to generate a calling index of 3, but in a specific survey some frogs were not calling, so an index value of only 1 was observed). Sampling yields (index) observations y_{it} and, conditional on C_i, the sampling distribution of y_{it} is a multinomial with C_i cells (or $C_i + 1$ if $C_i = 0$ is a possible state) where the cell probabilities are functions of various detection probability parameters. Of course, C_i (the true abundance state for site i) is generally unobserved, so inference is based on the unconditional likelihood, having removed C_i from the likelihood by integration. The resulting unconditional likelihood for y_{it} is a compound multinomial that contains, in addition to detection probability parameters, a set of parameters that describes the distribution of C_i across sites.

Multiple-Season Model

In many cases, the closure assumption used in single-season modeling will not be reasonable for the complete duration of the study. In the case of multiple-season or multiple-year studies, it may be reasonable to consider the situation as a sequence of closed seasons (common across all sites), where changes in the occupancy state of sites may occur between seasons, but not within seasons (similar to the sampling framework for Pollock's robust design in mark-recapture).

The detection history for a multiple season occupancy study may therefore be expressed as a sequence of K single-season detection histories. For example, the detection history $\mathbf{H}_i = 11\ 00\ 01$ would represent a three-season study in which the species was detected in the first and second surveys in season 1, was never detected in season 2, and was detected only in the second survey in season 3. Similar to the single-season situation, due to imperfect detectability, we do not know whether the species was present but undetected in season 2 or instead was genuinely absent.

Likelihood

MacKenzie et al. (2003) extended the single season model of MacKenzie et al. (2002) to multiple seasons by introducing two dynamic parameters that can be used to model changes in the occupancy state of sites over time. Let ε_t be the probability that a site occupied in season t is unoccupied by the species in season $t + 1$ (local extinction), and γ_t be the probability that an unoccupied site in season t is occupied by the species in season $t + 1$ (colonization). Therefore, a matrix of the probability of transitioning between occupancy states (between seasons) may be defined as:

$$\boldsymbol{\phi}_t = \begin{bmatrix} 1-\varepsilon_t & \varepsilon_t \\ \gamma_t & 1-\gamma_t \end{bmatrix}, \tag{8.6}$$

where rows of $\boldsymbol{\phi}_t$ represent the occupancy state of the site at t (state 1 = occupied; state 2 = unoccupied) and columns represent the occupancy state at t + 1. For completeness, a row vector $\boldsymbol{\phi}_0$ may be defined as $\boldsymbol{\phi}_0 = [\psi_1\ 1 - \psi_1]$, where ψ_1 is the probability that the site is occupied in the first season ($t = 1$).

To incorporate detection probabilities into the model, we define a column vector $\mathbf{p}_{\mathbf{H},t}$, which denotes the probability of observing the portion of the detection history $\mathbf{H}_i$ relevant to season t, conditional upon occupancy state. For instance,

$$\mathbf{p}_{110,t} = \begin{bmatrix} p_{t,1}p_{t,2}\left(1-p_{t,3}\right) \\ 0 \end{bmatrix}, \text{ and } \mathbf{p}_{000,t} = \begin{bmatrix} \prod_{j=1}^{3}\left(1-p_{t,j}\right) \\ 1 \end{bmatrix}, \tag{8.7}$$

where $p_{t,j}$ denotes the detection probability for visit j in season t. Note that whenever the species is detected at least once during a season, the second element of $\mathbf{p}_{\mathbf{H},t}$ will be zero because it is impossible to observe such a history if the site is in the unoccupied state. Similarly, the second element of $\mathbf{p}_{0,t}$ will always be 1, because the all-zero history is the only observable outcome if the site is unoccupied.

For any given detection history, the probability of observing such an outcome can be expressed as $\Pr(\mathbf{H}_i = \boldsymbol{\phi}_0\prod_{t=1}^{K-1}[D(\mathbf{p}_{\mathrm{H},t})\boldsymbol{\phi}_t]\mathbf{p}_{\mathrm{H},K}$, where $D(\mathbf{p}_{\mathrm{H},t})$ is a 2×2 diagonal matrix with the elements of $\mathbf{p}_{\mathrm{H},t}$ along the main diagonal (top left to bottom right), zero otherwise. For example, consider again the detection history $\mathbf{H}_i$ = 11 00 01. The probability of observing this would be

$$\Pr\left(\mathbf{H}_i = 11\ 00\ 01\right) = \boldsymbol{\phi}_0 D\left(\mathbf{p}_{11,1}\right)\boldsymbol{\phi}_1 D\left(\mathbf{p}_{00,2}\right)\boldsymbol{\phi}_2\mathbf{p}_{01,3}$$

$$= \begin{bmatrix}\psi_1 & 1-\psi_1\end{bmatrix}\begin{bmatrix}p_{1,1}p_{1,2} & 0 \\ 0 & 0\end{bmatrix}\begin{bmatrix}1-\varepsilon_1 & \varepsilon_1 \\ \gamma_1 & 1-\gamma_1\end{bmatrix}$$

$$\times \begin{bmatrix}\prod_{j=1}^{2}\left(1-p_{2,j}\right) & 0 \\ 0 & 1\end{bmatrix}\begin{bmatrix}1-\varepsilon_2 & \varepsilon_2 \\ \gamma_2 & 1-\gamma_2\end{bmatrix}\begin{bmatrix}\left(1-p_{3,1}\right)p_{3,2} \\ 0\end{bmatrix}$$

$$= \psi_1 p_{1,1}p_{1,2}\left\{\left(1-\varepsilon_1\right)\prod_{j=1}^{2}\left(1-p_{2,j}\right)\left(1-\varepsilon_2\right)+\varepsilon_1\gamma_2\right\}\left(1-p_{3,1}\right)p_{3,2}\ . \tag{8.8}$$

The central term in brackets on the final line represents the two possibilities for the species during the second season when it was not detected. Either the species (1) did not go locally extinct between seasons 1 and 2, was undetected in the three surveys during the second season, and continued to occupy the site into season 3 (with probability $(1 - \varepsilon_1)\prod_{j=1}^{2}(1 - p_{2,j})(1 - \varepsilon_2)$); or (2) went locally extinct between seasons 1 and 2 and recolonized the site between seasons 2 and 3 (with probability $\varepsilon_1\gamma_2$).

Once the probability of observing each detection history has been established, the model likelihood can be formed as usual, that is, $L(\psi_1, \boldsymbol{\varepsilon}, \boldsymbol{\gamma}, \mathbf{p}|\mathbf{H}_1,\ldots,\mathbf{H}_U) = \prod_{i=1}^{U}\Pr(\mathbf{H}_i)$. Missing observations and covariate information can also be included in the model in the same manner as described above for the single-season case.

Parameters and Estimation

As with any likelihood-based model, the parameters may be estimated either by maximum likelihood techniques or by Bayesian methods in which a posterior distribution for the parameters is computed from the data after first specifying an appropriate prior distribution. Regardless of how the estimation proceeds, note that the probability of occupancy at time t may be derived recursively by the expression:

$$\psi_t = \psi_{t-1}(1-\varepsilon_{t-1}) + (1-\psi_{t-1})\gamma_{t-1}. \tag{8.9}$$

Alternatively, in some instances the rate of change in occupancy may be of interest, which could be defined as:

$$\lambda_t = \frac{\psi_{t+1}}{\psi_t}. \tag{8.10}$$

Using these relationships, the model could be reparameterized so that these parameters (e.g., ψ or λ) are estimated directly by rearranging equations (8.9) and (8.10) to make either ε_{t-1} or γ_{t-1} the subject. Estimation of such a parameter, λ_t, is a frequent objective of monitoring programs, and reduced-parameter models (e.g., $\lambda = \lambda_t$ for all t) can be used to provide estimates of "trend" in occupancy. In addition, the potential effects of covariates on these quantities could be assessed directly. However, by reparameterizing the model, there are some constraints that must be imposed upon allowable parameter values that we have found can make numerical procedures unstable.

Case Study 1: Occupancy Analyses Based on Avian Point Counts

Here we consider data derived from point counts of gray catbird (*Dumetella carolinensis*) along a North American Breeding Bird Survey route located in New Hampshire. Point counts were made at 50 locations along the route, and these were repeated 11 times within a 1-month period. For purposes here, the data were reduced to detection/nondetection. Royle and Nichols (2003) provide more detailed analyses of these data and similar data for other species, and Link et al. (1994) provide additional description and background for the data.

We consider the site-occupancy model in which p is constant over the course of the study. In this case, the sufficient statistic is the detection frequency distribution (number of sites where only one detection occurred, exactly two detections occurred, etc.). For the catbird data, the sufficient statistics are

Detection:	0	1	2	3	4	5	6	7	8	9	10	11
Frequency:	31	6	4	5	2	0	2	0	0	0	0	0

Thus, there were 31 stops at which the species was not detected at any survey, six stops at which the species was detected on only one survey, etc.

The naive estimate of the proportion of sites occupied is 19/50 = 0.38. Applying the model of MacKenzie et al. (2002) described above ($p_t = p$) yields estimates $\hat{p} = 0.230$ ($\hat{SE}(\hat{p}) = 0.022$) and $\hat{\psi} = 0.403$ ($\hat{SE}(\hat{\psi}) = 0.053$). Although the estimated detection probability is small, the fact that there are 11 sample periods implies that nondetection at an occupied site is small ($(1 - 0.230)^{11} = 0.06$). Thus, the estimated site occupancy in this instance is similar to the proportion of sites where detection occurred (the naive estimate).

This example highlights an important point—that there may be little gain from statistical estimation under intensive sampling (or when p is large), but in practice one must typically compromise and trade off temporal replication for spatial replication in order to reduce the variance of $\hat{\psi}$. This leads to higher rates of nondetection for the sample sites, in which case formal estimation becomes more important because the naive estimator of site occupancy will be more biased. For point counts of breeding birds, the possibility exists that variable numbers of pairs (or breeding males) may inhabit the vicinities of the point count locations, inducing heterogeneity in p as discussed above. To investigate this potential heterogeneity, we fit the Poisson mixture model described above (also Royle and Nichols 2003) to the gray catbird data, yielding $\hat{\mu} = 0.546$ (birds/sample site) and $\hat{r} = 0.179$. The relative difference in AIC was 2.58 in favor of the heterogeneity model, indicating slightly more support for this model. Also, the estimated μ implies that about 58% of sites are unoccupied (have zero birds), 32% of sites are occupied by one bird, 8% by two birds, and about 2% by three or more birds (as derived from the Poisson distribution function). As catbirds are more frequently detected by their call, these estimates may roughly equate to breeding males.

Note that r under the heterogeneity model is the *individual* bird detection probability, whereas p from the model without heterogeneity is the probability of detecting one or more birds. When converted to a site-level detection probability to be consistent, $\hat{r} = 0.179$ at the individual level equates to a site-level detection probability of $\hat{p} = 0.222$, roughly in line with the estimate obtained previously. The estimate of μ implies a site occupancy rate of $1 - e^{-\mu} = 0.421$, which is only about 5% larger than the previous estimate (i.e., neglecting heterogeneity) and about 11% larger than the naive estimate.

Although heterogeneity due to variable abundance does not appear to be substantial for the catbird data (or rather, does not appear to have a large effect on estimated occupancy), it may be that an estimate of mean abundance has more direct relevance for demographic modeling purposes, for example, if interests focus on understanding gray catbird demographic processes.

Case Study 2: Occupancy Study Design for Monitoring Mahoenui Giant Weta Populations

Weta are one of the more unique and specialized groups of New Zealand insects. More than 70 endemic species of weta survive in New Zealand today. Even more ancient than species such as the tuatara, weta remain almost unchanged from their ancestors of 190 million years ago. The indigenous Maori people call them "devils of the night" because of their nocturnal habit and their rather peculiar body shape, spiny legs, and curved tusks. All weta are flightless and relatively large, but members of a subgroup called the giant weta are among the largest insects in the world. A pregnant female of one species locally known as the wetapunga weighed in at 71 g. Many of the giant weta species are endangered and survive only on offshore islands without introduced mammalian predators.

A schoolteacher discovered the Mahoenui giant weta (*Deinacrida mahoenui*) in a patch of gorse (*Ulex europaeus*) in 1962. Gorse is a perennial pest plant with sharp spiny stems and bright yellow flowers that can form dense tickets; it was originally introduced to New Zealand as a hedging plant by the early European settlers. The weta species occurs naturally in only two locations in New Zealand (Sherley 1998). One of these locations, Mahoenui, is a 240-ha scientific reserve near the town of Te Kuiti on the North Island of New Zealand. A key part of the current management program for this species is monitoring its abundance in the reserve. Such monitoring is difficult because this weta species is so rare, its gorse habitat is dense and prickly, and the reserve has a large number of steep-sided gullies.

Previous monitoring of weta in the Mahoenui Reserve has involved counting weta on two to four transect lines. This index is used to infer changes in the local weta population size. However, such an inference from an index to a population is always difficult because of confounding factors such as a change in the ability of observers to detect the weta. Therefore, an alternative monitoring scheme based on site occupancy models is being considered.

Some of the advantages of the occupancy modeling approach are: (1) it is designed to deal with variation in detection probability; (2) it allows sampling effort to be varied among sites; and (3) it enables managers to include habitat characteristics as covariates in the model to enhance prediction. This final factor is particularly important because weta appear to have strong habitat preferences even within the scientific reserve (Sherley and Hayes 1993), and this information could be useful in explaining changes in weta abundance. We consider this proposed study here to discuss some of the practical aspects in developing a monitoring program based upon the occupancy models detailed above.

Practical Considerations

There are a number of practical limitations that need to be considered in designing a survey for weta in Mahoenui Reserve. Most importantly, the thickness of the gorse cover and the steepness of the terrain within the reserve make it difficult to walk through the area, limiting the number of sites that could be searched per day. Moreover, dead gorse cover cannot be removed to facilitate movement through the reserve, because the weta tend to prefer this type of habitat. In a previous study, Sherley and Hayes (1993) reported that 75% of weta were on dead gorse foliage. Thus, gorse bushes have to be searched in a way that minimizes damage, even if the overall effect is that some weta fail to be detected. Moreover, surveys have to be restricted to observing weta on gorse bushes rather than removing the insects from bushes (e.g., to mark individuals) to avoid excessive habitat damage.

Another practical consideration is that the gorse habitat is not static, and thus the vegetation must also be monitored to compare changes in the weta population with changes in the structure of available habitat. Finally, some flexibility is required in the sampling regime because the ability to detect a weta appears to be related to weather conditions. Some evidence suggests that there is a higher chance of detecting weta when conditions are cool and humid (P. Bradfield and G. Sherley, Department of Conservation, New Zealand, pers. comm.).

Proposed Survey Design

The proposed survey will be limited to a central block within the reserve due to the level of resources that are expected to be available. Similarly, to ensure that sites are reasonably accessible, monitoring sites will be randomly located near the edge of the central block, with the center of each site within 30 m of the gorse edge. Restricting the placement of sites in this manner constrains statistical inferences to the monitoring region only. We cannot make inferences at a larger scale (e.g., to the entire block or the reserve).

One annual survey will be conducted in midsummer (January/February), when weta numbers tend to be at their highest (Sherley and Hayes 1993), increasing the chances of detecting weta should they be present at a site. Given the current project budget, a maximum of three searches of each site could be conducted per annual survey. Organizers will aim to conduct searches on consecutive days, but they may be spaced more than 1 day apart to avoid days of high temperature and low humidity (when weta are less visible). All three searches should be within 1 week to meet the assumption of closure with respect to occupancy of the sites by weta.

Seventy-five sites will be searched in the proposed monitoring scheme. To maximize the spatial coverage of the survey, not all sites will be searched an equal number of times. By rotating which sites are searched on each of the 3 days of monitoring, 30 (randomly chosen) sites will be searched three times and 45 sites will be searched only twice. Simulations suggest that such a design should provide reasonable results provided that the probability of detecting weta in a search of a site (given presence) is greater than or equal to 0.5. Because of uncertainties about the likely detection probability that will be encountered in the field, it is prudent to have a substantial proportion of sites searched three times. For the same level of effort, only 60 sites could be monitored with an equal number of searches (three searches each). Sites will be permanently marked and resurveyed each year.

A site will consist of a circular area of 3 m radius ($\approx$ 28 m^2), randomly placed within the defined monitoring region within the reserve. This size of unit was chosen (with field staff input) as a compromise between making the unit large enough to have a reasonable chance of containing at least one individual and small enough that searches would not be overly time-consuming. Rectangular sites (15 m $\times$ 2 m) were considered, but it was thought that the elongated shape would introduce more habitat variation within each site. Assuming that habitat will be more homogeneous within a circular site, this method should facilitate the characterization and modeling of occupancy as a function of small-scale habitat structure. Circular sites are also advantageous because if they are randomly located, they are less likely to extend beyond the boundary of the gorse block than elongated rectangles.

Consistency in survey methods is important both for repeated searches of sites within a season and for repeat annual surveys. Within a site, all gorse bushes will be searched for weta, using consistent survey effort. Given the differences in the number of bushes within a site, a simple field rule will be to search sites with less then 1/3 gorse cover for 3 min, to search sites with 1/3–2/3 gorse cover for 6 min, and to search sites with more than 2/3 gorse cover for 9 min.

Each search of a site should occur at a different time of the day, because the surveyor's ability to detect weta may vary during the day due to changes in weta activity levels. If sites are always searched at similar times of the day, and the detectability of weta varies in response to activity levels, the study design may be introducing a form of heterogeneity in detection probabilities. Conceivably, sites that would always be searched during peak activity times may have higher detection probabilities than sites surveyed during periods of reduced activity, hence violating one of the model assumptions. By rotating the order in which sites are searched (thereby varying the time of searching a particular site) we may have more success at identifying and modeling changes in detectability due to time of day (i.e., by breaking any site × time correlation structure). Similarly, it is recommended that observers be rotated around sites to ensure that the same observer does not always search the same site, as there are likely to be differences between observers in their ability to locate weta.

Data recorded during searches will include temperature, humidity, time of day, observer name, and characteristics of the gorse habitat, such as bush size and height. For each observed weta, the surveyor will record its observed height, location on the gorse branch (i.e., near the tip or stem), the color and size of the branch on which it is found, and its gender and age.

Final Design

The challenge in designing a monitoring scheme for Mahoenui giant weta based on site-occupancy modeling is that so little is known about their occupancy and detection probabilities and the effect of small-scale habitat differences on these rates. Thus, the recommended monitoring design will be used for 3 years as a pilot study before a full review of the methodology is carried out. This review will assess whether the design provides information of sufficient quality and quantity for management of the weta population. This initial 3-year monitoring period will provide valuable information on current levels of occupancy, detection probabilities, and required field effort.

Design factors that will be reviewed include the optimum number of sites and the number of searches within a survey. Data collection procedures will also be reviewed. The modeling above suggests that the minimum acceptable probability of detection is greater than or equal to 0.5. If the detection rate is lower than this threshold, it might be improved by increasing the length of time spent at each site, by changing the observation method to more actively search within each gorse bush, or by having larger sites. However, if weta numbers are very low, refining the data collection process will not improve detection rates. The review will also assess the habitat variables being collected and the training needs of field staff.

References

Azuma, D. L., J. A. Baldwin, and B. R. Noon. 1990. *Estimating the Occupancy of Spotted Owl Habitat Areas by Sampling and Adjusting for Bias*. USDA Forest Service, Pacific Southwest Research Station, General Technical Report PSW-124, Berkeley, California.

Buckland, S. T., D. R. Anderson, K. P. Burnham, J. L. Laake, D. L. Borchers, and L. Thomas. 2001. *Introduction to Distance Sampling*. Oxford University Press, Oxford, U.K.

Burnham, K. P., and D. R.Anderson. 1998. *Model Selection and Inference: A Practical Information-theoretic Approach*. Springer-Verlag, New York.

Burnham, K. P., and W. S. Overton. 1978. Estimation of the size of a closed population when capture probabilities vary among animals. *Biometrika* 65:625–633.

Chao, A., and S.-M. Lee. 1992. Estimating the number of classes via sample coverage. *Journal of the American Statistical Association* 87:210–217.

Chao, A., S.-M. Lee, and S. L. Jeng. 1992. Estimation of population size for capture-recapture data when capture probabilities vary by time and individual animal. *Biometrics* 48:201–216.

Cole, L. C. 1949. The measurement of interspecific association. *Ecology* 30:411–424.

Coull, B. A., and A. Agresti. 1999. The use of mixed logit models to reflect heterogeneity in capture-recapture studies. *Biometrics* 55:294–301.

Diamond, J. M. 1975. Assembly of species communities. pp. 342-444 in M. L. Cody and J. M. Diamond, eds., *Ecology and Evolution of Communities*. Harvard University Press, Cambridge, Massachusetts.

Dorazio, R. M., and J. A. Royle. 2003. Mixture models for estimating the size of a closed population when capture rates vary among individuals. *Biometrics* 59:351–364.

Fienberg, S. E., M. S. Johnson, and B. W. Junker. 1999. Classical multilevel and Bayesian approaches to population size estimation using multiple lists. *Journal of the Royal Statistical Society of London, Series A* 163:383–405.

Gaston, K. J., and T. M. Blackburn. 1996. Global scale macroecology: Interactions between population size, geographic range size and body size in the Anseriformes. *Journal of Animal Ecology* 65:701–714.

Gaston, K. J., R. M. Quinn, S. Wood, and H. R. Arnold. 1996. Measures of geographic range size: The effects of sample size. *Ecography* 19:259–268.

Geissler, P. H., and M. R. Fuller. 1987. Estimation of the proportion of area occupied by an animal species. *Proceedings of the Section on Survey Research Methods of the American Statistical Association* 1996:533–538.

Gotelli, N. J. 2000. Null model analysis of species co-occurrence patterns. *Ecology* 81:2606-2621.
Hanski, I. 1992. Inferences from ecological incidence functions. *American Naturalist* 139:657–662.
_____. 1994. A practical model of metapopulation dynamics. *Journal of Animal Ecology* 63:151–162.
_____. 1997. Metapopulation dynamics: From concepts and observations to predictive models. pp. 69–91 in I. A. Hanski and M. E. Gilpin, eds., *Metapopulation Biology: Ecology, Genetics, and Evolution*. Academic Press, New York.
_____. 1999. *Metapopulation Ecology*. Oxford University Press, Oxford, U.K.
Kendall, W. L. 1999. Robustness of closed capture-recapture methods to violations of the closure assumption. *Ecology* 80:2517–2525.
Kéry, M. 2002. Inferring the absence of a species—a case study of snakes. *Journal of Wildlife Management* 66:330–338.
Lancia, R. A., J. D. Nichols, and K. H. Pollock. 1994. Estimating the number of animals in wildlife populations. pp. 215–253 in T. Bookhout, ed., *Research and Management Techniques for Wildlife and Habitats*. The Wildlife Society, Bethesda, Maryland.
Lande, R. 1987. Extinction thresholds in demographic models of territorial populations. *American Naturalist* 130:624–635.
_____. 1988. Demographic models of the northern spotted owl (*Strix occidentalis caurina*). *Oecologia* 75:601–607.
Levins, R. 1969. Some demographic and genetic consequences of environmental heterogeneity for biological control. *Bulletin of the Entomological Society of America* 15:237–240.
Levins, R. 1970. Extinction. pp. 77–107 in M. Gustenhaver, ed., *Some Mathematical Questions in Biology. Volume II*. American Mathematical Society, Providence, R.I.
Link, W. A., R. J. Barker, J. R. Sauer, and S. Droege. 1994. Within-site variability in surveys of wildlife populations. *Ecology* 74:1097–1108.
MacKenzie, D. I., and L. Bailey. 2004. Assessing the fit of site occupancy models. *Journal of Agricultural, Biological and Environmental Statistics* 9:300–318.
MacKenzie, D. I., and W. L. Kendall. 2002. How should detection probability be incorporated into estimates of relative abundance? *Ecology* 83:2387–2393.
MacKenzie, D. I., J. D. Nichols, J. E. Hines, M. G. Knutson, and A. B. Franklin. 2003. Estimating site occupancy, colonization and local extinction probabilities when a species is detected imperfectly. *Ecology* 84:2200–2207.
MacKenzie, D. I., J. D. Nichols, G. B. Lachman, S. Droege, J. A. Royle, and C. A. Langtimm. 2002. Estimating site occupancy when detection probabilities are less than one. *Ecology* 83:2248–2255.
McCoy, E. D., and K. L. Heck, Jr. 1987. Some observations on the use of taxonomic similarity in large-scale biogeography. *Journal of Biogeography* 14:79–87.
Moilanen, A. 2002. Implications of empirical data quality for metapopulation model parameter estimation and application. *Oikos* 96: 516–530.
Nichols, J. D., and K. U. Karanth. 2002. Statistical concepts; assessing spatial distribution. pp. 29–38 in K. U. Karanth and J. D. Nichols, eds., *Monitoring Tigers and Their Prey*. Centre for Wildlife Studies, Bangalore, India.
Norris, J. L., III, and K. H. Pollock. 1996. Nonparametric MLE under two closed capture-recapture models with heterogeneity. *Biometrics* 52:639–649.

Otis, D. L., K. P. Burnham, G. C. White, and D. R. Anderson. 1978. Statistical inference from capture data on closed animal populations. *Wildlife Monographs* 62:1–135.

Pielou, E. C. 1977. *Mathematical Ecology*. Wiley, New York.

Pledger, S. 2000. Unified maximum likelihood estimates for closed capture-recapture models for mixtures. *Biometrics* 56:434–442.

Robbins, C. S., D. K. Dawson, and B. A. Dowell. 1989. Habitat area requirements of breeding forest birds of the middle Atlantic states. *Wildlife Monographs* 103:1–34.

Royle, J. A. 2004a. Modeling abundance index data from anuran calling surveys. *Conservation Biology* 18:1–8.

———. 2004b. N-mixture models for estimating population size from spatially replicated counts. *Biometrics*. 60:108–115.

Royle, J. A., and J. D. Nichols. 2003. Estimating abundance from repeated presence absence data or point counts. *Ecology* 84:777–790.

Scott, J. M., P. J. Heglund, M. L. Morrison, J. B. Haufler, M. G. Rafael, W. A. Wall, and F. B. Samson, eds. 2002. *Predicting Species Occurrences: Issues of Accuracy and Scale*. Island Press, Washington, D.C.

Seber, G. A. F. 1982. *Estimation of Animal Abundance and Related Parameters*. 2nd ed. Griffin, London.

Sherley, G. H. 1998. *Threatened Weta Recovery Plan*. Biodiversity Recovery Unit, Department of Conservation, Wellington, New Zealand.

Sherley, G. H., and L. M. Hayes. 1993. The conservation of a giant weta (*Deinacrida* n.sp. Orthoptera: Stenopelmatidae) at Mahoenui, King Country: Habitat use, and other aspects of its ecology. *New Zealand Entomologist* 16:55–68.

Smith, E.P., and G. van Belle. 1984. Nonparametric estimation of species richness. *Biometrics* 40:119–129.

Stauffer, H. B., C. J. Ralph, and S. L. Miller. 2002. Incorporating detection uncertainty into presence-absence surveys for marbled murrelet. pp. 357–365 in J. M. Scott, P. J. Heglund, M. L. Morrison, J. B. Haufler, M. G. Rafael, W. A. Wall, and F. B. Samson, eds., 2002. *Predicting Species Occurrences: Issues of Accuracy and Scale*. Island Press, Washington, D.C.

Thomas, M. D., J. A. Brown, F. W. Maddigan, and L. A. Sessions. 2003. Comparison of trap-catch and bait interference methods for estimating possum densities. *New Zealand Plant Protection Society* 56:81–85.

Thompson, W. L., G. C. White, and C. Gowan. 1998. *Monitoring Vertebrate Populations*. Academic Press, San Diego.

Travis, J. M. J. 2003. Climate change and habitat destruction: A deadly anthropogenic cocktail. *Proceedings of the Royal Society of London B* 270:467–473.

White, G. C., and K. P. Burnham. 1999. Program MARK: Survival rate estimation from both live and dead encounters. *Bird Study* 46:S120–S139.

White, G. C., K. P. Burnham, and D. R. Anderson. 2001. Advanced features of program MARK. pp. 368–377 in R. Field, R. J. Warren, H. Okarma, and P. R. Sievert, eds., *Wildlife, Land, and People: Priorities for the 21st Century*. The Wildlife Society, Bethesda, Maryland.

Williams, B. K., J. D. Nichols, and M. J. Conroy. 2002. *Analysis and Management of Animal Populations*. Academic Press, San Diego.

Yoccoz, N. G., J. D. Nichols, and T. Boulinier. 2001. Monitoring of biological diversity in space and time. *Trends in Ecology and Evolution* 16:446–453.

9

A Bayesian Approach to Estimating Presence When a Species Is Undetected

James T. Peterson and Peter B. Bayley

The conservation and management of animal populations requires knowledge of the locations of existing populations coupled with a credible assessment of the suitability of the environment to support biota. Many species, however, are difficult to sample due to morphological and behavioral characteristics or are difficult to locate because they are relatively rare (e.g., Rabinowitz et al. 1986). This often results in a large number of samples yielding no observations. In these instances, populations are more expensive to study per unit of useful information, because deductions and inferences are needed from a greater proportion of samples in which they were absent. This presents sampling and interpretational challenges that are further exacerbated when the biologist cannot be sure that the zero sample (i.e., nondetection of the species) indicates that the species was absent.

Here we describe a Bayesian approach to estimating the probability of species presence for zero samples. Although our examples concentrate on the quantifying presence of rare or difficult-to-sample fish, the principles we describe apply to any animal population whose individuals have a probability of being observed that is less than 1. Although the term "rarity" of a species is conventionally applied to the size of its population, the methods we outline apply to any population in which the most appropriate sampling protocol and design produces zero observations.

Definitions

Observability (q) is the probability of observing (or capturing) a single individual in a given area. The proportion of N individuals in an area that are observed equals the observability, providing that all individuals have the same observability. This is typically termed catchability in fisheries. The qualification, "in a given area," means that as conditions affecting observability change in space or time, observability will change. Therefore, we treat observability (q) as a random variable that is frequently conditional on sampling conditions, as well as on the protocol used for observation or capture.

Detectability (d) is the probability of detecting one or more individuals in a given area. Therefore, if there are N individuals of a species with equal observability, q, detectability is:

$$d = 1 - (1 - q)^N, \qquad (9.1)$$

provided that the individuals respond independently to the observation process. Detectability is very sensitive to abundance and observability (Figure 9.1).

Other probability measures are defined as follows. When one or more individuals are present (= *conditional* event F) in a given area, the proba-

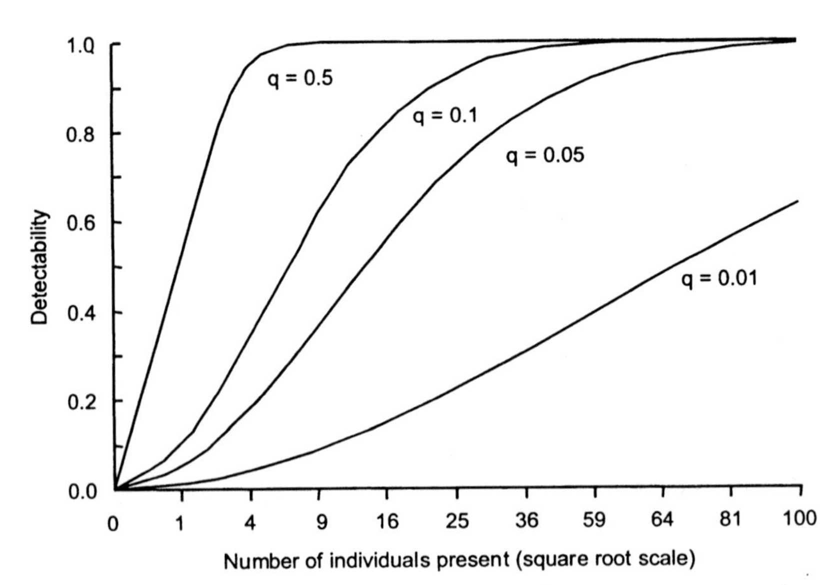

Figure 9.1. Detectability (d in equation [9.1]) of a species versus numbers of individuals present (N) for selected values of observability (q).

bility of observing zero individuals (= event *Co*) is given by $P(Co \mid F)$, which is the complement of detectability, *d*. Conversely, when zero individuals have been observed (= *conditional* event *Co*) in a given area, the probability of presence (event = *F*) is denoted $P(F \mid Co)$.

Previous Approaches: Advantages and Limitations

The most common approach to quantifying the spatial distribution of a species is to presume that all individuals present are observed in each sampled quadrat, that is, observability, $q = 1$. Unfortunately, this is not true for mobile species in natural conditions, such as fish, when sampled using the most affordable or legal methods available. Complete detectability based on visual observation methods may be realistic for some terrestrial species, but the challenge in quantifying such observations then consists of defining and measuring a consistent quadrat size to be sampled so that comparable samples can be taken and inferences can be made to larger scales.

Biologists have attempted to minimize the influence of incomplete observation on estimates of species presence using what can be roughly categorized as sampling effort or inferential approaches (Bayley and Peterson 2001). Sampling effort approaches implicitly recognize incomplete detectability and rely on multiple samples to ensure species detection over a region of interest. When the region of interest is a sample site (sample unit), sites are repeatedly sampled over time to ensure a probability of detection above some arbitrary threshold (Stauffer et al. 2002). The required number of repeated samples is determined using species detection probabilities as estimated via a binomial model fit to repeated sampling data from multiple sites. Another sampling effort approach takes a regional (larger scale) focus rather than a site-specific focus to assess species presence (Bonar et al. 1997). A statistical function describing the spatial distribution of the species in the region is presumed, and a predetermined number of random samples is taken to ensure the desired probability of detecting the species, assuming a density above some arbitrary threshold.

Neither site-specific nor regional sampling effort approaches produce estimates of probability of presence because they require the adoption of arbitrary thresholds. Also, neither approach explicitly accounts for variation in detectability among sample units or through time. Therefore, it is valid to apply these approaches only when detectability is constant, which requires that either (1) abundance per unit area (*N* from equation (9.1)) and observability (q) of individuals in each sampled area are constant, or

(2) abundance and observability are inversely related in such a manner that constant detectability is fortuitously maintained (Figure 9.1). Either of these conditions would rarely be satisfied under most sampling circumstances, especially when the sampling unit needs to be sufficiently consistent to make valid statistical inferences.

Inferential approaches explicitly account for variable detectability and can be used with existing data that are consistent with a specific sample design. For instance, MacKenzie et al. (2002) recently developed a method for estimating the site occupancy rates for zero observations that incorporate covariates, such as sample unit characteristics. Similarly, Geissler and Fuller (1987) developed a method that accounts for incomplete detectability in estimating the proportion of sample sites occupied. These methods require the collection of data from a number of sites that are sampled on multiple occasions and the assumption that sites are "closed" to changes in occupancy (MacKenzie et al. 2002; see also Chapter 8, this volume). That is, a species cannot colonize or abandon sample sites during the study period. These relatively rigorous restrictions, however, may limit the usefulness of these and similar approaches. Repeated visits may not be feasible or affordable, particularly for investigations that require a relatively large number of study sites. Closure assumptions also are likely to be violated for relatively mobile organisms, such as fishes (e.g., Gowan et al. 1994). Therefore, a more flexible approach may be desirable. Below we describe an approach to quantifying the spatial distribution of species that uses empirical estimates of species detection probabilities to estimate probabilities of species presence, given a zero sample.

Estimating Sampling Detection Probabilities

The probability of detecting a species depends on an estimate of observability and an estimate of the number of individuals in a sampling site (Figure 9.1). Using empirical estimates of observability and abundance, one can estimate detection probabilities as 1 minus the probability of capturing no (zero) animals:

$$\hat{d} = 1 - \hat{P}(Co \mid N), \tag{9.2}$$

where $\hat{P}(Co \mid N)$ is the estimated probability of failing to observe any individuals when N are present and where N is defined above. Direct use of equation (9.2) is difficult because the number of vulnerable animals, N, is unknown and varies among samples, which changes detectability. The estimate of this number for any sample is more realistically represented by a probability distribution function (PDF) of likely numbers. Therefore, the empirically derived $P(F_N)$ is expressed as a PDF for the unconditional probability of N animals being present, where N varies from zero to some reasonably large number. Equation (9.2) can therefore be reexpressed as the sum of joint probabilities corresponding to each possible N:

$$\hat{d} = 1 - \sum_{0} \hat{P}(Co \mid F_N) \cdot \hat{P}(F_N) . \tag{9.3}$$

Terms in the summations in equation (9.3) approach zero for increasing N, and the rate of approach of their product is greater for higher catchability. In what follows, we discuss methods for estimating $P(Co \mid F_N)$ and $P(F_N)$.

Modeling Fish Catchability

An unbiased estimator of catchability (observability) is required to obtain a reliable estimator of detectability. Fish catchability can be estimated site by site using a variety of population estimation techniques, such as capture-recapture methods (Williams et al. 2002). Because meaningful studies require standardized protocols and numerous samples, we believe it is more efficient to develop protocol-based models from field calibrations that predict catchability (observability) based on species, individual size, and habitat characteristics that use mark-recapture to estimate the vulnerable population. These models have employed quasi-likelihood (e.g., Bayley 1993; Peterson and Rabeni 2001; Bayley and Austen 2002) or beta-binomial regression (Peterson et al. 2004). Assuming that capture is a binomial process and that individuals respond independently and have the same catchability, $P(Co \mid F_N)$ is estimated as:

$$\hat{P}(Co_i \mid F_{N_i}) = (1 - \hat{q}_i)^{\hat{N}_i} , \tag{9.4}$$

where $\hat{q}_i$ and $\hat{N}_i$ are estimators of catchability and number of individuals in the ith sample unit, respectively.

The prevalence of overdispersion (variance in excess of a presumed binomial distribution) in many catchability and mark and recapture models is at least partly due to fish not responding independently to capture (Bayley and Dowling 1993; Bayley and Herendeen 2000; Peterson and Rabeni 2001; Bayley and Austen 2002). These violations of binomial assumptions can result in the systematic underestimation of detection probabilities (Peterson 1999). Thus, a more realistic version of equation (9.4) that accounts for overdispersion is:

$$\hat{P}(Co_i \mid F_{N_i}) = (1-\hat{q}_i)^{\hat{N}_i^*}, \tag{9.5}$$

where $\hat{N}_i^*$, the effective number of groups for sample unit i, is estimated by $\hat{N}_i \bullet [\hat{N}_i \bullet \hat{s}^2 \bullet \hat{q}_i \bullet (1-\hat{q}_i) + 1]^{-1}$, and $\hat{s}^2$ is an estimate of extrabinomial variance in catchability that represents overdispersion (Bayley and Peterson 2001). Alternatively, overdispersion can be incorporated via a beta-binomial distribution as:

$$\hat{P}(Co_i \mid F_{N_i}) = \left[\frac{\Gamma(\hat{N}_i+1)\Gamma(\hat{a}_i)\Gamma(\hat{a}_i+\hat{b}_i)\Gamma(\hat{N}_i+\hat{b}_i)}{\Gamma(\hat{N}_i+1)\Gamma(\hat{a}_i+\hat{b}_i+\hat{N}_i)\Gamma(\hat{a}_i)\Gamma(\hat{b}_i)}\right], \tag{9.6}$$

where $\hat{a}_i$ and $\hat{b}_i$ are shape parameters (Peterson et al. 2002) for the ith sample unit estimated from a beta-binomial regression, in which $\hat{a}_i = \frac{\hat{q}_i}{\gamma}$ and $\hat{b}_i = 1 - \frac{1-\hat{q}_i}{\gamma}$, where γ is a dispersion parameter.

Modeling Fish Distribution and Abundance

The number of fish, in the form of a PDF, $P(F_N)$, must be estimated from field samples. Because a sufficient number of samples with population estimates will not typically be available or affordable, unbiased abundance estimates (e.g., estimated using catchability models) and all zero estimates resulting from zero catches must be included. The best approach is to model species-specific abundance as a function of ecologically plausible habitat variables. The error distribution function assumed in the species-specific predictive model is used to estimate the PDF, $P(F_N)$. The main challenge is in obtaining good empirical support for a function (or mixed function, e.g., Welsh et al. 1996) that adequately predicts real absences and low numbers at the scale of the sample unit. Accurate prediction of zeros and small abundances is more important than for larger numbers because detectability becomes larger and less sensitive as abundance increases (Figure 9.1).

Because fish are discrete units (i.e., a sample site cannot contain a fraction of a fish), the abundance PDF, $P(F_N)$, is best modeled using a discrete statistical distribution. Two commonly used discrete distributions are the Poisson and the negative binomial. Both distributions can be used to model organismal abundance as a function of covariates (e.g., habitat characteristics) via generalized linear models (McCullagh and Nelder 1989). However, they fundamentally differ in how they model variance and approximate species distributions. The Poisson distribution prescribes a variance (equal to the mean) and assumes that individuals are randomly distributed after accounting for the effects of covariates in a regression. In previous work, we found that the Poisson was inadequate for modeling the abundance and distribution for several species of stream-dwelling fishes (Peterson 1999; Bayley and Peterson 2001) and believe that the Poisson assumptions are likely violated for many other animal populations. Violations of the assumptions can result in the systematic underestimation of detection probabilities (Peterson 1999). We currently model fish abundance using a negative binomial distribution, which allows for variance in excess of that prescribed by its Poisson cousin. The probability that the ith site contains individuals is estimated by the negative binomial as:

$$\hat{P}(F_{\hat{N}_i}) = \left(1 + \frac{\hat{m}_i}{\hat{k}_i}\right)^{-\hat{k}_i} \cdot \frac{\Gamma(\hat{k}_i + \hat{N}_i)}{\Gamma(\hat{N}_i + 1) \cdot \Gamma(\hat{k}_i)} \cdot \left(\frac{\hat{m}_i}{\hat{m}_i + \hat{k}_i}\right)^{\hat{N}_i}, \quad (9.7)$$

where $\hat{m}_i$ is the estimator of mean abundance for sample unit i, and $\hat{k}$ is the estimator of the negative binomial dispersion parameter for sample unit i, which decreases as dispersion increases.

Incorporating Detection Probabilities into Presence-Absence Surveys and Inference

The interpretation of an event when a species is not detected requires the consideration of total probability (Bayley and Peterson 2001), which means that all occurrences that could cause a zero catch must be considered. At the broadest level, all occurrences would fall into either Situation A: the species was present but not caught, or Situation B: the species was absent from the particular area sampled. Using Bayesian terminology, the total probability of a zero catch (event Co) occurring, $P(Co)$, can be expressed as the sum of probabilities of Situations A and B:

$$P(Co) = P(Co \mid F) \cdot P(F) + P(Co \mid \sim F) \cdot P(\sim F) . \quad (9.8)$$

The first term is the probability of Situation A, which is the product (= joint probability) of $P(F \mid Co)$, the probability of missing a species (event Co) when it is present (event F) and $P(F)$, the prior estimate of probability of the species being present (the Bayes empirical prior). The second term is the probability of Situation B, which is the product of $P(Co \mid \sim F)$,the probability of not capturing a species when it is absent (= 1) and $P(\sim F)$, the probability of the species being absent (event $\sim F$). Logically, $P(F) + P(\sim F) = 1$.

Situation A also can be expressed as the joint probability that a species is present given a zero catch event, $P(F \mid Co)$, and that it is not captured, $P(Co)$. Therefore, this product is equivalent to the first term in equation (9.8):

$$P(F \mid Co) \cdot P(Co) = P(Co \mid F) \cdot P(F), \tag{9.9}$$

where equation (9.9) is known as conditional probability symmetry.

The standard Bayes' rule is derived by rearranging equation (9.9) and substituting for $P(Co)$ using equation (9.8). It describes the probability of a species being present (event F) given a zero capture (event Co):

$$P(F \mid Co) = \frac{P(Co \mid F) \cdot P(F)}{P(Co \mid F) \cdot P(F) + P(Co \mid \sim F) \cdot P(\sim F)}. \tag{9.10}$$

Thus, the numerator is the product of the unconditional probability of presence (the Bayesian prior, $P(F)$) adjusted by the likelihood of missing the species when present, $P(Co \mid F)$. This joint probability can be interpreted as being proportional to the number of possibilities of zero catches when present, whereas the denominator represents all possibilities of zero catches (Situations A and B, equation (9.8)) and thereby serves to standardize the range of conditional probabilities, $P(F \mid Co)$, between zero and one.

Site-specific posterior probabilities of presence are estimated using site-specific estimates of $P(Co \mid F_N)$ and $P(F_N)$. Reexpressing equation (9.10) as the sum of joint probabilities corresponding to each possible N, the posterior probabilities are estimated as:

$$\hat{P}(F_i \mid Co_i) = \frac{\sum_{1} \hat{P}(Co_i \mid F_{N_i}) \cdot \hat{P}(F_{N_i})}{\sum_{0} \hat{P}(Co_i \mid F_{N_i}) \cdot \hat{P}(F_{N_i})}, \tag{9.11}$$

where $P(Co_i \mid F_{N_i})$ and $P(F_{N_i})$ of the ith sample unit are estimated using equations (9.4), (9.5), or (9.6) and equation (9.7), respectively.

Sampling Considerations and Assumptions

Our approach to estimating detectability and probability of presence for zero catches can be used with any standardized sampling protocol, provided that observability (catchability) and animal abundance and distribution, $P(F_N)$, can be quantified. We assume that species-specific data used to fit models of observability and animal abundance and distribution are from independent samples in the region collected from a representative, preferably probabilistic, sampling design. Further, we assume that the conditions and sample protocol used to collect these data are similar to those encountered during normal sampling activities. This assumption is crucial because model-based predictions should be applied only to situations that are similar to those under which the models were parameterized (Mangel et al. 2001). For example, the $P(F_N)$ models fit to data collected from sample units specifically chosen to contain high population densities would likely overestimate animal abundance and, hence, detection probabilities.

When making inferences from a set of samples, it is essential to maintain relatively consistent sample unit dimensions (i.e., spatial extent). Although it is not always possible to maintain a constant sampled area, large variation compromises our ability to interpret catch data and generalize even when the species is encountered. With respect to stream fish species richness estimation, a popular approach is to expand the area sampled until some estimated proportion of the "species richness" has been encountered (Angermeier and Smogor 1995). We have argued that this compromises the interpretation of a data set with widely varying sampled areas because of the various mechanisms that contribute to the species-area phenomenon (Bayley and Peterson 2001). This argument also applies to single species. Habitat heterogeneity generally increases with sample unit size (Wiens 1989), which may affect the accuracy of animal abundance models. Sample variance also is influenced by sample unit size (Wiens 1989), which accordingly will influence dispersion parameter estimates, such as k in the negative binomial model. The influence of sample area would be particularly acute for species with low average densities and clustered distributions that respond to sparsely distributed resources, such as critical habitats.

Empirical Validation of Probability of Presence Estimations

Here we describe a direct validation of zero-catch samples in stream systems that do not contain listed or at-risk species. When the sample from a standard protocol is followed by intensive sampling in a closed area, including the use of an ichthyocide, the latter provides a validation of the presence or absence of the species. An estimate of its abundance also can be derived from the ichthyocide sampling data. An example from a series of boat electrofishing samples indicates that probability of presence, $P(F \mid Co)$, increases with population size for zero catches (Figure 9.2), but with considerable unexplained variance. This variance is largely due to unexplained error in the models estimating the prior distribution, $P(F_N)$. We describe below a categorical approach to measure the degree to which such variance affects our ability to correctly predict absence or presence across several samples and among four sampling protocols.

We compared counts of zero-catch events correctly or incorrectly predicted according to whether the $P(F \mid Co)$ estimate is greater or less than 0.5 (Figure 9.3A). The upper left and lower right frequencies in each 2 × 2 table

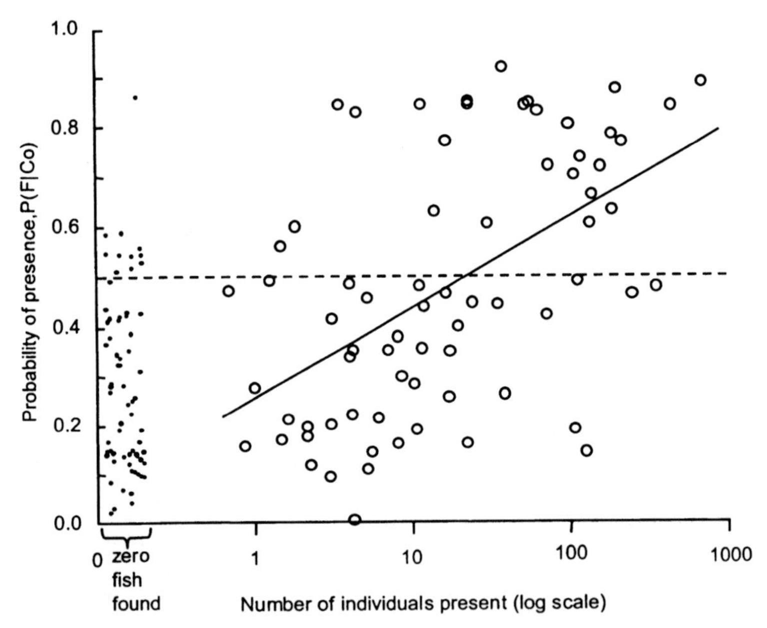

Figure 9.2. Probability of presence estimates ($P(F \mid Co)$, y-axis) versus numbers of individuals present (estimated from subsequent rotenone calibration) from samples in which a boat electrofisher caught zero fish of any one of 16 species among 12 blocked-off stream reaches (from Bayley and Peterson 2001). Open circles and solid regression line represent all zero-samples in which examples of that species were subsequently encountered. Dots (shown in jittered form) represent $P(F \mid Co)$ estimates for remaining zero-catch samples in which no individuals of the species were encountered. (Dotted line at $P(F \mid Co) = 0.5$ divides estimates into frequencies for presence and absence as shown in Figure 9.3A.)

A

Boat Electrofisher:

	PRES.	ABS.	Total
P(F|Co) ≥ 0.5	27	11	38
P(F|Co) < 0.5	44	54	98
Total	71	65	136

Pearson $\chi^2 = 7.5$, P=0.006

Electric Seine:

	PRES.	ABS.	Total
P(F|Co) ≥ 0.5	28	35	63
P(F|Co) < 0.5	166	438	604
Total	194	476	667

Pearson $\chi^2 = 7.9$, P=0.005

Backpack Electrofisher:

	PRES.	ABS.	Total
P(F|Co) ≥ 0.5	19	22	41
P(F|Co) < 0.5	26	59	85
Total	45	81	126

Pearson $\chi^2 = 3.0$, P=0.08

Net Seine:

	PRES.	ABS.	Total
P(F|Co) ≥ 0.5	6	2	8
P(F|Co) < 0.5	28	74	102
Total	34	76	110

Pearson $\chi^2 = 7.9$, P=0.005*

All Sampling Protocols:

	PRES.	ABS.	Total
P(F|Co) ≥ 0.5	80	70	150
P(F|Co) < 0.5	264	625	889
Total	344	695	1039

Pearson $\chi^2 = 32$, P<0.001

B

	Odds of being correct given:		
	P(F|Co)≥0.5	P(F|Co)<0.5	Either condition
Boat Electrofisher:	2.45	1.23	3.01
Electric Seine:	0.80	2.64	2.11
Backpack Electrofisher:	0.86	2.27	1.96
Net Seine:	3.00	2.64	7.93
All Sampling Protocols:	1.14	2.37	2.71

Figure 9.3. Validation of probability of presence estimates, $P(F \mid Co)$, for zero-catch events involving any one of 16 species from 121 field calibrations (methods in Bayley and Peterson 2001). The only information from each site used to estimate $P(F \mid Co)$ was the zero catch and habitat conditions. Presence/absence was determined independently using a combination of additional passes with the gear and an ichthyocide treatment that followed the standard protocol sample. (A) Frequency tables by sampling protocol show numbers of events in which species was present (PRES) or absent (ABS) (estimated independently from rotenone calibration) by numbers of corresponding estimates of $P(F \mid Co)$ that were greater or less than 0.5. "Net seine" protocol was based on two passes of a 20-foot (6.1-m) net, whereas protocols of all other gears involved one pass through the blocked reach. "Electric seine" combined results from 30-foot (9.1-m) and 50-foot (15.2-m) versions. (B) Odds of correctly predicting presence or absence given estimates of $P(F \mid Co)$ that were greater or less than 0.5, respectively, and the odds (odds ratio) of being correct for either condition (calculated from row proportions of tables shown in (A)). *Test was suspect because frequencies of > 20% of fitted cells were < 5.

represent correct predictions, and Pearson's chi-square test of interaction was used to estimate the significance of the frequency of correct predictions compared to the null hypothesis of equal cell proportions. All gears separately and collectively indicated strong significance ($P < 0.01$) except for the backpack electrofisher ($P = 0.08$), with the direction of the interaction consistently being in favor of correct predictions for all protocols.

We then used the ratios of row percentages of frequencies from Figure 9.3A to estimate the odds of making correct prediction of presence or absence (Figure 9.3B). For example, using a boat electrofisher, estimates of $P(F \mid Co) > 0.5$ were 2.4 times more likely to correctly predict presence than to incorrectly predict presence. Although the odds estimates of the electric seine and backpack shocker were less than 1 when the species was present, bias was still reduced compared with predicting zero when the catch was zero, and the odds of making correct predictions of absence (second column) were high. The third column in Figure 9.3B shows the odds ratio (= cross-product ratio), which is the product of the odds of being correct when the species was either present or absent. For example, using electric seines, respective estimates of $P(F \mid Co)$ were 2.1 times more likely to correctly predict presence or absence than to incorrectly predict absence or presence.

Recommendations and Future Directions

The previous analysis was intended to demonstrate the validity of the approach and not to stimulate a choice among alternative sampling methods. Rather than being alternatives, some methods were more appropriate in terms of catchability than others, depending on the width and depth of the stream reach, or previous studies may have involved a different gear or protocol. Therefore, the approach accommodates different sampling protocols because $P(F \mid Co)$ is independent of method. This principle applies to any set of animal observation protocols, but we recommend that the number of standardized protocols be kept to a minimum because if plausible observability (catchability) models are not available, which is sadly the case for the great majority of methods, such models need to be derived from field data for each protocol.

Although the site-by-site Bayesian approach for zero catches produces estimates with high variance, the important goal is to provide a set of predictions within the geographic range of the species or population thereof with much less bias than simply using the catch data. The problem with using catch data directly may not be merely a negative bias due to a large proportion of zero samples but that detectability is highly variable and is often negatively related to habitat complexity. Therefore, some of the best stream habitats, such as those providing refuge for fish from sampling gear, also have lowest catchability that can counteract the effect of higher animal density.

Biologists often are interested in making inferences to larger areas, such as habitat patches in a metapopulation dynamics context (e.g., Dunham and Rieman 1999). We briefly outline two approaches to making such inferences: (1) joint probability of presence, and (2) a direct patch-based approach.

Site-specific probabilities of presence, $P(F_i \mid Co_i)$, can be combined to make inferences to larger areas such as an ecologically defined patch from one or more calibrated protocols (Bayley and Peterson 2001). The joint probability of presence from u samples, each with an estimator of $P(F_i \mid Co_i)$, is given by:

$$\hat{P}(F_u \mid Co_u) = 1 - \prod_{i=1}^{u} (1 - \hat{P}(F_i \mid Co_i)), \qquad (9.12)$$

which represents the joint probability of presence across u random samples in a patch. To obtain an estimate of probability of presence in the whole patch, $P(F_i \mid Co_i)$ must be modeled as a function of ecological variables, and predictions for all possible areas in the patch must be used, not just those sites sampled as in equation (9.12). Naturally, this whole-patch estimate will be greater than the u-sample-based version equation (9.12), which in turn will be greater than the mean of the $P(F_i \mid Co_i)$. Also, any sample of known presence will, logically, produce a value of 1 for either joint estimate, which at least is true for the time of sampling.

The direct patch-based approach also requires a set of random samples within the defined patch, but with a consistent protocol (Peterson and Dunham 2003). In this case, the combined probability of missing the species when present from u samples (the equivalent of $P(F \mid Co)$ in equation (9.11)) is used in the following Bayesian relationship for a patch:

$$\hat{P}(F_u \mid Co_u) = \frac{\left(\prod_{i=1}^{u}\left(\sum_{1} \hat{P}(Co_i \mid F_{N_i}) \cdot \hat{P}(F_{N_i} \mid Oc)\right)\right) \cdot \hat{P}(Oc)}{\left(\prod_{i=1}^{u}\left(\sum_{1} \hat{P}(Co_i \mid F_{N_i}) \cdot \hat{P}(F_{N_i} \mid Oc)\right)\right) \cdot \hat{P}(Oc) + \hat{P}(\sim Oc)}, \qquad (9.13)$$

where $\hat{P}(F_{N_i} | Oc)$ is the conditional site-specific PDF estimator of the number of fish modeled from empirical data, $\hat{P}(Oc)$ is the estimator of prior probability of patch occupancy, and $\hat{P}(\sim Oc)$ is the estimator of probability that the patch is unoccupied, $P(\sim Oc) = 1 - P(Oc)$. The product ($\prod$) function in the numerator and denominator of equation (9.13) represents the joint probability of missing the species when present from all u samples, and therefore, the definition of the probability of presence is the same as in equation (9.12).

It has not yet been determined if the joint probability of presence or a direct patch-based approach provides a more efficient estimator of patch-based probability of presence for comparable numbers of samples. The patch-based probability of presence approach requires a consistent protocol and a prior estimate of $P(Oc)$, but it may be more efficient in regions that contain patches with very low probabilities of occurrence. Effort and costs can be minimized by assuming that the prior is unknown (i.e., an uninformative prior) or could be based on expert opinion. The latter, however, should be based on objective, supportable criteria.

Although the Bayesian approach outlined here has logical fundamentals, there is much room for improvement even with estimates at the site-scale equation (9.11). With moderate or high catchabilities, variation in $P(F \mid Co)$ is largely due to unexplained variation in the models predicting the vulnerable population size at sites with zero catches. Use of an unrealistic model for the species abundance PDF also may be responsible. Although there remain other models to be explored, the bottom line is to have sufficient samples with abundance estimates in the system being studied. This requirement will obviously become more difficult as the population becomes rarer and/or observability becomes very small, so there is no free lunch.

Acknowledgments

We thank the many individuals who assisted during the field portion of this research in Illinois. The manuscript was improved with suggestions from M. Conroy, M. Freeman, and C. Moore.

References

Angermeier, P. L., and R. A. Smogor. 1995. Estimating number of species and relative abundances in stream-fish communities: effects of sampling effort and discontinuous spatial distribution. *Canadian Journal of Fisheries and Aquatic Sciences* 52:936–949.

Bayley, P. B. 1993. Quasi-likelihood estimation of marked fish recapture. *Canadian Journal of Fisheries and Aquatic Sciences* 50:2077–2085.

Bayley, P. B., and D. J. Austen. 2002. Capture efficiency of a boat electrofisher. *Transactions of the American Fisheries Society* 131:435–451.
Bayley, P. B., and D. C. Dowling. 1993. The effect of habitat in biasing fish abundance and species richness estimates when using various sampling methods in streams. *Polish Archives in Hydrobiology* 40:5–14.
Bayley, P. B., and R. A. Herendeen. 2000. The efficiency of a seine net. *Transactions of the American Fisheries Society* 129:901–923.
Bayley, P. B., and J. T. Peterson. 2001. An approach to estimate probability of presence and richness of fish species. *Transactions of the American Fisheries Society* 130:620–633.
Bonar, S. A., M. Divens, and B. Bolding. 1997. *Methods for Sampling the Distribution and Abundance of Bull Trout/Dolly Varden*. Research Report RAD97-05. Washington Department of Fish and Wildlife, Olympia, Washington.
Dunham, J. B., and B. E. Rieman. 1999. Metapopulation structure of bull trout: Influences of habitat size, isolation, and human disturbance. *Ecological Applications* 9:642–655.
Geissler, P. H., and M. R. Fuller. 1987. Estimation of the proportion of area occupied by an animal species. *Proceedings of the Section on Survey Research Methods of the American Statistical Association* 1986:533–538.
Gowan, C., M. K. Young, K. D. Fausch, and S. C. Riley. 1994. Restricted movement in resident stream salmonids: A paradigm lost? *Canadian Journal of Fisheries and Aquatic Sciences* 51:2626–2637.
MacKenzie, D. I., J. D. Nichols, G. B. Lachman, S. Droege, J. A. Royle, and C. A. Langtimm. 2002. Estimating site occupancy rates when detection probabilities are less than one. *Ecology* 83:2248–2255.
Mangel, M., O. Fiksen, and J. Giske. 2001. Theoretical and statistical models in natural resource management and research. pp. 57–72 in T. M. Shenk and A.B. Franklin, eds., *Modeling in Natural Resource Management*. Island Press, Washington, D.C.
McCullagh, P., and J. A. Nelder. 1989. *Generalized Linear Models*. 2nd ed. Chapman and Hall, London.
Peterson, J. T. 1999. *On the Estimation of Detection Probabilities for Sampling Stream-Dwelling Fishes*. BPA Report DOE/BP-25866-7. Bonneville Power Administration, Portland, Oregon.
Peterson, J. T., and J. Dunham. 2003. Combining inferences from models of sampling efficiency, detectability, and suitable habitat to classify landscapes for conservation of threatened bull trout, *Salvelinus confluentus*. *Conservation Biology* 17:1070–1077.
Peterson, J. T., J. Dunham, P. Howell, R. Thurow, and S. Bonar. 2002. *Protocol for Detecting Bull Trout Presence*. Western Division American Fisheries Society. Available from http://www.wdafs.org/committees/bull_trout/bull_trout_committee.htm (accessed June 15, 2004).
Peterson, J. T., and C. F. Rabeni. 2001. Evaluating the efficiency of a one-square-meter quadrat sampler for riffle-dwelling fish. *North American Journal of Fisheries Management* 21:76–85.
Peterson, J. T., R.F. Thurow, and J. Guzevich. 2004. An evaluation of multi-pass electrofishing for estimating the abundance of stream-dwelling salmonids. *Transactions of the American Fisheries Society* 133:462–475.
Rabinowitz, D., S. Cairns, and T. Dillon. 1986. Seven forms of rarity and their frequency in the flora of the British Isles. pp. 182–204 in M. E. Soulé, ed.,

Conservation Biology: The Science of Scarcity and Diversity. Sinauer Associates, Sunderland, Massachusetts.

Stauffer, H. B., C. J. Ralph, and S. L. Miller. 2002. Incorporating detection probabilities into presence-absence surveys for marbled murrelet. pp. 357–375 in J. M. Scott, P. J. Heglund, F. Samson, J. Haufler, M. Morrison, M. Raphael, and B. Wall, eds., *Predicting Species Occurrences: Issues of Accuracy and Scale*. Island Press, Washington, D.C.

Welsh, A. H., R. B. Cunningham, C. F. Donnelly, and D. B. Lindenmayer. 1996. Modelling the abundance of rare species: statistical models for counts with extra zeros. *Ecological Modelling* 88:297–308.

Wiens, J. A. 1989. Spatial scaling in ecology. *Functional Ecology* 3:385–397.

Williams, B. K., J. D. Nichols, and M. J. Conroy. 2002. *Analysis and Management of Animal Populations*. Academic Press, New York.

10

Searching for New Populations of Rare Plant Species in Remote Locations

Elizabeth L. Poon and Chris R. Margules

Limited resources often do not permit biologists to conduct extensive searches for every species listed as rare or threatened or to undertake surveys over large areas to determine species locations (Nicholls 1989; Margules and Austin 1994; Wessels et al. 1998). Further, it is inefficient to depend upon "accidental discoveries" (Good and Lavarack 1981) to estimate species distributions. A more systematic and structured approach is needed. One such approach, the method detailed in this chapter, is to stratify the region of interest using environmental variables, note strata where populations of rare species are known, and search for new populations in similar environments in the same general location and in the same environments at different geographic locations. In this way limited resources of labor, time, and funds can be used efficiently.

This approach is analogous to the way experienced field botanists would deal with the problem of searching for new populations of rare species. They create mental models of suitable habitat or environments and then look in the places those models suggest. We propose a systematic and explicit formalization of those models to make more available the sort of knowledge involved in constructing them. Using two plant species from Cape York Peninsula in northern Australia as case studies, this chapter details a protocol for searching for new populations of rare plants and explains how to use a systematic sampling approach to determine whether a species is truly rare or whether it is perceived to be rare as a result of inadequate sampling.

Study Area and Two Rare Species

Cape York Peninsula (CYP) occupies more than 137,000 km^2. It is sparsely populated; most people live in aboriginal or mining settlements or on large cattle stations (ranches). The road network is not extensive; it is unpaved and in poor condition for most of the year. Many parts of the Peninsula are inaccessible during the wet season. This study was located in the south-eastern part of CYP, between 143.0°E and 145.0°E and 14.0°S and 16.0°S (Figure 10.1). Rivers and creeks generally flow north to Princess Charlotte Bay, are relatively short in distance, and flow intermittently. The study area encompasses the Normanby, Kennedy, Laura, and Morehead river catchments. Eucalypt and melaleuca woodlands are the most extensive vegetation types within the study area, although grasslands, mangroves, closed forests, and wetlands are also present (Neldner et al. 1995). Most of the study area is within the Laura Basin, which has extensive lowlands characterized by areas of residual sands and large alluvial plains. Uplands in the southeast consist of conglomerate, coarse sandstone and leached shale of Lower Cretaceous age. Soils to the west are rocky and shallow and are derived from metamorphic and intrusive rocks (Horn et al. 1995).

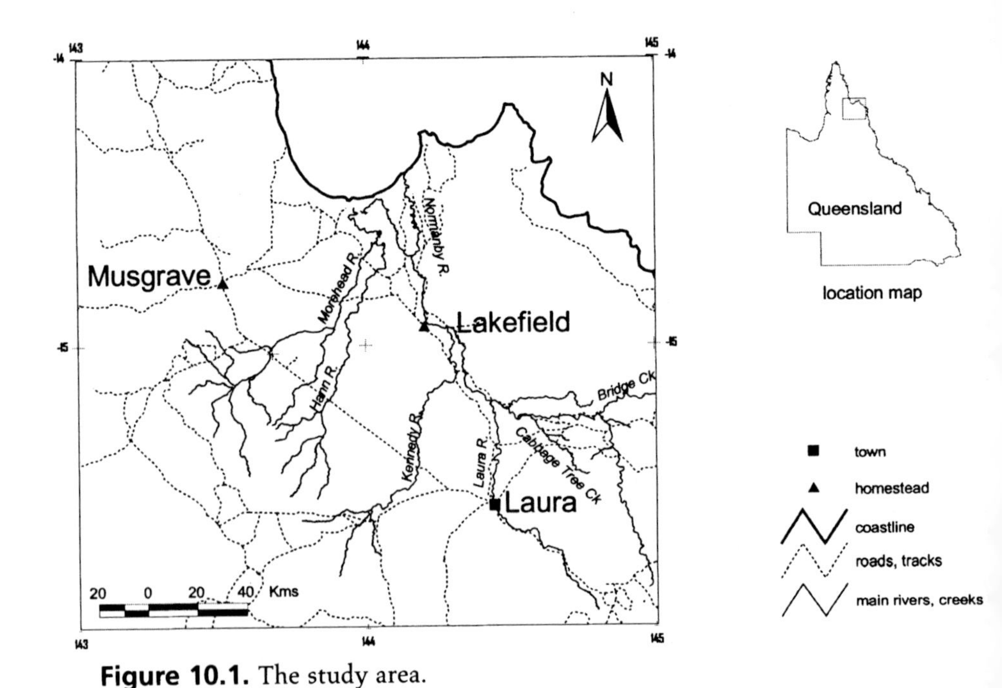

Figure 10.1. The study area.

Coix gasteenii Simon

Coix gasteenii is a tropical perennial member of the family Poaceae that grows to 2 m in height. It is the only species of *Coix* known to occur in Australia. The other five species in this genus occur throughout Asia, from India to Japan to as far south as Papua New Guinea. Known populations of *C. gasteenii* grow on clay-loam soils of Cainozoic origin, which are found in active floodplains, streams, and channels (Simon 1989).

Although *C. gasteenii* is closely associated with drainage lines, it has not been found in areas of high water flow such as wide, main-river systems or where riverbanks are steep and channels are narrow. Instead it occurs in areas where still waters persist after flooding and gradually recede with very little flow, such as the backwash channels of creeks. *C. gasteenii* grows in the dappled shade of gallery forests that fringe these areas. The palm *Corypha utan* is often present either at the site or nearby. At four of its locations, *C. gasteenii* is growing in close proximity to cane grass (*Mnesithea rottboellioides*). In all locations, vegetation surrounding the drainage lines is open eucalypt woodland or grassland with *Melaleuca* spp. Rainforest species also are present along the watercourse at the type location (Fox 2000). *C. gasteenii* occurs in areas where the mean annual rainfall ranges from 1,137 to 1,340 mm and mean annual temperature ranges between 24.4°C and 24.8°C (estimates derived from climatic surfaces generated by ESOCLIM; Houlder et al. 2001).

To date, five populations of *C. gasteenii* have been found growing in the southeastern part of CYP, along the Normanby River and Cabbage Tree Creek. Populations vary in size from a few scattered plants in localized patches to locally abundant patches of several hundred to possibly thousands of individuals. This species is currently listed as Endangered under State legislation (DEH 2003a).

Jedda multicaulis Clarkson

Jedda multicaulis is a shrub in the family Thymelaeaceae that grows to approximately 2 m in height. The genus is monospecific. Its closest relatives are lianas in South America and Asia. *J. multicaulis* grows on flat to gently undulating country, with well-drained, red sandy loam soils in open eucalypt woodland dominated by *Eucalyptus tetrodonta* and *Corymbia tessellaris*/*C. clarksoniana* (Clarkson 1986).

To date, there are only two known populations and they occur within approximately 80 km of each other in a north-south direction. The southern population occurs on a cattle ranch to the northwest of the town of Laura and the northern population occurs on another cattle ranch to the west of the CYP Development Road. The bushes are distinctive in appearance with a dense weeping habit, dark green foliage, and large fruit 6–7 cm long × 4–6 cm wide. The bushes grow in scattered clusters or waves, their growth habit at times radiating out from a central point, which may be a consequence of vivipary. Mean annual rainfall ranges from 1,028 mm at the southern population to between 1,070 and 1,083 mm at the northern population. Mean annual temperature ranges from 24.8°C to 25.0–25.2°C, respectively (estimates derived from climatic surfaces generated by ESOCLIM; Houlder et al. 2001).

J. multicaulis is currently listed nationally as Vulnerable, under State and Commonwealth legislation (DEH 2003a), because there are only two known populations and they both occur on grazing properties, with neither of the populations yet under protection in reserves or under conservation agreements.

The status of all plant species from CYP listed as rare and threatened is presently under review in order to update current management strategies. Verifying the true rarity status will help direct scarce management resources to those species and areas most threatened.

Materials and Methods

The sampling protocol consists of the following steps: (1) map all existing field records in environmental space (see Figure 10.3); (2) identify gaps in the coverage of environmental space; (3) conduct field surveys to fill those gaps; (4) reassess the rarity or commonness of the species; (5) construct spatial models from all available data to predict likely locations; (6) conduct field surveys to test predictions; and (7) confirm rarity or otherwise. The fourth step is an opportunity to reconsider the status of the species. If field sampling reveals new populations in environments that have not been sampled previously or have been only poorly sampled, depending on the size and extent of those populations, it might be concluded that the species is not rare and the whole protocol can stop. If the species is confirmed rare at Step 4, Step 5 is used to focus the search for new populations to determine the level of rarity and to find additional populations to aid in managing the species to ensure persistence.

Here we describe the data and analyses required for the protocol (also see Poon 2002).

Species Data

We obtained field locations of the two species from records in the HERBRECS vegetation database, which is maintained by the Queensland Herbarium. This database contains the field locations of all collections of plants held as specimens by the Herbarium since 1874. Of two field locations for *C. gasteenii*, one was from the type location and the other was from a cultivated source in Mareeba. We identified three additional locations for *C. gasteenii* through consultation with Queensland Parks and Wildlife Service (QPWS) staff who had been involved in searches for rare plants on CYP. *C. gasteenii* had been collected at these sites, but specimens had not yet been lodged at the Queensland Herbarium. One other location for *C. gasteenii* was deleted, as it was determined to be inaccurate. The field locations for *J. multicaulis* were accurate to within 50 m. Many early records from herbaria are unreliable because locations were often vague and maps were poor. This level of inaccuracy was expected.

Knowledge of where species do not occur, or have not been detected, also is useful. In general, herbaria do not record species absences so it was necessary to create "surrogate absences." These were derived from two other vegetation databases, CORVEG and SAVMON. The CORVEG database is the result of botanical surveys begun in 1979 and extensive vegetation surveys that were carried out across CYP by John Neldner and John Clarkson for the Cape York Peninsula Land Use Strategy (CYPLUS) during 1989–1994. All vascular plant species within 50 m × 10 m quadrats were recorded in three strata: ground, middle, and canopy. Two hundred fifteen CORVEG sites were located within the study area. We assumed that if a plant species had not been recorded at these sites, then it was truly absent. The data were considered to be reliable, as experienced botanists had conducted the surveys, although it also was recognized that even experienced botanists could miss some species, particularly if they were cryptic species.

Gay Crowley and Stephen Garnett established the SAVMON database in 1999 as part of a project designed to monitor the effects of different fire and grazing regimes on savannah vegetation. All plant species present in quadrats measuring 50 m × 4 m were recorded, and those not recorded were assumed to be absent. Absences recorded during field sampling as

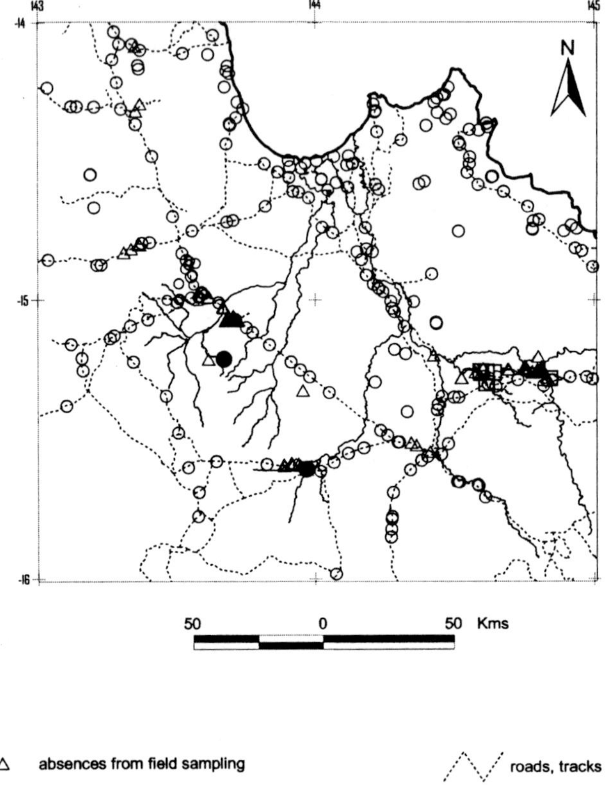

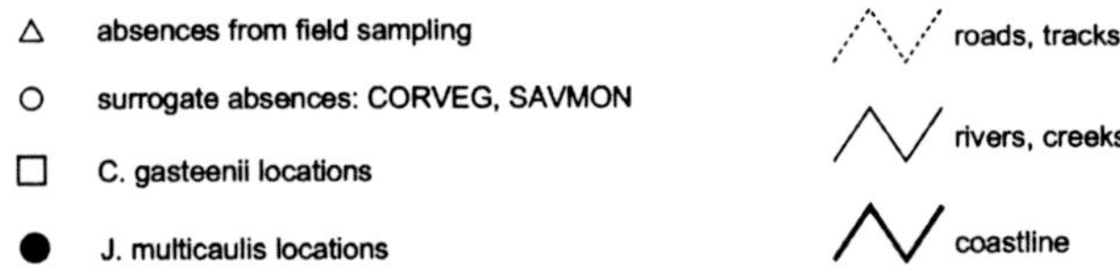

Figure 10.2. Geographic locations of known-species locations, surrogate absences, and absences from field sampling within the study area.

part of this project also were added to the database as they became available after Step 3 of the protocol (Figure 10.2).

Environmental Data

In selecting environmental variables to help predict species distributions, the emphasis should be on those variables that modulate physical processes and biological responses (Richards et al. 1990). Temperature, water, and nutrient regimes have been used widely to estimate species distribution patterns (e.g., Mackey et al. 1989; Cawsey et al. 2002). The water regime on CYP is highly seasonal, with most areas experiencing a water deficit for a quarter to half of the year (Nix 1982). Therefore, rainfall seasonality or rainfall of the wettest and driest quarters would have reflected the significant influence that water limitation has on vegetation more accurately than annual mean rainfall. Mackey (1994) used the rainfall of the driest quarter of the year as a major factor limiting the distribution of tropical rainforests in northern Queensland because it severely affects the soil moisture available to rainforest plants. Neldner (1996) used the rainfall of the wettest quarter of the year as a primary predictor of the distribution patterns of *Eucalyptus tetrodonta* and *E. miniata* in northern CYP; these species are susceptible to waterlogging. The extremes of temperature within the year can reflect limitations of plant growth; for example, if a species is susceptible to frost, its distribution may be limited to areas that are frost-free (Nix 1983). Temperature, combined with rainfall and geology, also can provide a measure of evaporation and an estimate of plant-available moisture. We used geological substrate to estimate nutrient regime as a surrogate for more detailed soils data. Thus, based on previous studies and field experience, the environmental variables used as predictors of the distribution patterns of species in this study were mean rainfall of the wettest quarter of the year, mean temperatures of the hottest and coldest quarters, and geology. They reflect the climatic seasonality within the study area and the interactions among water, climate, and terrain, which all affect plant growth (Nix 1982).

Climate Data

Climate surfaces for the whole study area were generated using the ESOCLIM module of ANUCLIM 5.1 (Houlder et al. 2001), which is based on an Australian Department of Defence digital elevation model (DEM). With a resolution of 10 m, this was one of the more detailed resolution DEMs available for any area in Australia. The steps taken to generate climate surfaces are explained in Houlder et al. (2001). From the climate surfaces, we calculated estimates of the mean temperature of the hottest quarter of the year (t_hq: October, November, December); mean temperature of the coldest quarter (t_cq: June, July, August) and mean rainfall of the wettest quarter (rf_wq: January, February, March). We derived annual mean temperature and annual mean rainfall for all field sites (presences and absences) as described by Austin et al. (2000).

Geological Data

We obtained geological data from the Department of Natural Resources and Mines at a scale of 1:250,000, as raster images in .tiff format scanned from

1:250,000 sheet maps. Geological coverage also was obtained from Geoscience Australia as fully attributed vector files for one 1:250,000 sheet, and a complete coverage at 1:1 million was available in WGS 1984 format.

When the complete digital coverage was compared with the partial coverage, we observed some loss in detail in the former because several smaller geological units in the latter had been generalized. Austin (2002) argued that complete digital coverage of a study area was preferable to partial coverage, so the decision was made to sacrifice some detail to gain full coverage. The geological data could be improved or updated as more detailed maps of the area become available. Currently, 1:100,000 geological maps are available for part of the study area.

Data Treatment

For Step 2 of the protocol, the field records of known populations of the two study species, plus surrogate absences, were plotted in environmental space, which was defined by overlaying climate (Figure 10.3) and geology (not shown). This plot was inspected for adjacent or very similar combinations of environ-

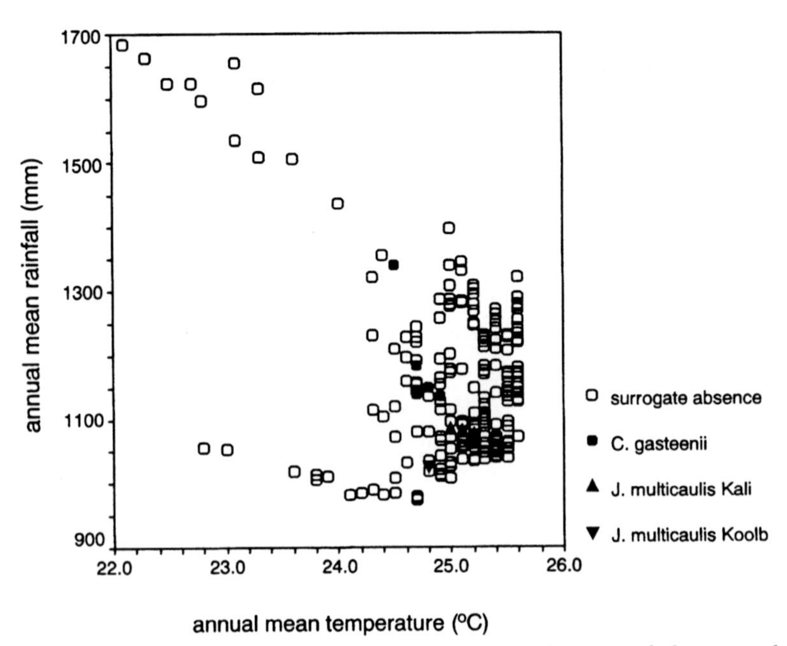

Figure 10.3. Surrogate absences and known populations of the two plant species plotted in environmental space. Note the large gaps in coverage of this space. "Kali" and "Koolb" refer to locations of *J. multicaulis* populations.

mental variables that had not been sampled, that is, had no records of the two species and no surrogate absences. These environments were chosen for the first field sampling survey in Step 3. We selected 96 field sites to span the unsampled or poorly sampled environmental space similar to that of known records of the two rare species. We assumed that there was an equal probability of the species occurring throughout this environmental space. Therefore, random sampling was considered unnecessary, and for the sake of efficiency and to minimize time and effort, we deliberately chose field sample sites to intersect with or be within 800 m of a road or track. Fortunately, examples of all unsampled or poorly sampled environments could be found within these constraints.

Spatial Distribution Models

There are basically two kinds of spatial models: those that measure the similarity of environments at known species locations with environments in other parts of the region and those that enumerate how often species are represented in different environments in order to calculate a relationship. Statistical models are of this second kind, and if they can be used, they are more reliable in the sense that they provide probabilities of occurrence with error and variance estimates. However, they require multiple observations for valid results. Because there were so few field locations of the two study species, modeling in Step 5 of the protocol required methods of the first kind. Both BIOCLIM (Busby 1991) and DOMAIN (Carpenter et al. 1993; CIFOR 2002) are examples of such methods. These two were compared as part of a wider study not reported here. Although both methods have their advantages and disadvantages, BIOCLIM was preferred for this study because it successfully predicted a previously unknown population of *C. gasteenii* (see Results).

The input to BIOCLIM was latitude, longitude, and elevation in order to generate, from the climate surfaces, a climate profile for each known species location. We compared this profile to those at grid points throughout the region to identify similar climate profiles. The degree of similarity can be used to estimate different degrees of "suitability" of climate for the species. The levels of suitability could be ranked based on pairs of percentiles, so grid points that fall outside the range of climates sampled could be considered "unsuitable," those that fall within the range but are outside the 5–95 percentile range at the extremes could be considered "marginal," and those within the 10–90 percentile range could be considered to be of a "most suitable climate."

There are several assumptions associated with biological data that are inputted into BIOCLIM (Nix 1982; Busby 1991; Houlder et al. 2001). If the assumptions are not true, then the accuracy of the BIOCLIM predictions may be adversely affected. It is assumed that: (1) known species locations are representative of their actual distributions; (2) geocoding of species locations is accurate; (3) identification of species is accurate; (4) anomalous data points have been checked; and (5) there is no taxonomic uncertainty regarding the species. Each of these assumptions was discussed in greater detail by Nix (1982) and Busby (1991).

We used the ESOCLIM module of the software package ANUCLIM 5.1 and a fine-resolution, digital elevation model to generate climate estimates based on geographic location and elevation for points throughout the study area. We then generated a climate profile for the species based on known species locations. These climate profiles then were overlaid on the geology coverage because in the field, particularly during the first and second field surveys, we observed that the species occurred on certain geology. We excluded the predicted climatically suitable areas that occurred on geology where it was known that the species did not occur. This enabled us to conduct a more focused search.

We used the above models in Step 6 to target new field sample sites, chosen to span the geographic range of areas determined as most suitable by BIOCLIM. In addition, we restricted our sampling for *C. gasteenii* to within 500 m of drainage lines because experienced field botanists believed this species was restricted to drainage lines (John Clarkson and Jill Landsberg, Queensland Parks and Wildlife Service, pers. comm.) and the first field survey (Step 2) supported this observation. We sampled areas outside drainage lines during the first field survey but no new populations were found.

Results

Here we present the step-by-step results of our search for populations of *C. gasteenii* and *J. multicaulis*. Our spatial models were described in the previous section.

Step 1

Figure 10.3 maps the known locations of the two species and the surrogate absences in rainfall and temperature space across the study area. [Note: The environmental predictors in this case are shown as mean annual rainfall and mean annual temperature, for ease of illustration.] *C. gasteenii* occurred over a wide rainfall range but a narrow temperature range, whereas *J. multicaulis* had a similar narrow temperature range and a more restricted rainfall range. Note also that there were large gaps in the coverage of environmental space by existing field sample sites, whether they included known occurrences of the two species or were surrogate absences. This may have been because these environmental combinations did not exist within the study area, they may have occurred in inaccessible terrain, or the areas may have never been visited by botanists. The two species and surrogate absences also were plotted against geology to identify gaps in the coverage of geology, but this is not shown here.

Steps 2 and 3

Figure 10.4, in a more restricted environmental space, maps the surrogate absences, the known species locations, and the field sites that we sampled in Step 3 to extend the coverage of field records in similar unsampled or poorly sampled environmental space. None of these field sample sites contained new populations of either species. These records were added to the database as absences.

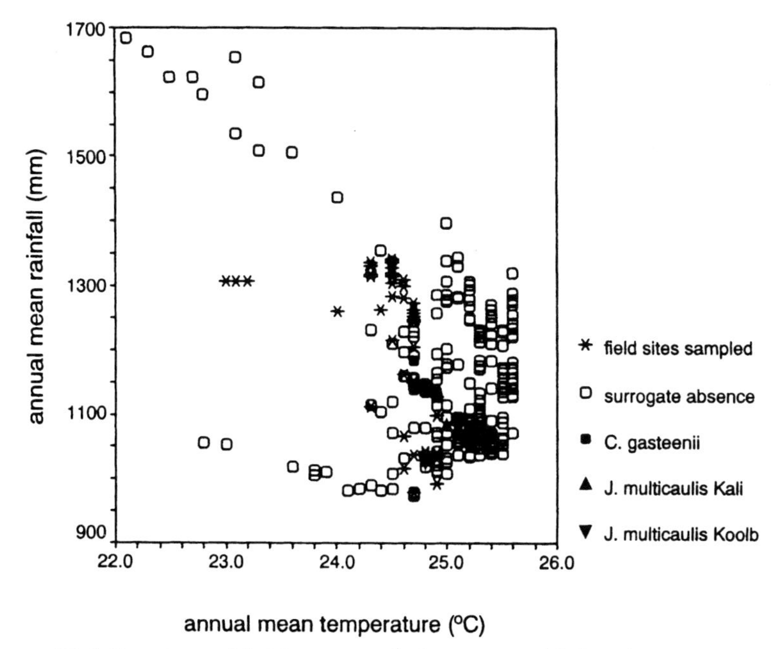

Figure 10.4. Locations of field sites sampled in Step 3 added to the same (slightly more restricted) plot in Figure 10.3. Note that they sample the gaps in coverage that were revealed in Figure 10.3.

Step 4

Both species were clearly rare and restricted in range. Their rarity was not an artifact resulting from inadequate sampling, so it was deemed appropriate to proceed to Step 5.

Step 5

Figure 10.5 maps the areas predicted by BIOCLIM to be most climatically suitable for *C. gasteenii*. Figure 10.6 shows the same map for *J. multicaulis*. These were the areas chosen for field survey in Step 6 below. Each figure also shows the known locations of the species.

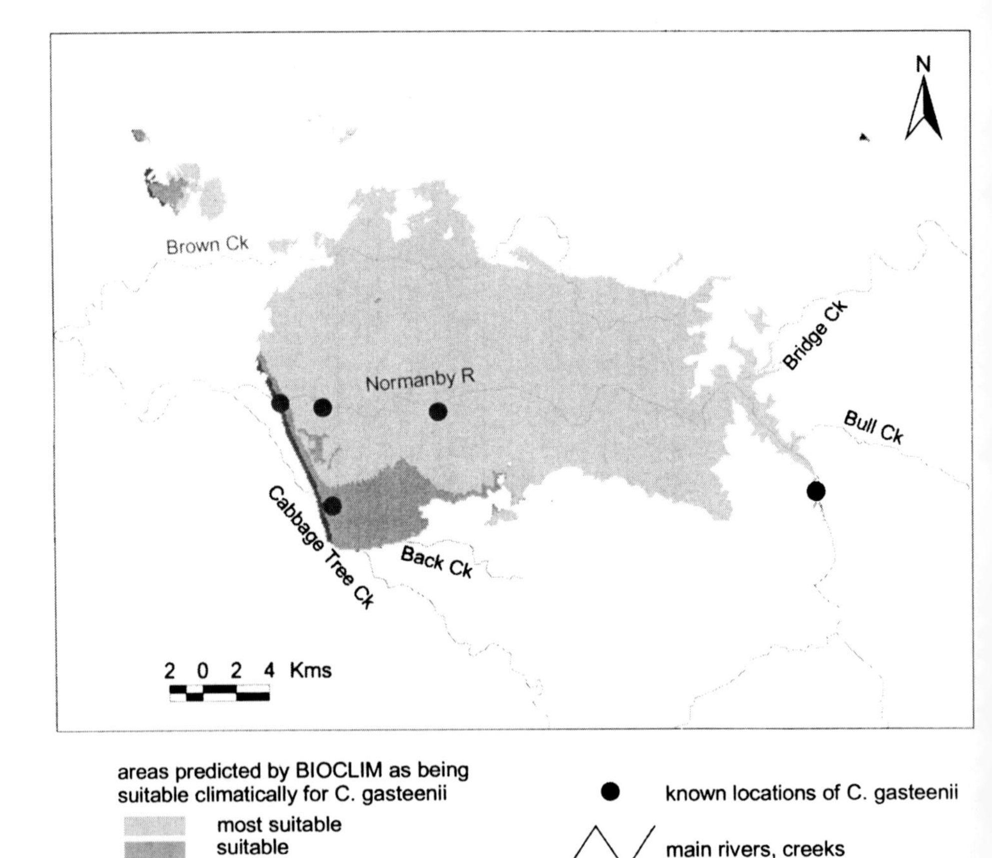

Figure 10.5. The area that is climatically similar to known locations of *C. gasteenii*.

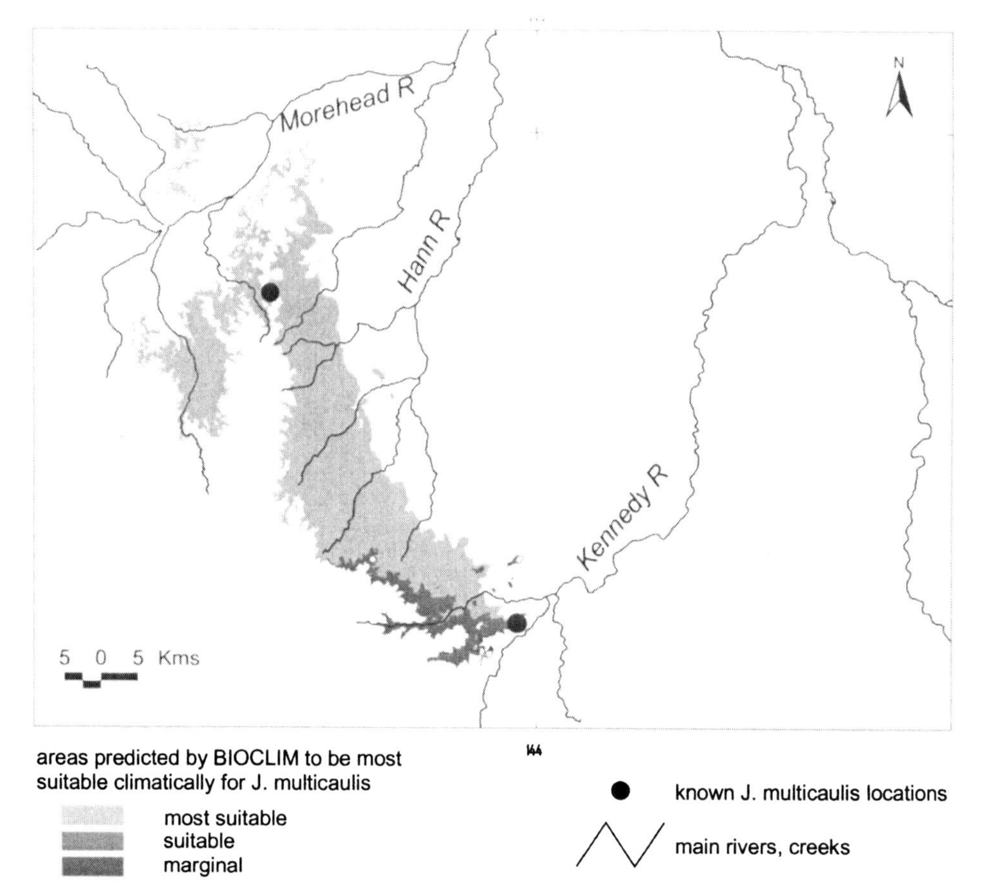

Figure 10.6. The area that is climatically similar to known locations of *J. multicaulis*.

Step 6

We designed a new field survey to search the climatically suitable areas for the two species. The search for *J. multicaulis* was limited by restricted access to private property, and no new populations of the species were found. However, we discovered one new population of *C. gasteenii* (Figure 10.7). Thus, the narrower focus made possible by spatial modeling was successful because it led to the discovery of a new population of this rare species.

Discussion

We were able to focus our search for previously undetected populations of two rare plants by combining a systematic search of poorly sampled or

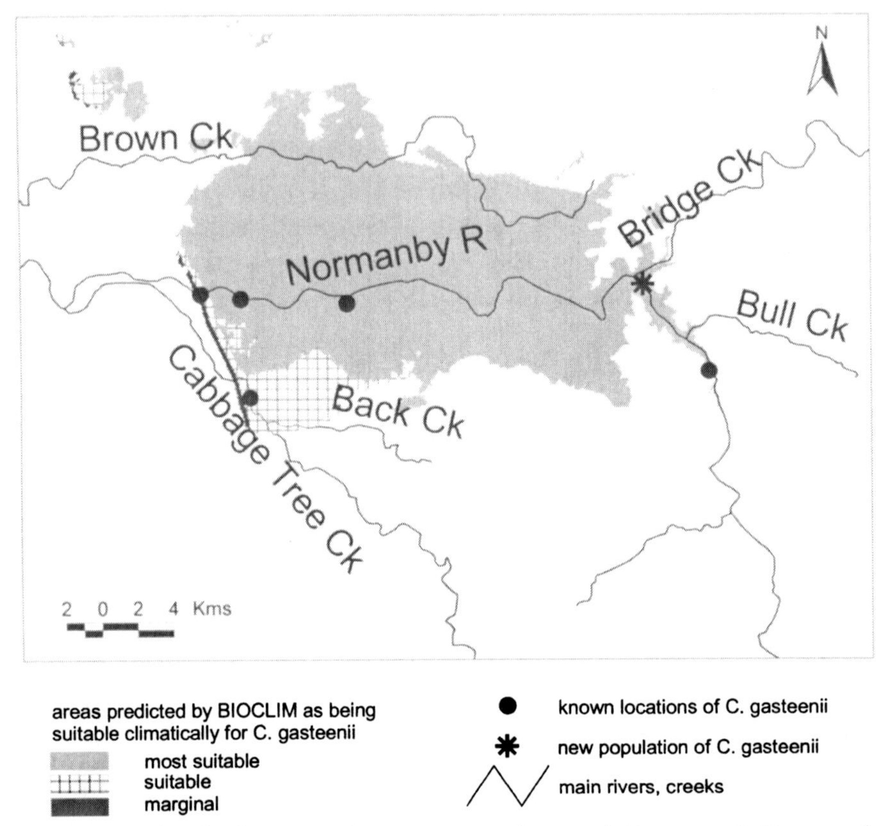

Figure 10.7. The location of one new population of *C. gasteenii* discovered during the course of the field survey in Step 6.

unsampled environmental space, defining the environmental envelope occupied by the two species, and predicting potential locations of new populations with spatial modeling. Ensuring as much of the environmental space within the study area as possible was sampled was important to reduce uncertainty in the spatial models. This process led to the discovery of one new population of one of those species. Figure 10.8 is a flow chart of the steps that constitute this search protocol. Step 6A was not used in the study reported here, but it represents an option to successively improve the spatial modeling in Step 5 by incorporating new field data records.

One possible way to improve this procedure would be to implement random sampling within the environmental space that species are predicted to occupy. Austin and Heyligers (1989, 1991) employed random

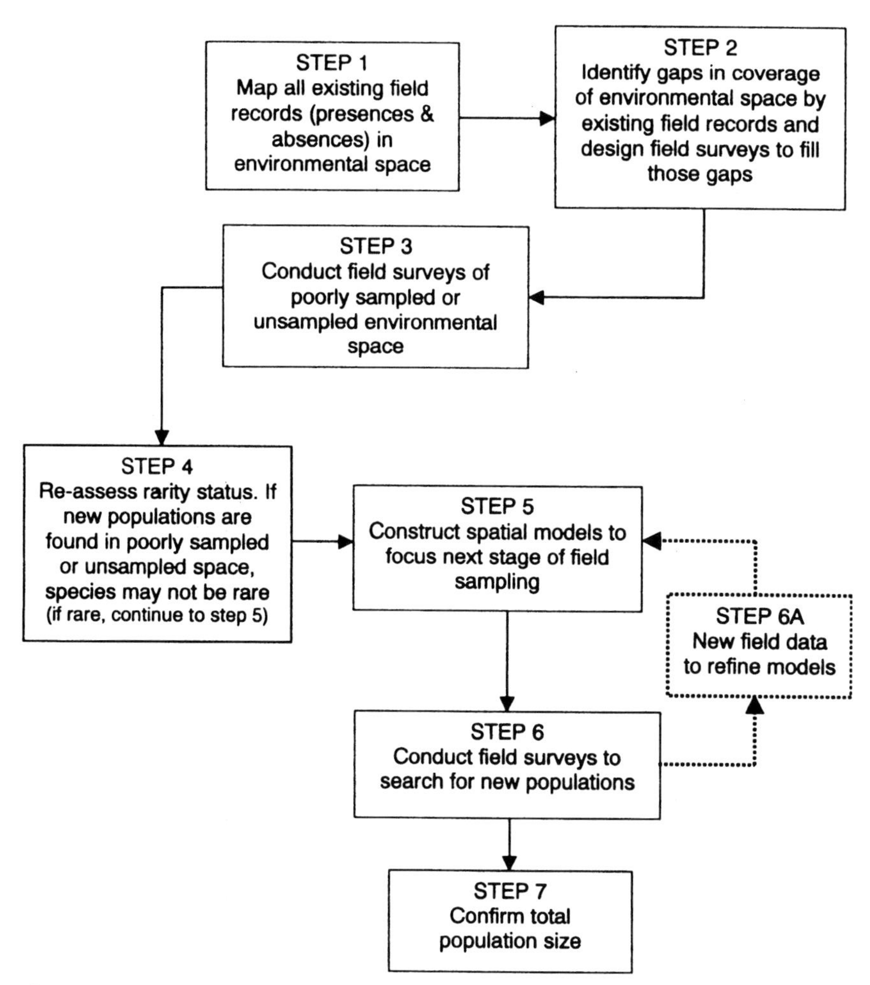

Figure 10.8. Major steps in a protocol that can be used to search for new populations of rare plants. In Step 4, one must decide whether to proceed to Step 5. Step 6A is an option to use more field sampling to refine spatial models derived during Step 5.

sampling within environmental cells following an environmental stratification for a field survey of trees in the coastal forests of northern New South Wales. However, they still found it necessary to adjust sample-site location to the nearest possible access point. Random sampling in the present study would simply have required more field time than was available. Because we assumed that the target species had the same chance of being encountered anywhere in the space identified, choosing sample sites for ease of access was appropriate.

Searching as much of the geographical area representing environmental space as time and resources allow is preferable because unknown or unrecognized factors may determine local distribution patterns. For example, factors such as the gradient of slope or the width of a drainage line determining the subsequent energy of water passing through it might limit possible locations for *C. gasteenii*. Other combinations of very localized conditions, which are too specialized or unique to map on a broad scale, might limit the distribution patterns of other species. Alternatively, ecological or evolutionary history may play a role, as it does for species with relic distributions or different species occupying similar but isolated environments. In such cases, it may not be possible to model distribution patterns adequately. However, procedures of the kind proposed here should be informed by experience, intuition, common sense, and local ecological knowledge. Indeed, the protocol in Figure 10.8 is our attempt to formalize and make explicit the kind of process that experienced field botanists would go through in their own minds if they were asked to look for new populations of rare plants. Moreover, our protocol adds some scientific rigor by making the search repeatable. Our protocol also aims to make the procedure accessible to people who may not have acquired local experience and ecological knowledge and may not have the option to do so.

We have employed this approach in a tropical savannah, but there is no reason to suppose that it will not work equally well in other biomes. Environmental stratification has been employed widely in the design of field surveys (e.g., Austin and Heyligers 1989 in temperate forests; Wessels et al. 1998 in subtropical savannahs; Austin et al. 2000 in temperate woodlands), and it is currently being used in the tropical rainforests of northeastern Queensland to guide field surveys of vertebrates, selected invertebrates, and plants. The same general protocol applies: plot existing field survey sites in environmental space, identify poorly sampled or unsampled space, and target geographical locations representing these environmental spaces with new surveys. There have been numerous range extensions of both plants and animals discovered in this way in the past 2 years (Andrew Ford, CSIRO Sustainable Ecosystems [CSE], Tropical Forest Research Centre [TFRC], and Steve Williams, James Cook University, pers. comm.)

In the same biome, spatial modeling was used to predict new locations of the magnificent brood frog, *Pseudophryne covacevichae*, which is a species listed as Vulnerable (DEH 2003b). Searches of these locations revealed one new population; unfortunately this success has never been published (David Hilbert, CSE, TFRC, pers. comm.). Spatial modeling also has been used to predict new locations of the endangered northern bettong, *Bettongia tropica*. The model has been used by the Queensland Parks and Wildlife Service to guide ongoing surveys; however, currently no new populations have been found (Hilbert et al. 2000).

Our basic argument is that a protocol should be developed that first evaluates the possibility that rarity is an artifact of inadequate sampling and then successively narrows down the search for new populations in the systematic way described above. Such an approach is especially relevant when available resources—time, funds, and staff—are scarce. It is also relevant to an area such as Cape York, which has a very sparse human population scattered over a vast area that is still only partly traversed by roads. This is an area that few field botanists have visited, where existing field records of all plant species are few and the coverage of field sites is seriously inadequate. Resources for botanical field surveys are always scarce, and many parts of the world remain poorly known from a botanical perspective. Perhaps similar protocols can be used in other regions and for other species to search for new populations of rare plants.

Acknowledgments

We would like to thank Dave Gillieson and Jill Landsberg for their guidance and support during the CYP study. We also thank John Clarkson and an anonymous reviewer for their comments on this manuscript. The generosity of CYP landowners for granting permission to work on their land is also greatly appreciated.

References

Austin, M. P. 2002. Case studies of the use of environmental gradients in vegetation and fauna modelling: Theory and practice in Australia and New Zealand. pp. 73–82 in J. M. Scott, P. J. Heglund, M. L. Morrison, J. B. Haufler, M. G. Raphael, W. A. Wall, and F. B. Samson, eds., *Predicting Species Occurrence: Issues of Accuracy and Scale*. Island Press, Washington, D.C.

Austin, M. P., E. M. Cawsey, B. L. Baker, M. M Yialeloglou, D. J. Grice, and S. V. Briggs. 2000. *Predicted Vegetation Cover in the Central Lachlan Region*. Final report of the Natural Heritage Trust Project AA 1368.97, CSIRO, Canberra, Australia.

Austin, M. P., and P. C. Heyligers. 1989. Vegetation survey design for conservation: Gradsect sampling of forests in north-eastern New South Wales. *Biological Conservation* 50:13–32.

______. 1991. New approaches to vegetation survey design: Gradsect sampling. pp. 31–36 in C. R. Margules and M. P. Austin, eds. *Nature Conservation: Cost Effective Biological Surveys and Data Analysis*. CSIRO, Australia.

Busby, J. R. 1991. BIOCLIM—A bioclimate analysis and prediction system. pp. 64–68 in C. R. Margules and M. P. Austin, eds., *Nature Conservation: Cost Effective Biological Surveys and Data Analysis*. CSIRO, Australia.

Carpenter, G., A. N. Gillison, and J. Winter. 1993. DOMAIN: A flexible modelling procedure for mapping potential distributions of plants and animals. *Biodiversity and Conservation* 2:667–680.

Cawsey, E.M., Austin, M.P. and Baker, B.L. 2002. Regional vegetation mapping in Australia: a case study in the practical use of statistical modelling. *Biodiversity and Conservation* 11: 2239–2274.

CIFOR (Centre for International Forestry Research) 2002. DOMAIN. Available from http://www.cifor.cgiar.org/research_tools/domain/index.htm (accessed August 5, 2002).

Clarkson, J. R. 1986. *Jedda*, a new genus of Thymelaeaceae (sub-tribe Linostomatinae) from Australia. *Austrobaileya* 2:203–210.

Department of the Environment and Heritage (DEH). 2003a. Environment Protection and Biodiversity Conservation Act 1999. Available from http://www.deh.gov.au/cgi-bin/sprat/public/publicthreatenedlist.pl?wanted=flora (accessed Nov. 21, 2003).

_____. 2003b. Environment Protection and Biodiversity Conservation Act 1999. Available from www.deh.gov.au/cgi-bin/sprat/public/publicthreatenedlist.pl?wanted=fauna (accessed Nov. 21, 2003).

Fox, I. 2000. *A Conservation Assessment of* Coix gasteenii (*Poaceae*). Queensland Herbarium, Mareeba, Australia.

Good, R. B., and P. S. Lavarack. 1981. The status of Australian plants at risk. pp. 81–92 in H. Synge, ed., *The Biological Aspects of Rare Plant Conservation*. Wiley, London.

Hilbert, D. W, A. W. Graham, and T. A. Parker. 2000. *Forest and Woodland Habitats of the Northern Bettong* (Bettongia tropica) *in the Past, Present and Future: A Report Prepared for Queensland Parks and Wildlife Service*. Cooperative Research Centre for Tropical Rainforest Ecology and Management, CSIRO Tropical Forest Research Centre, Atherton, Queensland.

Horn, A. M., E. A. Derrington, G. C. Herbert, R. W. Lait, and J. R. Hillier. 1995. *Groundwater Resources of Cape York Peninsula*. Cape York Peninsula Land Use Strategy, Office of the Co-ordinator General of Queensland, Brisbane, Department of Environment, Sports and Territories, Canberra, Queensland Department of Primary Industries, Brisbane and Mareeba, and Australian Geological Survey Organisation, Mareeba.

Houlder, D., M. Hutchinson, H. Nix, and J. McMahon. 2001. *ANUCLIM 5.1 User's Guide*. Australian National University/Centre for Resource and Environmental Studies (CRES), Canberra.

Mackey, B. G. 1994. Predicting the potential distribution of rainforest structural characteristics. *Journal of Vegetation Science* 5:43–54.

Mackey, B. G., H. A. Nix, J. A. Stein, S. E. Cork, and F. T. Bullen. 1989. Assessing the representativeness of the Wet Tropics of Queensland World Heritage Property. *Biological Conservation* 50:279–303.

Margules, C. R., and M. P. Austin. 1994. Biological models for monitoring species decline: The construction and use of databases. *Philosophical Transactions of the Royal Society of London B*. 344:69–75.

Neldner, V. J. 1996. *Improving Vegetation Survey: Integrating the Use of Geographic Information Systems and Species Modelling Techniques in Vegetation Survey. A*

Case Study Using the Eucalypt Dominated Communities of Cape York Peninsula. Ph.D. thesis, Australian National University, Canberra.

Neldner, V. J., D. C. Crossley, and M. Cofinas. 1995. Using geographic information systems (GIS) to determine the adequacy of sampling in vegetation surveys. *Biological Conservation* 73:1–17.

Nicholls, A. O. 1989. How to make biological surveys go further with Generalised Linear Models. *Biological Conservation* 50:51–75.

Nix, H. A. 1982. Environmental determinants of biogeography and evolution in Terra Australia. pp. 47–66 in W. R. Baker and P. J. M. Greenslade, eds., *Evolution of the Flora and Fauna of Arid Australia.* Peacock Publishers, Adelaide, Australia.

Nix, H. A. 1983. Climate of tropical savannas. pp. 37–62 in F. Bourlière, ed., *Tropical Savannas.* Elsevier, Amsterdam.

Poon, E. L. 2002. *Searching for New Populations of Rare Plant Species. Two Case Studies from Cape York Peninsula.* Honours thesis, James Cook University, Cairns, Australia.

Richards, B. N., R. G. Bridges, R. A. Curtin, H. A. Nix, K. R. Shepherd, and J. Turner. 1990. *Biological Conservation of the South-eastern Forests.* Report of the Joint Scientific Committee, Australian Government Publishing Service, Canberra.

Simon, B. K. 1989. A new species of *Coix* L. (Poaceae) from Australia. *Austrobaileya* 3:1–5.

Wessels, K. J., A. S. van Jaarsveld, J. D. Grimbeek, and M. J. van der Linde. 1998. An evaluation of the gradsect biological survey method. *Biodiversity and Conservation* 7:1093–1121.

Part IV

ESTIMATING ABUNDANCE, DENSITY, AND OTHER PARAMETERS

Because abundance and density are commonly used in population monitoring, Part IV of this volume is devoted to methods for estimating these and other population parameters for rare or elusive species or populations. Chapters in Part IV generally have a larger emphasis on the second stage of a two-stage design (counting or estimation methods) compared to chapters in Part II. Nonetheless, both stages are important to obtaining reliable parameter estimates.

Recent scientific and technological advances have provided biologists with a means to sample populations without actually handling individuals. These noninvasive methods are particularly relevant to sampling rare or elusive species. Therefore, the first four chapters in Part IV describe noninvasive methods for estimating abundance of rare or elusive species in terrestrial and marine habitats: genetic methods (Waits, Chapter 11); photographic sampling (Karanth et al., Chapter 12); methods based on snow tracks (Becker et al., Chapter 13); and hydroacoustic sampling (Hanselman and Quinn, Chapter 14 in part). This general class of methods will likely continue to gain importance with further advances in science and technology (see Chapter 17).

Bats as a group have been notoriously difficult to sample effectively to obtain reliable estimates of abundance. In Chapter 15, O'Shea et al. argue that it is more feasible to obtain reliable survival estimates for bats than meaningful abundance estimates. Therefore, they address the issue of banding bats, review and critique previous survival studies of bats, and describe their own research in estimating survival rates of bats.

Ganey et al. conclude Part IV with a comprehensive evaluation of the effectiveness of a monitoring protocol for detecting trends in abundance and in finite rate of population growth (λ) of Mexican spotted owls. All preceding chapters in this volume could be viewed, and perhaps should be viewed, within this monitoring context. As with other chapters in this volume, concepts discussed by Ganey et al. are not limited in application to one species or taxon.

11

Using Noninvasive Genetic Sampling to Detect and Estimate Abundance of Rare Wildlife Species

Lisette P. Waits

One exciting new approach for detecting and estimating abundance of rare wildlife species is noninvasive genetic sampling. This type of sampling is particularly appealing to wildlife biologists because they are able to obtain critical data without capturing, handling, or even observing the animals. Noninvasive genetic sampling was first reported as a method that used hair (Taberlet and Bouvet 1992) and feces (Höss et al. 1992) to obtain genetic samples from small, elusive brown bear (*Ursus arctos*) populations in Europe. Shortly after, Morin and Woodruff (1992) demonstrated the effectiveness of the technique using shed hairs as a source of DNA to study social structure in chimpanzees (*Pan troglodytes*). Since the initiation of this new field of research, a number of different noninvasive sources of DNA have been used, including hair (Taberlet and Bouvet 1992), feces (Höss et al. 1992), urine (Valiere and Taberlet 2000), snake skins (Bricker et al. 1996), regurgitates (Valiere et al. 2003), sloughed whale skin (Valsecchi et al. 1998), feathers (Taberlet and Bouvet 1991), egg shells (Strausberger and Ashley 2001), and skulls in owl pellets (Taberlet and Fumagalli 1996). Currently, hair and fecal samples are the most commonly used noninvasive DNA sources for detecting and estimating abundance of rare wildlife species.

Noninvasive genetic sampling methods have been used to address a wide range of research questions in wildlife biology. Classical conservation genetic evaluations of genetic diversity and population structure have been performed using noninvasive sources of DNA from brown bears (Taberlet and Bouvet 1992; Kohn et al. 1995; Taberlet et al. 1997) and elephants (*Elephas maximus* and *Loxodonta cyclotis*) (Fernando et al. 2000; Eggert et al. 2003). Noninvasive genetic sampling has been used to evaluate migration and gene flow in humpback whales (*Megaptera novaeangliae*) (Palsbøll et al. 1997), Italian wolves (*Canis lupus*) (Lucchini et al. 2002), and bonobos (*Pan paniscus*) (Gerloff et al. 1999). Fecal DNA analysis has provided important information about paternity and social structure in gibbons (*Hylobates muelleri*) (Oka and Takenaka 2001), chimpanzees (*Pan troglodytes*) (Morin et al. 1994; Constable et al. 2001), langurs (*Presbytis entellus*) (Borries et al. 1999), bonobos (Gerloff et al. 1999), and rhinoceros (*Diceros bicornis*) (Garnier et al. 2001). Fecal DNA sampling has been used as a tool for detecting hybrids and coyotes (*Canis latrans*) in efforts to control hybridization in red wolves (*Canis rufus*) (Adams et al. 2003). Morphologically based criteria for species identification of scats have been evaluated and updated using fecal DNA analysis (Farrell et al. 2000; Davison et al. 2002). Predators of livestock (Farrell et al. 2000) or endangered species (Ernest et al. 2002; Banks et al. 2003a) have been identified using fecal DNA analysis. These studies highlight the power and utility of noninvasive genetic sampling for addressing research questions in wildlife populations. The remainder of this chapter will focus on the use of noninvasive genetic sampling to detect and estimate abundance of rare wildlife species.

Obtaining and Amplifying DNA

There are two main types of DNA in animal cells: mitochondrial DNA (mtDNA) and nuclear DNA (nDNA) (Avise 1994). MtDNA is a double-stranded, circular molecule contained in the mitochondria, which are maternally inherited in most animals (Birky et al. 1989; Avise 1994). MtDNA is present in hundreds to thousands of copies per cell and is the DNA molecule generally used for species identification of noninvasive samples. Nuclear DNA chromosomes are found in the nucleus of the cell and have a biparental mode of inheritance. Nuclear DNA is used for individual identification and sex identification of noninvasive genetic samples. To perform species identification or individual identification, the DNA must be amplified using a technique called the polymerase chain reaction (PCR). PCR allows researchers to make copies of specific target regions of mtDNA and nDNA (Mullis et al. 1986). When extracting DNA from noninvasive sources, both mtDNA and nDNA are obtained, but mtDNA mol-

Table 11.1.

Mitochondrial DNA (mtDNA) and nuclear DNA (nDNA) amplification success rates for noninvasive genetic sampling studies.

Species	DNA source	mtDNA	nDNA	Citation
Chimpanzee	Single hair		50%	Gagneux et al. (1997)
Brown bear	Single hair		50%	Taberlet et al. (1997)
Brown bear	Multiple hairs	90%	77%	Poole et al. (2001)
Wombat	Single Hair		96%	Banks et al. (2003a)
Elephant	Feces	72%	60%	Eggert et al. (2003)
Coyote	Feces	79%	48%	Kohn et al. (1999)
Brown bear	Feces	96%	88%	Murphy et al. (2003b)
Baboon	Feces	100%	86%	Frantzen et al. (1998)
Reindeer	Feces		93%	Flagstad et al. (1999)
Wombat	Feces		83%	Banks et al. (2002)
Otter	Feces		20%	Dallas et al. (2003)
Lynx	Feces	99%		Palomeres et al. (2002)
Seal	Feces	72%	60%	Reed et al. (1997)
Wolf	Feces	84%	64%	Lucchini et al. (2002)

ecules are present at much higher copy numbers than nDNA molecules. Thus, PCR amplification success rates for species identification of noninvasive samples are higher than success rates for individual identification (Table 11.1).

Extraction of DNA from hair samples uses cells that are attached to the root of the hair. DNA extraction is generally performed using two methods: chelex protocols (Walsh et al. 1991; Goossens et al. 1998; Woods et al. 1999; Banks et al. 2003b) or commercially available silica-binding extraction kits (Poole et al. 2001; Riddle et al. 2003). Fecal DNA is obtained from sloughed intestinal epithelial cells. Fecal DNA extraction methods have included chelex protocols (Paxinos et al. 1997; Palomares et al. 2002), phenol-chloroform (Ernest et al. 2000; Fernando et al. 2000; Oka and Takenaka 2001), diatomaceous earth/guanidine-thiocyanate (Gerloff et al. 1995; Kohn et al. 1995; Lucchini et al. 2002), magnetic beads (Flagstad et al. 1999) and commercially available silica-binding extraction kits (Farrell et al. 2000; Goossens et al. 2000; Constable et al. 2001; Creel et al. 2003). Fecal DNA extracts can contain high concentrations of PCR inhibitors, and extraction methods are designed to minimize inhibitors while maximizing

DNA yield. A few studies have evaluated the effectiveness of different extraction methods (Kohn et al. 1995; Reed et al. 1997; Paxinos et al. 1997; Flagstad et al. 1999; Lathuilliere et al. 2001; Frantz et al. 2003), but there is no clear consensus on an optimal method for all species. Currently, the most commonly used method for extracting DNA from fecal samples is silica-binding extraction kits.

Preservation methods of hair and fecal samples collected in the field can have important impacts on DNA amplification success rates (Wasser et al. 1997; Murphy et al. 2000, 2003b; Roon et al. 2003). Hair samples can be preserved by storing at room temperature in ethanol (Oka and Takenaka 2001), by desiccating with silica beads (Gagneux et al. 1997), or by freezing (Constable et al. 2001). Roon et al. (2003) recently compared the effectiveness of silica desiccation and –20°C freezing for brown bear samples stored for up to 1 year before extraction. When evaluating mtDNA amplification, they reported that there were no declines in success rates over the storage period and no significant differences between preservation methods. However, when evaluating nDNA amplification performance, success rates decreased significantly (~20%) between 6 months and 1 year of storage, regardless of preservation method. Also, nDNA amplification success rates were slightly higher (1–10%) for samples preserved at –20°C.

Several studies have evaluated effectiveness of different fecal DNA preservation methods (Wasser et al. 1997; Frantzen et al. 1998; Murphy et al. 2000, 2003b; Frantz et al. 2003). Using bear fecal samples, Wasser et al. (1997) qualitatively evaluated freezing, freeze-drying, silica desiccation, 100% ethanol, and 30 other methods for preserving DNA over a 6-month period. Silica desiccation was recommended as the optimal method due to strong performance and ease of use in the field. In contrast, Murphy et al. (2000, 2003b) reported that silica desiccation performed poorly for brown bear fecal samples and recommended preservation in 90–100% ethanol. When preserving baboon fecal DNA samples, Frantzen et al. (1998) reported that samples preserved in DETs buffer had the highest nDNA amplification rates compared to air-drying, freezing, and 70% ethanol. In contrast, 70% ethanol preservation outperformed DETs buffer and –20°C freezing for badger (*Meles meles*) fecal samples (Frantz et al. 2003). The variation in results among studies may be due to species-specific effects or interactions between storage methods and extraction methods (Frantz et al. 2003). The majority of recently published fecal DNA studies used ethanol for sample preservation (Bayes et al. 2000; Dallas et al. 2000; Fernando et al. 2000; Goossens et al. 2000; Constable et al. 2001; Lucchini et al. 2002; Dallas et al. 2003; Eggert et al. 2003).

Species Identification

Detecting the presence of rare or endangered species is one of the important uses of noninvasive genetic sampling (Foran et al. 1997). The use of noninvasive genetic sampling to identify species was initiated by the development of a fecal DNA test to differentiate the endangered San Joaquin kit fox (*Vulpes macrotis mutica*) from other sympatric canids (Paxinos et al. 1997) and a fecal DNA test to differentiate grey seals (*Halichoerus grypus*) and harbor seals (*Phoca vitulina*) (Reed et al. 1997). A growing number of research projects are using hair or fecal samples to document the presence of a target species. In the United States, the Forest Service has initiated a large-scale hair survey as part of an initiative to gather data on the current distribution of Canada lynx (*Lynx canadensis*) (McKelvey et al. 1999; McDaniel et al. 2000). Detection of lynx hair samples is accomplished via PCR amplification of two portions of mtDNA followed by restriction enzyme analysis (Mills et al. 2000b). In Europe, fecal DNA analysis is being used in surveys to determine the distribution of the Iberian lynx (*Lynx pardinus*) (Palomares et al. 2002). In France and Switzerland, Valiere et al. (2003) used mtDNA sequencing of hair and fecal samples to differentiate dog (*Canis familiaris*), wolf, and fox (*Vulpes vulpes*) samples and to document the expansion of the Italian wolf population. Restriction enzyme-based techniques have been developed to differentiate mtDNA of otter (*Lutra lutra*), American mink (*Mustel vison*) and polecat (*Mustela putoris*) in Europe (Hansen and Jacobsen 1999) and fisher (*Martes pennanti*), wolverine (*Gulo gulo*), marten (*Martes americana*), and striped skunk (*Mephitis mephitis*) in North America (Riddle et al. 2003). Also, a length variation in the mtDNA control region has been used extensively to differentiate brown bears and black bears (*Ursus americanus*) in noninvasive genetic sampling projects in North America (Woods et al. 1999; Mowat and Strobeck 2000; Murphy et al. 2000; Poole et al. 2001).

Individual Identification and Population Estimation

After identifying the presence of a rare or endangered species in a particular geographic area, researchers and managers will want to estimate the size and density of the population. Noninvasive genetic sampling provides a method of obtaining mark-recapture population estimates without capturing or handling individuals. The first step in developing noninvasive genetic techniques to estimate abundance of animals is to develop methods to identify individuals. Individual identification is achieved by using PCR to amplify nDNA microsatellite markers and by collecting genotype data at four to twelve microsatellite loci, depending on population variation (Paetkau 2003). Recently, noninvasive genetic sampling has been used to obtain minimum number alive or mark-recapture population estimates in brown bears (Taberlet et al. 1997; Woods et al. 1999; Mowat and Strobeck 2000; Poole et al. 2001), black bears (Woods et al. 1999; Paetkau 2003), cougars (*Puma concolor*) (Ernest et al. 2000, 2002), wolves (Lucchini et al. 2002; Creel et al. 2003), humpback whales (Palsbøll et al. 1997), coyotes (Kohn et al. 1999), hairy-nosed wombats (*Lasiorhinus krefftii*) (Banks et al. 2002; Banks et al. 2003b), otters (Dallas et al. 2003), seals (Reed et al. 1997), badgers (*Meles meles*) (Frantz et al. 2003), pine martens (*Martes americana*) (Mowat and Paetkau 2002), and elephants (Eggert et al. 2003). However, several studies have highlighted the challenges of obtaining accurate population estimates using the low quantity, low quality DNA obtained from hair and feces (Taberlet et al. 1996; Gagneux et al. 1997; Goossens et al. 1998; Taberlet and Waits 1998; Taberlet et al. 1999; Mills et al. 2000a; Waits and Leberg 2000; Creel et al. 2003). The following sections will review these potential pitfalls, suggest possible solutions, and review results from population estimation studies.

Overestimation Challenges

Multiple studies have documented that errors in multilocus genotypes can lead to overestimates in population size (Taberlet et al. 1996, 1997, 1999; Gagneux et al. 1997; Goossens et al. 1998; Waits and Leberg 2000; Miller et al. 2002; Creel et al. 2003; Paetkau 2003). There are three main types of errors that may be present in multilocus genotypes: (1) laboratory scoring or recording errors; (2) contamination; or (3) PCR amplification errors. Paetkau (2003) provided a recent review of methods for detecting and minimizing genotyping errors. Laboratory scoring errors can be minimized through rigorous protocols such as independent analyses of gels, rigorous training of new personnel, and minimizing manual recording of numbers (Paetkau 2003). Contamination errors can be minimized by establishing facilities dedicated to the extraction and PCR setup of noninvasive (low quantity) DNA sources (Taberlet et al. 1996, 1997, 1999). PCR amplification errors are more challenging to avoid and detect (Taberlet et al. 1999; Paetkau 2003).

The two main types of PCR amplification errors are false alleles and false homozygotes (Figure 11.1). A false allele error occurs when the recorded genotype includes an allele that is not present in the correct (true) genotype of the individual (Taberlet et al. 1996). False allele errors generally occur at lower rates than false homozygote errors (Table 11.2) and can frequently be detected when three different alleles are observed for a heterozygous sample. A false homozygote error (also known as allelic dropout) occurs when only one allele is amplified at a locus yet the true genotype of the individual contains two different alleles (a heterozygote) (Gerloff et al. 1995; Taberlet et al. 1996; Gagneux et al. 1997). PCR amplification error rates increase as DNA concentration decreases (Taberlet et al. 1996; Morin et al. 2001); thus, PCR amplification errors are particularly problematic for low quantity and quality DNA sources like single hairs and feces. Multiple studies have shown that PCR amplification error rates from noninvasive genetic samples can be greater than 20% (Table 11.2) and will greatly inflate population estimates if undetected. Waits and Leberg (2000) used computer simulations to demonstrate that abundance estimates from mark-recapture data could be > 200% the true size of the population if genotyping error rates were 5% per locus and 7–10 loci were used to identify individuals.

A number of different methods have been proposed for minimizing the impacts of genotyping errors; the multiple-tubes approach was the first standardized technique (Taberlet et al. 1996). This conservative approach involves repeating amplifications of homozygous loci an average of eight times and heterozygous loci an average of three times to achieve 99% con-

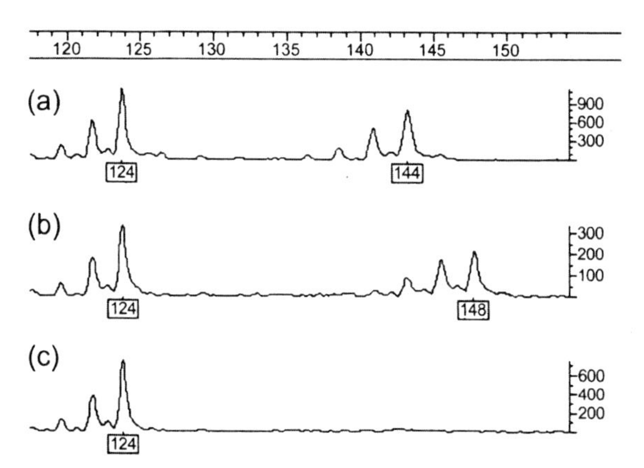

Figure 11.1. Examples of genotyping errors for a heterozygous genotype: (a) true genotype; (b) false allele error (148); and (c) false homozygote error.

Table 11.2.

Frequency of false allele (FA) and false homozygote (FH) genotyping errors in noninvasive genetic sampling studies.

Species	DNA source	% FA	% FH	Citation
Badger	Feces	8	27	Frantz et al. (2003)
Coyotes	Feces	2.5	2.5	Kohn et al. (1999)
Chimpanzee	Shed hair	5.5	31	Gagneux et al. (1997)
Marmot	Single hair	4	14	Goossens et al. (1998)
Marmot	Ten hairs	0	0.5	Goossens et al. (1998)
Otters	Feces	1	2	Dallas et al. (2003)
Wombat	Hair	< 0.5	< 0.5	Sloane et al. (2000)
Orangutan	Feces	3	4	Goossens et al. (2000)
Baboon	Feces	NR	48	Smith et al. (2000)
Baboon	Feces	1	8	Bayes et al. (2000)
Brown bear	Feces	3	6	Murphy et al. (2003b)

fidence in the accuracy of a genotype (Taberlet et al. 1996). This approach produces accurate genotypes but can be extremely expensive and time-consuming to implement. A more efficient maximum likelihood-based approach has been recommended that assesses genotype reliability and strategically reamplifies loci that are most likely to contain errors (Miller et al. 2002). Frantz et al. (2003) also reported an adaptation to the multiple-tubes approach, called the comparative multiple-tubes approach, that requires fewer amplifications and minimizes amplification errors. The use of quantitative PCR techniques to determine DNA quantity in each sample and allow researchers to focus PCR replication efforts also has been suggested as a method for optimizing success rates and minimizing error rates (Morin et al. 2001). Currently, the need for specialized equipment and species-specific optimization has prevented extensive use of quantitative PCR methods.

Noninvasive genetic sampling projects that use DNA extracts from multiple hairs per sample have lower genotyping error rates (Goossens et al. 1998; Sloane et al. 2000; Paetkau 2003) (Table 11.2) and less comprehensive data filtering methods have been developed (Woods et al. 1999; Mowat and Strobeck 2000; Banks et al. 2003b). These methods generally focus on detecting and reamplifying unique genotypes that appear only once within a data set and/or pairs of unique genotypes that closely resemble one another, such as differing at a single locus. Criteria for which genotypes are reanalyzed, which loci are reamplified, and how many reamplifications are carried out vary from study to study (Kohn et al. 1999; Woods et al. 1999; Mowat and Strobeck 2000; Poole et al. 2001; Banks et al. 2003b). The effectiveness of these data filtering methods at removing bias from mark-recapture estimates was recently evaluated using simulations, and population inflation was less than 5% when per locus error rates were less than or equal to 1% (Roon et al. in press). However, when per locus error rates were greater than or equal to 5%, population estimates often were inflated by greater than 10%.

Underestimation Challenges

If the genetic markers used lack the variability to distinguish individuals, then two distinct individuals in a population may by chance carry the same genotype. This can artifically inflate the number of recaptures, thereby decreasing the minimum count of individuals in a population. The probability of this occurrence, or probability of identity ($P_{(ID)}$), depends on allelic diversity, number of loci analyzed, and the percentage and degree of related individuals in a population (Waits et al. 2001). Waits and Leberg (2000) and Mills et al. (2000a) demonstrated that this "shadow effect" could lead to a negative bias in mark-recapture estimators. Capture probability and population size affected the magnitude of this bias, but negative biases up to 20% were observed at $P_{(ID)} = 0.20$ (Mills et al. 2000a). The potential for large negative bias can be minimized by using a sufficient number of highly polymorphic loci and a conservative threshold, such as $P_{(ID)}$ (sibs) < 0.05 (Woods et al. 1999; Waits et al. 2001; Paetkau 2003).

Population Estimation Studies

Noninvasive genetic sampling methods are increasingly being used to survey populations (Schwartz et al. 1998, 1999). Most sampling protocols of noninvasive projects are not designed to detect all individuals, thus mark-recapture models are applied to estimate population size. As with standard mark-recapture studies, noninvasive genetic sampling studies must meet the assumptions of mark-recapture models for accurate estimators. When using a mark-recapture model that assumes demographic closure (White et al. 1982; Boulanger and McLellan 2001), violations of this assumption can be minimized by collecting samples over a short time period and sampling to maximize demographic closure (Boulanger et al. 2002). Variation in capture probability among individuals is another potential problem. Causes of variation leading to unequal trappability may include behavioral differences among animals, trap response, or time variation (Otis et al. 1978; Thompson et al. 1998). In noninvasive genetic sampling studies, a compounding factor in individual capture variation is the quality of DNA extracted from samples collected in the field (Farrell et al. 2000; Goossens et al. 2000; Murphy et al. 2003b; Paetkau 2003). For example, environmental conditions such as differences in humidity or exposure to shade can influence DNA amplification success rates (Farrell et al. 2000; Goossens et al. 2000; Paetkau 2003), and diet has been shown to affect fecal DNA success rates in bears (Murphy et al. 2003a). To address variation in capture probability, different mark-recapture models have been developed that relax the assumption of equal catchability and model individual heterogeneity in capture probability (Otis et al. 1978; Burnham and Overton 1979; Huggins 1989, 1991; Chao et al. 1992; Lee and Chao 1994; Pledger 2000). Software packages, such as MARK (White and Burnham 1999), are available to implement these models. To illustrate the potential of noninvasive genetic sampling for population estimation, I will review three recent studies in badgers, coyotes, and bears. One striking feature of all three studies is the ability to sample 50–80% of the population in a very short period of time in the field (10–60 days).

Frantz et al. (2003) developed a rapid and efficient method for estimating badger population size in the United Kingdom. They collected 53 fecal samples from badger latrines of three different social groups over a 10-day period. To ensure collection of fresh feces, all samples were collected prior to 10 A.M. and old droppings were dusted with builder's chalk. To test the accuracy of microsatellite genotyping and mark-recapture estimation, they collected fecal samples from 36 known individuals and independently estimated population size from extensive trapping and video observation. Genotypic data were collected from seven microsatellite loci using the comparative multiple-tubes approach to minimize genotyping errors. The authors obtained complete genotypes for 39 samples (74%) and identified 20 different individuals (16 matching known individuals). Mark-recapture estimates of abundance based on the M_h-Jackknife estimator, a model in which capture probabilities vary by individual animal, were 26 (22–40, 95% CI) and well within the range estimated from independent methods (24–34 individuals).

Kohn et al. (1999) developed an efficient and accurate surveying method based on fecal DNA to estimate abundance of a coyote population in the

Santa Monica Mountains of California. They gathered 651 carnivore feces along six trail-transects of a 15 km^2 study area over a 2-week period. To reduce effort and costs, they analyzed fecal samples at random, and stopped processing of samples when only one new genotype was discovered in 30 consecutively analyzed samples. Coyote samples were genotyped at three highly variable microsatellite loci with a single replication of all samples and a minimum of three replicates for ambiguous genotypes to minimize genotyping errors. Kohn et al. (1999) obtained 30 unique multilocus genotypes from 115 coyote samples and estimated population size using a rarefaction curve and a mark-recapture model that included individual capture heterogeneity (Burnham and Overton 1979). Both estimators provided very similar population estimates—rarefaction 38 (36–40, 95% CI) and mark-recapture 41 (38–45, 95% CI)—that were within the size range predicted by concurrent field studies. Thus, approximately 80% of the population was sampled, and estimates using the rarefaction curve indicated that another 220 samples would need to be analyzed to sample 90% of the population (Kohn et al. 1999).

Others have applied a more comprehensive sampling approach over large study areas by using barbed-wire hair traps for bear populations (Woods et al. 1999; Mowat and Strobeck 2000; Poole et al. 2001; Paetkau 2003). In British Columbia, Poole et al. (2001) used noninvasive genetic sampling across an 8,527 km^2 study area to estimate grizzly bear population size. They systematically divided the study area into 103 cells of 9 × 9 km and placed five different hair traps in each cell over a 60-day trapping period. They installed one barbed-wire hair trap, baited with a liquid scent lure, in each cell for approximately 12 days and then moved the trap to a new, randomly selected location over five trapping sessions. Poole et al. (2001) collected more than 2,000 hair samples and extracted 1,139 samples that were analyzed to species using an mtDNA test. Of the 1,023 samples with sufficient DNA, 544 were grizzly bears, 453 were black bears and 25 were wolves. All grizzly bear samples were genotyped using six microsatellites. Ninety-eight unique grizzly bear genotypes were obtained from the 420 samples with sufficient DNA. For mark-recapture modeling, a bear must be caught in two or more different capture sessions to be considered a recapture. Poole et al. (2001) observed significant variation in capture success among sessions; thus, Darroch's time varying model was used to obtain the mark-recapture estimate of 148 (124–182, 95% CI). This estimator was believed to be biased high due to geographic closure violations, so the population estimate was adjusted to 138 (114–172, 95% CI) to address this bias. This was the first estimate of population size for this study area; however, density estimates were within the range of other brown bear populations in similar habitats (Pearson 1975).

Sex Identification

Noninvasive genetic sampling also can be used to provide critical data about the sex ratio of a population when researchers are able to obtain a representative sample from the population. Sex-identification methods have been optimized for chimpanzees and gorillas (*Gorilla gorilla gorilla*) (Bradley et al. 2001), canids (Kohn et al. 1999; Lucchini et al. 2002) otters (Dallas et al. 2000), felids (Ernest et al. 2000), red deer (*Cervus elaphus*) (Huber et al. 2002), bears (Taberlet et al. 1993), hairy-nosed wombats (Sloane et al. 2000), seals (Reed et al. 1997), elephants (Eggert et al. 2003), and whales (Palsbøll et al. 1997). Sex identification is generally performed by amplifying either a section of the amelogenin locus (Ennis and Gallagher 1994; Poole et al. 2001; Bradley et al. 2001) or the SRY gene (Reed et al. 1997; Woods et al. 1999; Dallas et al. 2000; Ernest et al. 2000; Mowat and Strobeck 2000; Eggert et al. 2003; Lucchini et al. 2002). When using fecal DNA, it is possible for sex-identification primers to amplify the prey rather than the predator that deposited the sample (Ernest et al. 2000; Murphy et al. 2003a). Thus, researchers working with carnivore fecal samples must first test the accuracy of sex identification with fecal samples of known sex (Ernest et al. 2000). If coamplification of prey DNA is a problem, species-specific primers should be designed (Murphy et al. 2003a) or Y chromosome fragments should be sequenced to verify the species identification (Lucchini et al. 2002).

Conclusion

Noninvasive genetic sampling provides a valuable new tool for detecting the presence of rare species, estimating population sizes and estimating sex ratios. Species identification can be performed fairly quickly and efficiently with high success rates. Currently, the main limitation is designing field techniques to efficiently sample large geographic areas. Estimation of population size and sex ratio from noninvasive samples is more challenging because success rates can be low and genotyping errors can bias estimators. However, a growing number of studies are demonstrating that careful study design and application of error-detection protocols can produce accurate population estimators that are obtained more quickly and efficiently than with traditional methods.

Many rare species are difficult to capture, and noninvasive genetic sampling provides one of the few feasible methods that can document presence and estimate abundance of these animals. Additional research on populations of known population size with many radio-collared animals and independent estimates of population size are greatly needed to estimate capture probabilities, evaluate different population estimators, and refine sampling methods. The power of noninvasive genetic sampling for wildlife biology has just begun to be explored, and there is great potential for the future.

References

Adams, J. R., B. T. Kelly, and L. P. Waits. 2003. Using fecal DNA sampling and GIS to monitor hybridization between red wolves (*Canis rufus*) and coyotes (*Canis latrans*). *Molecular Ecology* 12:2175–2186.

Avise, J. C. 1994. *Molecular Markers, Natural History and Evolution*. Chapman and Hall, New York.

Banks, S. C., A. Horsup, A. Wilton, and A. C. Taylor. 2003a. Genetic marker investigation of the source and impact of predation on a highly endangered species. *Molecular Ecology* 12:1663–1667.

Banks, S. C., S. D. Hoyle, A. Horsup, P. Sunnucks, and A. C. Taylor. 2003b. Demographic monitoring of an entire species, the northern hairy-nosed wombat (*Lasiorhinus krefftii*), by genetic analysis of non-invasively collected material. *Animal Conservation* 6:101–107.

Banks, S. C., M. P. Piggott, B. D. Hansen, N. A. Robinson, and A. C. Taylor. 2002. Wombat coprogenetics: Enumerating a common wombat population by analysis of faecal DNA. *Australian Journal of Zoology* 50:193–204.

Bayes, M. K., K. L. Smith, S. C. Alberts, J. Altman, and M. W. Bruford. 2000. Testing the reliability of microsatellite typing from faecal DNA in the savannah baboon. *Conservation Genetics* 1:173–176.

Birky, C., P. Fuerst, and T. Maruyama. 1989. Organelle gene diversity under migration, mutation and drift: Equilibrium expectations, approach to equilibrium, effects of heteroplasmic cells, and comparison to nuclear genes. *Genetics* 121:613–627.

Borries, C., K. Launhardt, C. Epplen, J. T. Epplen, and P. Winkler. 1999. DNA analyses support the hypothesis that infanticide is adaptive in langur monkeys. *Proceedings of the Royal Society of London Series B* 266:901–904.

Boulanger, J., and B. N. McLellan. 2001. Closure violation in DNA-based mark-recapture population estimation of grizzly bears. *Canadian Journal of Zoology* 79:642–651.

Boulanger, J., G. C. White, B. N. McLellan, J. Woods, M. Proctor, and S. Himmer. 2002. A meta-analysis of grizzly bear DNA mark-recapture projects in British Columbia, Canada. *Ursus* 13:137–152.

Bradley, B. J., K. E. Chambers, and L. Vigilant. 2001. Accurate DNA-based sex identification of apes using non-invasive samples. *Conservation Genetics* 2:179–181.

Bricker, J., L. M. Bushar, H. K. Reinart, and L. Gelbert. 1996. Purification of high quality DNA from shed skin. *Herpetological Review* 27:133–134.

Burnham, K. P., and W. S. Overton. 1979. Robust estimation of population size when capture probabilities vary amongst animals. *Ecology* 60:927–936.

Chao, A., S.-M. Lee, and S.-L. Jeng. 1992. Estimating population size for capture-recapture data when capture probabilities vary by time and individual animal. *Biometrics* 48:201–216.

Constable, J. L., M. V. Ashley, J. Goodall, and A. E. Pusey. 2001. Noninvasive paternity assignment in Gombe chimpanzees. *Molecular Ecology* 10:1279–1300.

Creel, S., G. Spong, J. L. Sands, J. Rotella, J. Zeigle, L. Joe, K. M. Murphy, and D. Smith. 2003. Population size estimation in Yellowstone wolves with error-prone noninvasive microsatellite genotypes. *Molecular Ecology* 12:2003–2009.

Dallas, J. F., D. N. Carss, F. Marshall, K.-P. Koepfli, H. Kruuk, S. B. Piertney, and P. J. Bacon. 2000. Sex identification of the Eurasian otter *Lutra lutra* by PCR typing of spraints. *Conservation Genetics* 1:181–183.

Dallas, J. F., K. E. Coxon, T. Sykes, P. R. Chanin, F. Marshall, D. N. Carss, P. J. Bacon, S. B. Piertney, and P. A. Racey. 2003. Similar estimates of population genetic composition and sex ratio derived from carcasses and faeces of Eurasian otter *Lutra lutra*. *Molecular Ecology* 12:275–282.

Davison, A., J. D. S. Birks, R. C. Brookes, T. C. Braithwaite, and J. E. Messenger. 2002. On the origin of faeces: Morphological versus molecular methods for surveying rare carnivores from their scats. *Journal of Zoology* (London) 257:141–143.

Eggert, L. S., J. A. Eggert, and D. S. Woodruff. 2003. Estimating population sizes for elusive animals: The forest elephants of Kakum National Park, Ghana. *Molecular Ecology* 12:1389–1402.

Ennis, S., and T. Gallagher. 1994. PCR based sex determination assay in cattle based on the bovine Amelogenin locus. *Animal Genetics* 25:425–427.

Ernest, H. B., M. C. T. Penedo, B. P. May, M. Syvanen, and W. M. Boyce. 2000. Molecular tracking of mountain lions in the Yosemite Valley region in California: Genetic analysis using microsatellites and faecal DNA. *Molecular Ecology* 9:433–442.

Ernest, H. B., E. Rubin, and W. M. Boyce. 2002. Fecal DNA analysis and risk assessment of mountain lion predation of bighorn sheep. *Journal of Wildlife Management* 66:75–85.

Farrell, L. E., J. Roman, and M. E. Sunquist. 2000. Dietary separation of sympatric carnivores identified by molecular analysis of scats. *Molecular Ecology* 9:1583–1590.

Fernando, P., M. E. Pfrender, S. E. Encalada, and R. Lande. 2000. Mitochondrial DNA variation, phylogeography and population structure of the Asian elephant. *Heredity* 4:362–372.

Flagstad, Ø., K. Røed, J. E. Stacy, and S. K. Jakobsen. 1999. Reliable noninvasive genotyping based on excremental PCR of nuclear DNA purified with a magnetic bead protocol. *Molecular Ecology* 8:879–884.

Foran, D. R., S. C. Minta, and K. S. Heinemeyer. 1997. DNA-based analysis of hair to identify species and individuals for population research and monitoring. *Wildlife Society Bulletin* 25:840–847.

Frantz, A. C., L. C. Pope, P. J. Carpenter, T. J. Roper, G. J. Wilson, R. J. Delahay, and T. Burke. 2003. Reliable microsatellite genotyping of the Eurasian badger (*Meles meles*) using faecal DNA. *Molecular Ecology* 12:1649–1661.

Frantzen, M. A. J., J. B. Silk, J. W. Ferguson, R. K. Wayne, and M. H. Kohn. 1998. Empirical evaluation of preservation methods for fecal DNA. *Molecular Ecology* 7:1423–1428.

Gagneux, P., C. Boesch, and D. S. Woodruff. 1997. Microsatellite scoring errors associated with noninvasive genotyping based on nuclear DNA amplified from shed hair. *Molecular Ecology* 6:861–868.

Garnier, J. N., M. W. Bruford, and B. Goossens. 2001. Mating system and reproductive skew in the black rhinoceros. *Molecular Ecology* 10:2031–2041.

Gerloff, U., B. Hartung, B. Fruth, G. Hohmann, and D. Tautz. 1999. Intracommunity relationships, dispersal pattern and paternity success in a wild living community of Bonobos (*Pan paniscus*) determined from DNA analysis of faecal samples. *Proceedings of the Royal Society of London Series B* 266:1189–1195.

Gerloff, U., C. Schlötterer, K. Rassmann, and D. Tautz. 1995. Amplification of hypervariable simple sequence repeats (microsatellites) from excremental DNA of wild living bonobos (*Pan paniscus*). *Molecular Ecology* 4:515–518.

Goossens, B., L. Chikhi, S. S. Utami, J. de Ruiter, and M. W. Bruford. 2000. A multisamples, multi-extracts approach for microsatellite analysis of faecal samples in an arboreal ape. *Conservation Genetics* 1:157–162.

Goossens, B., L. P. Waits, and P. Taberlet. 1998. Plucked hair samples as a source of DNA: Reliability of dinucleotide microsatellite genotyping. *Molecular Ecology* 7:1237–1241.

Hansen, M. M., and L. Jacobsen. 1999. Identification of mustelid species: otter (*Lutra lutra*), American mink (*Mustela vison*) and polecat (*Mustela putorius*), by analysis of DNA from faecal samples. *Journal of Zoology* (London) 247:177–181.

Höss, M., M. Kohn, S. Pääbo, F. Knauer, and W. Schröder. 1992. Excrement analysis by PCR. *Nature* 359:199.

Huber, S., U. Bruns, and W. Arnold. 2002. Sex determination of red deer using polymerase chain reaction of DNA from feces. *Wildlife Society Bulletin* 30:208–212.

Huggins, R. M. 1989. On the statistical analysis of capture experiments. *Biometrika* 76:133–140.

———. 1991. Some practical aspects of a conditional likelihood approach to capture experiments. *Biometrics* 47:725–732.

Kohn, M., F. Knauer, A. Stoffela, W. Schröder, and S. Pääbo. 1995. Conservation genetics of the European brown bear—A study using excremental PCR of nuclear and mitochondrial markers. *Molecular Ecology* 4:95–103.

Kohn, M., E. C. York, D. A. Kamradt, G. Haught, R. M. Sauvajot, and R. K. Wayne. 1999. Estimating population size by genotyping feces. *Proceedings of the Royal Society of London Series B* 266:657–663.

Lathuilliere, M., N. Menard, A. Gautier-Hion, and B. Crouau-Roy. 2001. Testing the reliability of noninvasive genetic sampling by comparing analyses of blood and fecal samples in Barbary macaques (*Macaca sylvanus*). *American Journal of Primatology* 55: 151–158.

Lee, S.-M., and A. Chao. 1994. Estimating population size via sample coverage for closed capture-recapture models. *Biometrics* 50:88–97.

Lucchini, V., E. Fabbri, F. Marucco, S. Ricci, L. Boitani, and E. Randi. 2002. Noninvasive molecular tracking of colonizing wolf (*Canis lupus*) packs in the western Italian Alps. *Molecular Ecology* 11:857–868.

McDaniel, G. W., K. S. McKelvey, J. R. Squires, and L. Ruggiero. 2000. Efficacy of lures and hair snares to detect lynx. *Wildlife Society Bulletin* 28:119–123.

McKelvey, K. S., J. Claar, G. W. McDaniel, and G. Hanvey. 1999. *National Lynx Detection Protocol.* Unpublished report. Rocky Mountain Research Station, USDA Forest Service, Missoula, Montana.

Miller, C. R., P. Joyce, and L. P. Waits. 2002. Assessing allelic drop-out and genotype reliability using maximum likelihood. *Genetics* 160:357–366.

Mills, L. S., J. J. Citta, K. P. Lair, M. K. Schwartz, and D. A. Tallmon. 2000a. Estimating animal abundance using non-invasive DNA sampling: Promise and pitfalls. *Ecological Applications* 10:283–294.

Mills, L. S., K. L. Pilgrim, M. K. Schwartz, and K. McKelvey. 2000b. Identifying lynx and other North American felids based on mtDNA analysis. *Conservation Genetics* 1:285–288.

Morin, P. A., K. E. Chambers, C. Boesch, and L. Vigilant. 2001. Quantitative polymerase chain reaction analysis of DNA from non-invasive samples for accurate microsatellite genotyping of wild chimpanzees (*Pan troglodytes verus*). *Molecular Ecology* 10:1835–1844.

Morin, P. A., J. J. Moore, R. Chakraborty, L. Jin, J. Goodall, and D. S. Woodruff. 1994. Kin selection, social structure, gene flow, and the evolution of chimpanzees. *Science* 265:1193–1201.

Morin, P. A., and D. S. Woodruff. 1992. Paternity exclusion using multiple hypervariable microsatellite loci amplified from nuclear DNA of hair cells. pp. 63–81 in R. D. Martin, A. F. Dixson, and E. J. Wickings, eds., *Paternity in Primates: Genetic Tests and Theories*. Karger, Basel, Switzerland.

Mowat, G., and D. Paetkau. 2002. Estimating marten *Martes americanus* population size using hair capture and genetic tagging. *Wildlife Biology* 8:201–208.

Mowat, G., and C. Strobeck. 2000. Estimating population size of grizzly bears using hair capture, DNA profiling, and mark-recapture analysis. *Journal of Wildlife Management* 64:183–193.

Mullis, K., F. Faloona, S. Scharf, R. Saiki, G. Horn, and H. Erlich. 1986. Specific enzymatic amplification of DNA in vitro: The polymerase chain reaction. *Cold Spring Harbor Symposia on Quantitative Biology* 51:263–273.

Murphy, M. A., L. P. Waits, and K. C. Kendall. 2000. Quantitative evaluation of fecal drying methods for brown bear DNA analysis. *Wildlife Society Bulletin* 28:951–957.

_____. 2003a. The influence of diet on faecal DNA amplification and sex identification in brown bears (*Ursus arctos*). *Molecular Ecology* 12:2261–2265.

Murphy, M. A., L. P. Waits, K. C. Kendall, S. K. Wasser, J. A. Higbee, and R. Bogden. 2003b. An evaluation of long-term preservation of brown bear (*Ursus arctos*) fecal DNA samples. *Conservation Genetics* 3:435–440.

Oka, T., and O. Takenaka. 2001. Wild gibbons' parentage tested by non-invasive DNA sampling and PCR-amplified polymorphic microsatellites. *Primates* 42:67–73.

Otis, D. L., K. P. Burnham, G. C. White, and D. R. Anderson. 1978. Statistical inference from capture data on closed animal populations. *Wildlife Monographs* 62:1–135.

Paetkau, D. 2003. An empirical exploration of data quality in DNA-based population inventories. *Molecular Ecology* 12:1375–1387.

Palomares, F., J. A. Godoy, A. Piriz, S. J. O'Brien, and W. E. Johnson. 2002. Faecal genetic analysis to determine the presence and distribution of elusive carnivores: Design and feasibility for the Iberian lynx. *Molecular Ecology* 11:2171–2182.

Palsbøll, P. J., J. Allen, M. Berub, P. J. Clapham, T. P. Feddersen, P. S. Hammond, R. R. Hudson, H. Jorgensen, S. Katona, A. H. Larsen, F. Larsen, J. Lien, D. K. Mattila, J. Sigurjonsson, R. Sears, T. Smith, R. Sponer, P. Stevick, and N. Oien. 1997. Genetic tagging of humpback whales. *Nature* 388:767–769.

Paxinos, E., C. McIntosh, K. Ralls, and R. Fleischer. 1997. A noninvasive method for

distinguishing among canid species: Amplification and enzyme restriction of DNA from dung. *Molecular Ecology* 6:483–486.
Pearson, A. M. 1975. *The Northern Interior Grizzly Bear,* Ursus arctos. Canadian Wildlife Services Report Series Number 43, Ottawa.
Pledger, S. 2000. Unified maximum likelihood estimates for closed capture-recapture models using mixtures. *Biometrics* 56:434–442.
Poole, K. G., G. Mowat, and D. A. Fear. 2001. DNA-based population estimate for grizzly bears *Ursus arctos* in northeastern British Columbia, Canada. *Wildlife Biology* 7:105–115.
Reed, J. Z., D. J. Tollit, P. M. Thompson, and W. Amos. 1997. Molecular scatology: The use of molecular genetic analysis to assign species, sex, and individual identity to seal faeces. *Molecular Ecology* 6:225–234.
Riddle, A. E., K. L. Pilgrim, L. S. Mills, K. S. McKelvey, and L. F. Ruggiero. 2003. Identification of mustelids using mitochondrial DNA and non-invasive sampling. *Conservation Genetics* 4:241–243.
Roon, D. A., L. P. Waits, and K. C. Kendall. 2003. A quantitative evaluation of two methods for preserving hair samples. *Molecular Ecology Notes* 3:163–166.
_____. In press. A simulation test of the effectiveness of several methods for error checking non-invasive genetic data. *Animal Conservation.*
Schwartz, M. K., D. T. Tallmon, and G. Luikart. 1998. Review of DNA-based census and effective population size estimators. *Animal Conservation* 1:1–7.
_____. 1999. Using genetics to estimate the size of wild populations: Many methods, much potential, uncertain utility. *Animal Conservation* 2:321–323.
Sloane, M. A., P. Sunnucks, D. Alpers, L. B. Beheregaray, and A. C. Taylor. 2000. Highly reliable genetic identification of individual northern hairy-nosed wombats from single remotely collected hairs: A feasible censusing method. *Molecular Ecology* 9:1233–1240.
Smith, K. L., S. C. Alberts, M. K. Bayes, M. W. Bruford, J. Altmann, and C. Ober. 2000. Cross-species amplification, non-invasive genotyping, and non-Mendelian inheritance of human STRPs in savannah baboons. *American Journal of Primatology* 51:219–227.
Strausberger, B. M., and M. V. Ashley. 2001. Eggs yield nuclear DNA from egg-laying female cowbirds, their embryos and offspring. *Conservation Genetics* 2:385–390.
Taberlet, P., and J. Bouvet. 1991. A single plucked feather as a source of DNA for bird genetic studies. *Auk* 108:959–960.
_____. 1992. Bear conservation genetics. *Nature* 358:197.
Taberlet, P., J.-J. Camarra, S. Griffin, E. Uhres, O. Hanotte, L. P. Waits, C. Paganon, T. Burke, and J. Bouvet. 1997. Non-invasive genetic tracking of the endangered Pyrenean brown bear population. *Molecular Ecology* 6:869–876.
Taberlet, P., and L. Fumagalli. 1996. Owl pellets as a source of DNA for genetics studies of small mammals. *Molecular Ecology* 5:301–305.
Taberlet, P., S. Griffin, B. Goossens, S. Questiau, V. Manceau, N. Escaravage, L. P. Waits, and J. Bouvet. 1996. Reliable genotyping of samples with very low DNA quantities using PCR. *Nucleic Acids Research* 24:3189–3194.
Taberlet, P., H. Mattock, C. Dubois-Paganon, and J. Bouvet. 1993. Sexing free-ranging brown bears *Ursus arctos* using hairs found in the field. *Molecular Ecology* 2:399–403.
Taberlet, P., and L. P. Waits. 1998. Non-invasive genetic sampling (correspondence). *Trends in Ecology and Evolution* 13:26–27.

Taberlet, P., L. P. Waits, and G. Luikart. 1999. Non-invasive genetic sampling: Look before you leap. *Trends in Ecology and Evolution* 14:323–327.

Thompson, W. L., G. C. White, and C. Gowan. 1998. *Monitoring Vertebrate Populations*. Academic Press, San Diego.

Valiere, N., L. Fumagalli, L. Gielly, C. Miquel, B. Lequette, M.-L. Poulle, J.-M. Weber, R. Arlettaz, and P. Taberlet. 2003. Long-distance wolf recolonization of France and Switzerland inferred from non-invasive genetic sampling over a period of 10 years. *Animal Conservation* 6:83–92.

Valiere, N., and P. Taberlet. 2000. Urine collected in the field as a source of DNA for species and individual identification. *Molecular Ecology* 9:2150–2154.

Valsecchi, E., D. Glockner-Ferrari, M. Ferrari, and W. Amos. 1998. Molecular analysis of the efficiency of sloughed skin sampling in whale population genetics. *Molecular Ecology* 7:1419–1422.

Waits, J. L., and P. L. Leberg. 2000. Biases associated with population estimation using molecular tagging. *Animal Conservation* 3:191–200.

Waits, L. P., G. Luikart, and P. Taberlet. 2001. Estimating probability of identity among genotypes in natural populations: Cautions and guidelines. *Molecular Ecology* 10:249–256.

Walsh, P. A., D. A. Metzger, and R. Higuchi. 1991. Chelex100 as a medium for simple extraction of DNA for PCR-based typing from forensic material. *BioTechniques* 10:506–513.

Wasser, S. K., C. S. Houston, G. M. Koehler, G. G. Cadd, and S. R. Fain. 1997. Techniques for application of faecal DNA methods to field studies of Ursids. *Molecular Ecology* 6:1091–1097.

White, G. C., D. R. Anderson, K. P. Burnham, and D. L. Otis. 1982. *Capture-recapture and Removal Methods for Sampling Closed Populations*. USDOE Report Number LA-8787-NERP, Los Alamos National Laboratory, Los Alamos, New Mexico.

White, G. C., and K. P. Burnham. 1999. Program MARK: Survival estimation from populations of marked animals. *Bird Study* 46:S120-S139.

Woods, J. G., D. Paetkau, D. Lewis, B. N. McLellan, M. Proctor, and C. Strobeck. 1999. Genetic tagging free ranging black and brown bears. *Wildlife Society Bulletin* 27:616–627.

12

Photographic Sampling of Elusive Mammals in Tropical Forests

K. Ullas Karanth, James D. Nichols, and
N. Samba Kumar

Tropical forests harbor much of the planet's terrestrial biodiversity (Terborgh 1992; WCMC 1992), including many threatened mammal species. However, many tropical forest mammals naturally occur at low densities because of traits such as large body size, specialized diets, or spatially dispersed social structures (Eisenberg 1981). Among these, rodents and carnivores tend to be especially elusive because of their nocturnal and secretive behaviors. Many tropical mammals have now become even more rare and elusive due to excessive hunting (Robinson and Bennett 2000) and other anthropogenic pressures (Karanth 2002). Therefore, understanding the ecology of elusive tropical mammals and monitoring their populations are critical conservation needs. In addressing these challenges, biologists have only recently started to employ modern animal population sampling methods.

Animal sampling programs throughout the world typically focus on estimation of one or more "system state variables" (e.g., population size; Williams et al. 2002) at different points in space and/or time. Monitoring programs are frequently developed with the intention of drawing inferences about variation of such quantities over time, space, and associated environmental and management variables.

In this chapter we outline a conceptual framework for animal sampling that includes discussion of underlying rationale (why sample), selection of state variables (what to sample), and general estimation principles (how to sample). We then discuss some of the special challenges presented by elusive tropical mammals and stress the importance of these questions in such sampling situations. Here, we particularly focus on the new remote photographic sampling techniques (Karanth and Nichols 1998; O'Brien et al. 2003) that are increasingly being employed in studies of tropical mammals.

Sampling Animal Populations: General Principles

Many existing programs for sampling animal populations, particularly those targeting rare tropical mammals, are not as useful as they might be because investigators do not devote adequate thought (see Jenelle et al. 2002; Karanth et al. 2003) to fundamental questions associated with establishment of sound sampling programs (see reviews by Thompson et al. 1998; Yoccoz et al. 2001; Pollock et al. 2002). In this section, we present a brief outline of the sort of thinking that we believe should precede and underlie sound animal sampling programs by focusing on three basic questions.

Why Sample?

Efforts to sample animal populations are generally associated with one of two main classes of endeavor: science or conservation. When animal sampling is a component of a scientific research program, estimates of state variables (e.g., abundance) provide the means of confronting model-based predictions with measures of true system response (Hilborn and Mangel 1997; Nichols 2001; Williams et al. 2002). The differences between estimates and model-based predictions then form the basis for rejecting hypotheses in a hypothesis-testing framework or for updating model weights in a multiple-hypothesis framework.

Estimates of state variables for animal populations and communities serve three distinct roles in the conduct of wildlife conservation. First, estimates of system state (e.g., the number of animals) are needed to make state-dependent management decisions (e.g., Williams et al. 2002). Second, system state is frequently contained in the objective functions (explicit statements of management objectives, usually expressed in mathematical form) for managing animal populations and communities, and evaluation of the objective function is an important part of management, addressing the question "to what extent are management objectives being met?" Finally, effective management of wildlife requires either a single model thought to be predictive of system response to management actions or a set of models with associated weights reflecting relative degrees of faith in the different models. The process of developing faith in a single model or weights for members of a model set involves the approaches described above as "science," in which model-based predictions are evaluated with respect to estimated changes in state variables.

In summary, there are some very good reasons for sampling animal populations and communities to estimate relevant state variables. Our suggestion is simply that these reasons be made explicit before commencing a study and that the estimation of state variables be viewed not as an end in itself, but as a component of a larger process of either science or management.

What to Sample?

Certainly the selection of what state variable(s) to estimate will depend on the scientific or management objectives of the study. When dealing with single species, the most commonly used state variable is abundance or population size (sometimes expressed as density). Estimation of abundance frequently requires substantial effort, but it is a natural choice for state variable in studies of population dynamics and management of single-species populations. Study and management of single-species populations also focus on the vital rates responsible for state variable dynamics, such as survival, reproduction, immigration, and emigration.

For some purposes, a useful state variable in single-species population studies is occupancy, defined as the proportion of area, patches, or sample units that is occupied by the species (Mackenzie et al. 2002). Vital rates associated with this state variable are rates of local (patch) extinction and colonization.

When scientific or conservation attention shifts to the community level of organization, many possible state variables exist. The basic multivariate state variable of community ecology is the species abundance distribution, specifying the number of individuals in each species in the community. Many derived state variables are obtained by attributing different values or weights to individuals of different species (Yoccoz et al. 2001). A commonly used state variable is simply species richness, the number of species within the taxonomic group of interest that is present in the community at any point in time or space. Vital rates determining changes in richness are simply local probabilities of extinction and colonization (e.g., Boulinier et al. 1998a, 2001).

The central point is that there is no single state variable that is preferred for the study of animal populations and communities. Instead, the selection of state variable should be closely tied to the objectives of the sampling programs, that is, the answer to the question, why sample?

How to Sample?

Reliable estimation of state variables and inferences about their variation over time and space require attention to two critical aspects of sampling animal populations—spatial variation and detectability (Lancia et al. 1994; Thompson et al. 1998; Yoccoz et al. 2001; Karanth and Nichols 2002). Spatial variation in animal abundance is relevant because investigators can seldom apply survey methods to every square meter of land in the area of interest. Instead, they must select a sample of locations to which survey methods are applied, and this selection must be done in such a way as to permit inferences about the locations that are not surveyed, and hence about the entire area of interest. Approaches to spatial sampling include simple random sampling, unequal probability sampling, stratified random sampling, systematic sampling, cluster sampling, double sampling, and various kinds of adaptive sampling (e.g., Cochran 1977; Thompson 1992).

Detectability refers to the fact that even in locations that are surveyed by investigators, it is very common for investigators to miss animals (i.e., animals go undetected). The investigator typically applies some survey method to each location that yields some sort of count statistic (number of animals seen, caught, harvested, photographed, etc.). Assume that the state variable of interest is abundance. Let N_{it} be the true number of animals associated with an area or sample unit of interest, i, at time t, and denote as C_{it} the associated count statistic. The count is best viewed as a random variable such that:

$$E(C_{it}) = N_{it} p_{it}, \tag{12.1}$$

where p_{it} is the detection probability (probability that a member of N_{it} appears in the count statistic, C_{it}). Estimation of N_{it} thus requires estimation of p_{it}:

$$\hat{N}_{it} = C_{it} / \hat{p}_{it}. \tag{12.2}$$

Typically, interest will not be in abundance itself but in relative abundance, the ratio of abundances at two locations ($\lambda_{ijt} = N_{it}/N_{jt}$), or in rate of population change, the ratio of abundances in the same location at two times (e.g., $\lambda_{it} = N_{it+1}/N_{it}$). Sometimes count statistics are treated as indices, and it is hoped that the ratio of count statistics can be used to estimate these abundance ratios. For example, consider the estimator $\hat{\lambda}_{it} = C_{it+1}/C_{it}$. The expectation of this estimator can be approximated using equation (12.1) as:

$$E(\hat{\lambda}_{it}) \approx \frac{N_{it+1}p_{it+1}}{N_{it}p_{it}}. \qquad (12.3)$$

As can be seen from equation (12.3), if the detection probabilities are very similar for the two sample times, then the estimator will not be badly biased, but when detection probabilities differ, then the estimator will be biased. If detection probability itself is viewed as a random variable, then we still require $E(p_{it}) = E(p_{it+1})$. Thus, we conclude that estimation of both absolute and relative abundance requires information about detection probability (also see Lancia et al. 1994; Karanth and Nichols 2002; Williams et al. 2002).

Sampling Tropical Forest Mammals: Ecological and Practical Issues

Because of their sensory acuity and evasive behavior, several tropical forest mammals (e.g., carnivores) usually cannot be surveyed by using methods based on visual detections like distance sampling (Buckland et al. 2001). Some other groups, such as ungulates or primates that are amenable to visual detection, may occur in dense cover or at low densities such that survey effort required (time invested or distances covered) to achieve adequate numbers of detections may be impractical. Often, because of prevailing social, environmental, and logistical constraints in tropical regions, investigators cannot employ potentially useful sampling approaches that involve animal-handling, such as radio-tagging (White and Garrott 1990) or traditional mark-recapture methods (Lancia et al. 1994; Thompson et al. 1998; Williams et al. 2002). Tropical biologists have tried to overcome these constraints by employing "camera-trapping" as an alternative, noninvasive sampling method for studying populations of rare and elusive animals.

Photographic Sampling of Animals

Cameras set in remote areas and activated by the animals themselves have been used to photograph mammals in tropical forests for many years

(Champion 1927; McDougal 1977). However, the use of animal-activated cameras for wildlife research is somewhat more recent (e.g., Gysel and Davis 1956; Pearson 1959, 1960) and has become very popular in developed countries (Cutler and Swann 1999). Camera-traps have been increasingly used in the scientific study of elusive tropical mammals (Griffiths and van Schaik 1993; Karanth 1995; Karanth and Nichols 1998, 2002; O'Brien et al. 2003; Trolle and Kéry 2003) in recent years. Many such investigations are currently underway in Asia, Africa, and Latin America.

Photographs as Count Statistics

As noted earlier on the "how" of animal sampling, inferences about animal populations and communities are virtually always based on some sort of count statistic. In situations where individual animals can be identified from photographs, camera-trap studies can be designed and analyzed using methods used for conventional capture-recapture sampling (e.g., Karanth 1995; Karanth and Nichols 1998, 2002; Trolle and Kéry 2003). Such studies provide estimates of the state variable used most frequently in wildlife studies, abundance (or density).

When interest is focused on single species in cases where individuals cannot be identified from photographs, one option is to use camera-trap data to estimate occupancy as a state variable (e.g., see MacKenzie et al. 2002). The count statistic in this case would be the number of sample units (areas sampled by camera-traps) at which the species had been photographed and identified, and the quantity to be estimated would be the proportion of these units actually occupied by the species.

Finally, we note that interest may instead be directed at the community level of organization. In this case, species richness of some group of mammals may be the state variable of interest (Cam et al. 2002). The count statistic would be the total number of species identified from camera-trapping, and inference would require estimation about the proportion of species in the community that was actually detected.

Basics of Camera-Trapping

A camera-trap consists of an automated device that is activated when the targeted animal moves into range and triggers one or more previously positioned cameras to take pictures of that animal. Usually several traps are deployed based on various design considerations. The equipment used can take a variety of forms ranging from cheap homemade pressure-pad devices to expensive, sophisticated commercial units (Cutler and Swann 1999; Karanth et al. 2002). The sampling process consists of deploying a number of camera-trap units in the surveyed area in a manner most conducive to obtaining photographs of the target species. Usually, the investigator periodically revisits and checks the traps to ensure their proper functioning and to replenish film or batteries.

Photographic Identification of "Captured" Animals

For community-level surveys of mammal species richness or for single species surveys designed to estimate habitat occupancy or to derive an index of relative abundance, the photographs obtained must be of adequate quality to unambiguously identify the animal species. The choice of trap sites and the positioning of cameras are governed by this need. In community-level surveys of mammals, because of inter-specific differences in size, grouping patterns, and behavior, the positioning and spacing of camera-traps involves a compromise among competing needs for optimally photo-capturing different species. Consequently, some species in the surveyed area may have capture probabilities that approach zero, a situation analogous to the "hole in the sampled area" problem (Karanth and Nichols 1998) encountered in single species capture-recapture studies. The design of community-level camera-trap studies should thus focus on attaining nonzero detection probabilities for all of the species in the community, guild, or taxonomic group of interest.

In studies that try to estimate abundance or density of a single species, camera-traps must yield high quality pictures that permit identification of individual animals. Naturally occurring marks on animals, such as the shape, arrangement, and patterns of stripes (tigers *Panthera tigris*), spots (cheetahs *Acionyx jubatus*), or rosettes (jaguars *Panthera onca*, leopards *Panthera pardus*, and ocelots *Felis pardalis*); the shape and configuration of body parts such as head, tusks, and ears (elephants *Elephas maximus*, *Loxodonta africana*, skin folds (Javan rhinos *Rhinoceros sondaicus*), and even injuries and scars (manatees *Trichechus* spp.) can be used to identify individuals. In a few cases, it may be possible to first physically capture the animals and artificially mark individuals for photographic identification in subsequent samples.

Because natural markings on animals are asymmetric, unambiguous individual identifications may necessitate photographs of both flanks, requiring the deployment of two or more cameras with each trap. Unfortunately, investigators sometimes deploy single cameras to cut costs, thereby losing scarce data as well as diminishing the ability to apply powerful capture-recapture analytic methods to photographic count statistics.

Equipment and Data Collection Protocols

Although homemade camera-traps can be constructed inexpensively, we do not recommend them for surveys of rare and elusive mammals because of their low reliability. Some of the commercial units are listed in publications (e.g., Karanth and Nichols 2002:187–188), and Web sites evaluate relative merits of different units (e.g., www. jesseshuntingpages.com/cams.html).

Most commercial camera-traps employ either "active" or "passive" tripping devices to fire the cameras. Active devices respond to an animal intercepting an electronic beam, whereas passive ones are triggered by the animal's body heat (Karanth and Nichols 2002). The more sophisticated (and expensive) camera-trap equipment permits the investigators to target their study species using several means: firing multiple cameras with a single tripping device; varying the period of beam-breakage to avoid smaller creatures; varying the interval between consecutive pictures; electronically "waking up" cameras that "sleep" in battery-saving mode; setting specific "time zones" for picture-taking to avoid undesirable species, and electronically storing the date and time for each tripping event.

Most currently available camera-traps use flashlight photography and capture images on ordinary film. However, new equipment that offers digital image capture, infrared photography that avoids flash, and even video-capture of images is now on the market. Whatever the type of equipment employed, it is critically important to ensure that each picture obtained on a film roll (or disk or videotape) is given a unique identification number and that subsequent data collection and film processing protocols permit clear, unambiguous identification of the time, date, and location for any photographic capture event. We recommend using predesigned data forms (Karanth and Nichols 2002:183) to ensure that different field personnel obtain capture records in a consistent manner.

Environmental and Social Factors Affecting Camera-Trap Surveys

In addition to ecology and behavior of study species, several environmental and social factors impose constraints on camera-trap surveys in the tropics. Commonly, rain and humidity restrict the work to certain seasons. Under humid conditions, camera-traps that rely on passive detection generally appear to perform more reliably than the more sensitive, active detection units (Kawanishi 2002).

In some cases, animal damage poses a threat to equipment. We found that elephants frequently damaged the equipment, and tigers did so occasionally. More often, human vandalism and theft are deterrents. Provision of a steel protective shell around the camera-trap (Karanth and Nichols 2002:184–186), locking devices, or cryptic hiding of the equipment are possible countermeasures against these problems.

Photographic sampling of rare mammals is usually conducted at landscape scales and over difficult terrain. Deployment of traps according to a predetermined survey design usually involves moving equipment over long distances, often on foot. This disadvantage is sometimes offset by the ready availability of inexpensive labor in the tropics. In many areas, camera-traps can be revisited only after several days. Locally hired labor may not have the skills necessary to record data or check the equipment, requiring the presence of the investigator even for routine revisits. The number of camera-traps deployed, the trap spacing used, the duration of the sampling periods, and consequently, the quality of the data obtained in camera-trap surveys, are thus influenced strongly by a variety of environmental and social factors.

Modeling and Estimation Using Photographic Data

This section describes how photographic count statistics on tropical forest mammals can be used to estimate state variables and rates of change in these variables. We will not present all of the relevant estimators or present their underlying rationale. Instead, we will point to literature with descriptions of these approaches and indicate how we believe these approaches might be used in camera-trap studies of tropical forest mammals.

Estimation of Abundance and Density

The appropriate methods for abundance estimation differ, depending on whether or not animals can be individually identified. For species and situations in which individuals cannot be identified, it may be possible to use the occupancy approach (Royle and Nichols 2003; also see next section) to draw inferences about abundance.

In some situations with no individual identification, it may be reasonable to use the count statistics as indices of relative abundance for comparing abundance at different times or locations. The reasonableness of such direct use of count statistics depends on the relationship between the counts and the true quantities of interest—abundances at the different times and places (Nichols and Karanth 2002). For example, if counts are related to abundance by a proportionality constant, such as detection probability in equation (12.1), then reasonable inference about relative abundance is possible only when that constant is very similar for the two times or locations being compared (see equation (12.3) and related discussion). Use of counts (trapping rates) as indices to abundance is thus based on restrictive, untested assumptions.

Given the above, when identification of individual animals is possible from photographs, capture-recapture models developed for closed populations provide a more robust approach to abundance estimation. Thus, there is little justification for conducting camera-trap surveys that generate only indices of abundance (e.g., some studies cited by Carbone et al. 2001). Instead it is preferable to compute estimates based on appropriate capture-recapture methods, because resources invested tend to be comparable in the two cases.

Under a capture-recapture sampling design, camera-traps are set throughout an area of interest, with attention devoted to eliminating holes—areas within the overall area of interest within which an animal might travel normally and never encounter a camera-trap (Karanth and Nichols 1998; Nichols and Karanth 2002; Karanth et al. 2004). If the investigators (or field assistants) have prior knowledge of habits and behavior of the target species, it is wise to use this knowledge in the deployment of camera-traps. For example, telemetry and sign studies clearly indicate the preference of tigers for traveling along trails and roads, so allocation of trap stations to trail or road systems is a reasonable means of sampling an area to get larger numbers of captures (Karanth et al. 2002). Placing camera-traps at mineral licks, water holes, animal latrines, bait stations, etc., may also increase capture probabilities and thereby improve the quality of the estimates.

If nothing is known about the habits of the target species, random allocation of traps, for example, using a grid system imposed on the study area, provides a reasonable means of sampling, although numbers of captures may be so low as to severely limit utility of results. The overall objective of the trap deployment should be that all individuals in the sampled area have nonzero probabilities (hopefully, similar across individuals) of encountering a camera-trap.

The most straightforward design involves setting camera-traps throughout the area of interest as discussed above and collecting photographs for a short period (say 5–45 consecutive days, depending on the species of interest). However, because camera-traps are expensive, enough units may not be available for this approach. Therefore, it sometimes may become necessary to move the traps around the area of interest with a different set of locations being sampled during each time interval (see designs suggested by Nichols and Karanth 2002).

Camera-trap designs for individually identifiable animals should yield capture histories of individuals. Each capture history (one per individual) is simply a record of whether or not the animal was caught at each sampling period. If we let "0" indicate no capture and "1" denote capture, then history 001010 indicates an animal caught only in periods 3 and 5 of a 6-period study. Capture histories for all individuals caught at least once provide the data needed to estimate abundance and hence, the number of animals exhibiting a capture history of all 0's (present in the sampled area but never caught).

The statistical models that have been most useful for such work are based on "closed" populations that do not change by birth, death, immigration, or emigration over the course of the sampling (Nichols and Karanth 2002). The closure assumption imposes the restriction that sampling be carried out over a sufficiently short time frame during which closure violations are not expected to occur. The various members of this class of models differ with respect to the incorporated sources of variation in detection probability. Otis et al. (1978) and White et al. (1982) provided classic descriptions of these models. Some more recent estimators and models also were summarized in Williams et al. (2002). Computer programs CAPTURE (Rexstad and Burnham 1991) and MARK (White and Burnham 1999) can be used to analyze capture history data and provide statistics useful in model selection as well as estimates of abundance.

In many cases density (number of animals per unit area), rather than abundance, is the quantity of interest. In such cases, it is necessary to estimate the area actually sampled by the camera-traps. This area will typically be larger than the area over which traps are actually spread. Estimation of the area sampled requires ancillary data from radio telemetry or distances between capture locations of camera-trapped animals (e.g., Wilson and Anderson 1985; Nichols and Karanth 2002; Karanth et al. 2004).

In some cases, it will be possible to sample an area for some period each year (e.g., 4 weeks of camera-trapping each summer) for a number of years. Capture-recapture designs that include sampling at two different time scales are referred to as "robust designs" (Pollock 1982; Pollock et al. 1990; Williams et al. 2002). Within the robust design, sampling periods separated by relatively long time intervals (e.g., 1 year) are referred to as primary periods, whereas periods separated by relatively short intervals (e.g., 1 day or 1 week) are referred to as secondary periods. Capture history data can be aggregated across secondary periods and used to estimate survival rates between primary periods. Capture history data over secondary periods within primary periods can then be used to estimate abundance. Finally, resulting survival rate and abundance estimates can be used together to estimate recruitment. We have conducted such analyses for camera-trap data on tigers collected between 1991 and 2000 at Nagarahole Reserve, India (Karanth et al. in prep.).

We emphasize that the duration of the survey and sampling periods, the location, placement, and spacing of traps, etc., must be dictated by the ecology of the animal. Recommending standardized protocols for camera-trapping, regardless of species and ecological context (e.g., Fonseca et al. 2003), is likely to lead to violations of major capture-recapture assumptions (Otis et al. 1978; White et al. 1982; Nichols and Karanth 2002) related to population closure, nonzero capture probabilities for all animals, and even to a single animal population being sampled in the first place!

Estimation of Habitat Occupancy

When individual animals cannot be identified, patch occupancy estimation can be used to draw inferences about target species. Depending on the species and their habitats, occupancy survey designs might involve placing camera-traps systematically or randomly throughout an entire area of interest or over habitat patches or appropriate habitat in an area of interest. If habitat is patchy, patches themselves can be used as the sample units. If habitat is not found in discrete patches, sample units must be selected. In many cases, it will be sensible to simply select a grid cell size, place a grid over the area of interest, and randomly select cells to be sampled. In such situations, consideration should be given to the size of the sample unit (grid cell) relative to the individual range size of the target species. For example, if occupancy is to be used as a state variable in a monitoring program, it would not be reasonable to set sample unit size so small that a single animal could occupy many sample units. In the case of territorial species, use of appropriately sized grid cells (e.g., approximately the size of the territory) might lead to occupancy estimates that could be interpreted as estimates of number of territorial animals.

In occupancy studies, identification of individuals is not assumed, and it is not necessary to deploy multiple cameras for unambiguous identification. Camera-traps should be deployed for a relatively short period of time (e.g., 2 weeks), as estimation of occupancy requires that the sampled locations be closed to changes in occupancy over the course of the sampling (i.e., animals do not move into the area and become established or depart the area permanently over the course of the sampling).

The data resulting from an occupancy study for a single season are detection histories (analogous to capture histories described above) for each sample unit. Detection histories are rows of 1's and 0's indicating days on which at least one individual of the species is or is not detected, respectively. For example, 0001000101, denotes a location at which the species was photographed on days 4, 8, and 10, by a camera left out for 10 days. Each sampled location has such a history. These detection histories differ from capture histories in that the number of locations at which no animals are detected (detection histories of all 0's) is known in occupancy studies. Detection probability is estimated from the patterns of detection and nondetection, at locations with at least one detection. The objective then becomes to estimate how many of these nondetection sites were actually occupied. The estimation thus explicitly accounts for the reality that nondetection does not equate to absence in so-called presence-absence (more properly "detection-nondetection") studies.

Detection history data are used to estimate the probability that a sample unit is occupied or, equivalently, the proportion of sample units occupied. This can be accomplished using a two-step approach that involves first estimating the number of sampled locations that are occupied and then dividing this estimate by the number of sampled locations (Nichols and Karanth 2002). A more efficient approach permits direct estimation of the occupancy parameter (MacKenzie et al. 2002; Royle and Nichols 2003; Chapter 8, this volume) in a single step. If the same locations are sampled with cameras each year, then the robust design approach can be used to estimate not only occupancy but also rate of change in occupancy over time and probabilities of local extinction and colonization of the sample units (Barbraud et al. 2003; MacKenzie et al. 2003). Programs PRESENCE (MacKenzie et al. 2002, 2003) and MARK (White and Burnham 1999) can be used to assist in model selection and to compute estimates of occupancy from detection history data.

Although this statistical approach to occupancy estimation is relatively new, Kawanishi (2002) has already successfully used it with camera-trap data on tropical forest mammals. For example, she divided her study sites in Taman Negara, Malaysia, into 9 km^2 grid cells for the purpose of estimating occupancy for several mammal species. Using camera-trapping and surveys of secondary animal signs to assess occupancy, she computed a naive occupancy estimate of 0.36 (number of cells known to be occupied divided by the total number of cells) for sambar deer, *Cervus unicolor*, at her Merapoh study site. Using the approach of Nichols and Karanth (2002), however, she estimated that 0.64 ($\widehat{SE} = 0.104$) of the grid cells were actually occupied by sambar. Although the differences between naive and estimated rates of occupancy computed by Kawanishi (2002) were not so large for all species, this example illustrates the potential importance of trying to properly account for detection probability in surveys of spatial distribution of tropical mammals.

Estimation of Species Richness

Instead of focusing on species-specific state variables such as occupancy and abundance, species richness within some group of mammals (e.g., ungulates, meso-carnivores) may be the target quantity for estimation in some studies. The sampling problem is that every species in the group may not be detected during survey efforts, and we would like a method that accounts for missed species. Spatial sampling and deployment of camera-traps will be similar to those used in occupancy studies. A key consideration in the design of community studies is that all of the species in the group of interest must have the potential to be detected. Single cameras are adequate, as animals must be identified to species only. Again, sampling should not extend over too long a period, because the mammal community is assumed to be closed over the period of sampling.

The data arising from a camera-trap study directed at species richness are detection histories for each species. Each detection history would indicate whether or not the species was detected at a sampling occasion. For example, assume that camera-traps were deployed for 10 consecutive nights. A history of 0011000101 would indicate a species that was detected on sample occasions 3, 4, 8, and 10, but not on other occasions. Each detected species has such a history. The different species are analogous to the different individuals in a standard capture-recapture setting. The patterns of detection and nondetection can thus be used to estimate species-level detection probability, and hence total number of species (including those not detected), using the models developed for closed populations (e.g., Otis et al. 1978). Because of differences in detection probabilities of animals of different species, and because of different abundances that contribute to variation in detectability at the species level, we suspect that models permitting heterogeneity in detection probabilities will be especially useful (Burnham and Overton 1979; Boulinier et al. 1998b).

In addition to estimating species richness, if the same locations are sampled over time (e.g., every year), as in the robust design (Pollock 1982), resulting data can be used to estimate rate of change in species richness and temporal variation in richness, as well as local extinction probabilities and turnover (proportion of species that is new) (Boulinier et al. 1998a; Nichols et al. 1998). The Web-based program COMDYN was developed to estimate richness and associated community-dynamic parameters (Hines et al. 1999). Examples of use of this approach in community investigations include Boulinier et al. (2001), Cam et al. (2002), and Doherty et al. (2003). Most of the published uses involve avian point count data, but camera-trap data on rare tropical mammals are also certainly suitable for employing this capture-recapture–based approach to estimating species richness.

Discussion

We argue that photographic sampling provides a logistically reasonable approach to monitoring elusive mammals in tropical forests. In the first section of this chapter, we posed three questions relevant to any monitoring program—why monitor, what should be monitored, and how should one conduct monitoring? With respect to the why, we emphasized that monitoring is not a stand-alone activity to be considered in isolation but should instead be viewed as a component of a larger process, usually either science or management. The choice of state variable to monitor depends very heavily on the reason for the monitoring. We emphasized that the "how" of monitoring involves at least two important sources of uncertainty, spatial sampling and detectability. Spatial variation in state variables of interest dictates that sample units to be surveyed must be selected in a manner that permits inference about the sample units not selected. Detectability refers to the usual inability to detect all individuals in a sample area that is surveyed. Count statistics must be collected in such a way that the associated detection probability can be estimated.

Because of the difficulty of working with and viewing animals in tropical forests, remote camera-traps provide an attractive means of sampling animals. If abundance of animals is the state variable of interest and if the animals possess natural marks permitting individual identification, full-fledged capture-recapture models can be used to estimate abundance. The count statistics are the numbers of different animals detected, and the detection histories of individual animals provide the data needed to draw inferences about detection probability, and hence abundance.

In some cases, it may be reasonable to use habitat occupancy as the state variable of interest. In this case, the count statistic is the number of patches or sample units at which the mammal species of interest is detected, and detection histories of the sample units are used to draw inferences about detection probability.

In community-level studies of mammals, the state variable of interest may be the number of animals (species richness) in a certain size class or guild or taxonomic group. In such studies, the count statistic is the number of different species detected, and detection probability is estimated using detection histories of the different species.

For each of the above three possible state variables, if sampling is conducted at approximately the same time for each of a number of years or seasons, it is possible to use robust design approaches (Pollock 1982) to estimate not only the state variable of interest but also the vital rates governing changes in the state variable.

Camera-trap surveys of elusive mammals usually involve heavy investments of resources and effort. However, many camera-trap survey protocols currently being implemented or recommended appear to be based on ad hoc considerations (e.g., Fonseca et al. 2003) that are unlikely to yield scientifically defensible results. We believe that scientific and conservation values of camera-trapping studies of elusive tropical mammals can be enhanced substantially by paying closer attention to issues covered above.

References

Barbraud, C., J. D. Nichols, J. E. Hines, and H. Hafner. 2003. Estimating rates of extinction and colonization in colonial species and an extension to the metapopulation and community levels. *Oikos* 101:113–126.

Boulinier, T., J. D. Nichols, J. E. Hines, J. R. Sauer, C. H. Flather, and K. H. Pollock. 1998a. Higher temporal variability of forest breeding bird communities in fragmented landscapes. *Proceedings of the National Academy of Sciences, U.S.A.* 95:7497–7501.

Boulinier, T., J. D. Nichols, J. E. Hines, J. R. Sauer, C. H. Flather, and K. H. Pollock. 2001. Forest fragmentation and bird community dynamics: inference at regional scales. *Ecology* 82:1159–1169.

Boulinier, T., J. D. Nichols, J. R. Sauer, J. E. Hines, and K. H. Pollock. 1998b. Estimating species richness: The importance of heterogeneity in species detectability. *Ecology* 79:1018–1028.
Buckland, S. T., D. R. Anderson, K. P. Burnham, J. L. Laake, D. L. Borchers and L. Thomas. 2001. *Introduction to Distance Sampling: Estimating Abundance of Biological Populations.* Oxford University Press, Oxford, United Kingdom.
Burnham, K. P., and W. S. Overton. 1979. Robust estimation of population size when capture probabilities vary among animals. *Ecology* 62:625–633.
Cam, E., J. D. Nichols, J. E. Hines, J. R. Sauer, R. Alpizar-Jara, and C. H. Flather. 2002. Disentangling sampling and ecological explanations underlying species-area relationships. *Ecology* 83:1118–1130.
Carbone, C., S. Christie, K. Conforti, T. Coulson, N. Franklin, J. R. Ginsberg, M. Griffiths, J. Holden, K. Kawanishi, M. Kinnaird, R. Laidlaw, A. Lynam, D. W. MacDonald, D. Martyr, C. McDougal, L. Nath, T. O'Brien, J. Seidensticker, D. Smith, M. Sunquist, R. Tilson, and W.N. Wan Shaharuddin. 2001. The use of photographic rates to estimate densities of tigers and other cryptic mammals. *Animal Conservation* 4:75–79.
Champion, F. W. 1927. *With a Camera in Tiger Land.* Chatto and Windus, London.
Cochran, W. G. 1977. *Sampling Techniques,* 3rd ed. Wiley, New York.
Cutler, T. L., and D. E. Swann. 1999. Using remote photography in wildlife ecology: A review. *Wildlife Society Bulletin* 27:571–581.
Doherty, P. F., Jr., G. Sorci, J. A. Royle, J. E. Hines, J. D. Nichols, and T. Boulinier . 2003. From the cover: Sexual selection affects local extinction and turnover in bird communities. *Proceedings of the National Academy of Sciences, U. S. A.* 100:5858–5862.
Eisenberg, J. F. 1981. *Mammalian Radiations.* University of Chicago Press, Chicago.
Fonseca, G., T. E. Lacher, P. Batra, J. Sanderson, S. Brandes, A. Espinel, C. Kuebler, A. Bailey, and J. Heath. 2003. *Tropical Ecology, Assessment, and Monitoring: Camera Trapping Protocol* (Electronically published report). Center for Applied Biodiversity Science, Conservation International, Washington, D.C. Available from http://teaminitiative.org (accessed October 15, 2003).
Griffiths, M., and C. P. van Schaik 1993. The impact of human traffic on the abundance and activity periods of Sumatran rainforest wildlife. *Conservation Biology* 7:623–626.
Gysel, L. W., and E. M. J. Davis. 1956. A simple automatic photographic unit for wildlife research. *Journal of Wildlife Management* 20:451–453.
Hilborn, R., and M. Mangel. 1997. *The Ecological Detective: Confronting Models with Data.* Princeton University Press, Princeton, New Jersey.
Hines, J. E., T. Boulinier, J. D. Nichols, J. R. Sauer, and K. H. Pollock. 1999. COMDYN: Software to study the dynamics of animal communities using a capture-recapture approach. *Bird Study* 46:S209–S217.
Jenelle, C. S., M. C. Runge, and D. I. MacKenzie. 2002. The use of photographic rates to estimate densities of tigers and other cryptic mammals: A comment on misleading conclusions. *Animal Conservation* 5:119–120.
Karanth, K. U. 1995. Estimating tiger *Panthera tigris* populations from camera-trap data using capture-recapture models. *Biological Conservation* 71:333–338.
_____. 2002. Nagarahole: Limits and opportunities in wildlife conservation. pp. 189–202 in J. Terborgh, C. P. Van Schaik, L. Davenport, and M. Rao, eds., *Making Parks Work: Strategies for Preserving Tropical Nature.* Island Press, Washington, D.C.

Karanth, K. U., N. S. Kumar, and J. D. Nichols. 2002. Field surveys: Estimating absolute densities of tigers using capture-recapture sampling. pp. 139–152 in K. U. Karanth and J. D. Nichols, eds., *Monitoring Tigers and Their Prey: A Manual for Researchers, Managers and Conservationists in Tropical Asia.* Centre for Wildlife Studies, Bangalore, India.

Karanth, K. U., and J. D. Nichols. 1998. Estimation of tiger densities in India using photographic captures and recaptures. *Ecology* 79:2852–2862.

_____, eds. 2002. *Monitoring Tigers and Their Prey: A Manual for Researchers, Managers and Conservationists in Tropical Asia.* Centre for Wildlife Studies, Bangalore, India.

Karanth, K. U., J. D. Nichols, N. S. Kumar, and J. E. Hines. In preparation. Assessing tiger population dynamics using photographic capture-recapture sampling.

Karanth, K. U., J. D. Nichols, N. S. Kumar, W. A. Link, and J. E. Hines. 2004. Tigers and their prey: Predicting carnivore densities from prey abundance. *Proceedings of the National Academy of Sciences, U.S.A.* 101:4854–4858.

Karanth, K. U., J. D. Nichols, J. Seidensticker, E. Dinerstein, J. L. D. Smith, C. McDougal, A. J. T. Johnsingh, R. Chundawat, and V. Thapar. 2003. Science-deficiency in conservation practice: The monitoring of tiger populations in India. *Animal Conservation* 6:141–146.

Kawanishi, K. 2002. *Population Status of Tigers in a Primary Rainforest of Peninsular Malaysia.* Ph.D. Dissertation, University of Florida, Gainesville.

Lancia, R. A., J. D. Nichols, and K. H. Pollock. 1994. Estimating the number of animals in wildlife populations. pp. 215–253 in T. Bookhout, ed., *Research and Management Techniques for Wildlife and Habitats.* The Wildlife Society, Bethesda, Maryland.

MacKenzie, D. I., J. D. Nichols, J. E. Hines, M. G. Knutson, and A. B. Franklin. 2003. Estimating site occupancy, colonization and local extinction when a species is detected imperfectly. *Ecology* 84:2200–2207.

MacKenzie, D. I., J. D. Nichols, G. B. Lachman, S. Droege, J. A. Royle, and C. Langtimm. 2002. Estimating site occupancy rates when detection probabilities are less than one. *Ecology* 83:2248–2255.

McDougal, C. 1977. *The Face of the Tiger.* Rivington Books, London.

Nichols, J. D. 2001. Using models in the conduct of science and management of natural resources. pp. 11–34 in T. M. Shenk and A. B. Franklin, eds., *Modeling in Natural Resource Management: Development, Interpretation, and Application.* Island Press, Washington, D.C.

Nichols, J. D., T. Boulinier, J. E. Hines, K. H. Pollock, and J. R. Sauer. 1998. Estimating rates of local extinction, colonization and turnover in animal communities. *Ecological Applications* 8:1213–1225.

Nichols, J. D., and K. U. Karanth. 2002. Statistical concepts: Estimating absolute densities of tigers using capture-recapture sampling. pp. 121–137 in K. U. Karanth and J. D. Nichols, eds., *Monitoring Tigers and Their Prey: A Manual for Researchers, Managers and Conservationists in Tropical Asia.* Centre for Wildlife Studies, Bangalore, India.

O'Brien, T., M. Kinnaird, and H. T. Wibisono. 2003. Crouching tigers, hidden prey: Sumatran tigers and prey populations in a tropical forest landscape. *Animal Conservation* 6:1–10.

Otis, D. L., K. P. Burnham, G. C. White, and D. R. Anderson.1978. Statistical inference from capture data on closed animal populations. *Wildlife Monographs* 62:1–135.

Pearson, O. P. 1959. A traffic survey of *Microtus-Reithrodontomys* runways. *Journal of Mammalogy* 40:169–180.

_____. 1960. Habits of *Microtus californicus* revealed by automatic photographic recorders. *Ecological Monographs* 30:231–249.

Pollock, K. H. 1982. A capture-recapture design robust to unequal probability of capture. *Journal of Wildlife Management* 46:757–760.

Pollock, K. H., J. D. Nichols, C. Brownie, and J. E. Hines. 1990. Statistical inference from capture-recapture experiments. *Wildlife Monographs* 107:1–97.

Pollock, K. H., J. D. Nichols, T. R. Simons, and J. R. Sauer. 2002. Large scale wildlife monitoring studies: Statistical methods for design and analysis. *Environmetrics* 13:1–15.

Rexstad, E., and K. P. Burnham. 1991. *User's Guide to Interactive Program CAPTURE.* Colorado Cooperative Fish and Wildlife Research Unit, Colorado State University, Fort Collins.

Robinson, J. G., and E. L. Bennett, eds. 2000. *Hunting for Sustainability in Tropical Forests.* Columbia University Press, New York.

Royle, J. A., and J. D. Nichols. 2003. Estimating abundance from repeated presence-absence data or point counts. *Ecology* 84: 777–790.

Terborgh, J. 1992. *Diversity and the Tropical Rainforest.* Freeman, New York.

Thompson, S. K. 1992. *Sampling.* Wiley, New York.

Thompson, W. L., G. C. White, and C. Gowan. 1998. *Monitoring Vertebrate Populations.* Academic Press, San Diego.

Trolle, M., and M. Kéry. 2003. Estimation of ocelot density in the Pantanal using capture-recapture analysis of camera-trapping data. *Journal of Mammalogy* 84:607–614.

White, G. C., D. R. Anderson, K. P. Burnham, and D. L. Otis. 1982. *Capture-recapture and Removal Methods for Sampling Closed Populations.* USDOE Report Number LA-8787-NERP, Los Alamos National Laboratory, Los Alamos, New Mexico.

White, G. C., and K. P. Burnham. 1999. Program MARK: Survival rate estimation from both live and dead encounters. *Bird Study* 46:S120–S139.

White, G. C., and R. A. Garrott. 1990. *Analysis of Wildlife Radiotracking Data.* Academic Press, San Diego.

Williams, B. K., J. D. Nichols, and M. J. Conroy. 2002. *Analysis and Management of Animal Populations.* Academic Press, San Diego.

Wilson, K. R., and D. R. Anderson. 1985. Evaluation of two density estimators of small mammal population size. *Journal of Mammalogy* 66:13–21.

WCMC (World Conservation Monitoring Center). 1992. *Global Biodiversity: Status of the Earth's Living Resources.* Chapman and Hall, London.

Yoccoz, N. G., J. D. Nichols, and T. Boulinier. 2001. Monitoring of biological diversity in space and time. *Trends in Ecology and Evolution* 16:446–453.

13

Using Probability Sampling of Animal Tracks in Snow to Estimate Population Size

Earl F. Becker, Howard N. Golden, and Craig L. Gardner

Sampling rare, elusive, or secretive species is difficult (Sudman et al. 1988; Becker et al. 1998; Chapter 2, this volume). For many of these species, it is easier to detect evidence of their presence than to see the animal (Caughley 1977). This is especially true with animal tracks in the snow (Thompson et al. 1989; Raphael 1994). Otis (1994) identified the need for population estimators for large-scale areas. He noted that techniques tend to break down for scarce populations, with additional problems arising if distribution patterns are fragmented or movements are dynamic.

Hayashi (1978, 1980) and Hayashi et al. (1979) used hare (*Lepus brachyurus angustidens*) tracks in the snow to estimate population size. Hayashi's (1980) estimator used a probability-sampling scheme, modeled after the Buffon needle problem (Ross 1976), in which the probability of observing a hare track in the snow from an aircraft is assumed to be constant. Reid et al. (1987) estimated river otter (*Lontra canadensis*) density from track counts in the snow along riverine and lake habitat. Their estimator required ground-based sampling and a high sample intensity (> 70%) to obtain precise results.

In this chapter, we discuss two techniques using aerial-based surveys of tracks in snow for estimating low-density mammal populations, focusing on wolves (*Canis lupus*) and wolverines (*Gulo gulo*). One technique employs line transects to intercept animal tracks and the other technique uses quadrat sampling to locate tracks. We present data from numerous surveys, discuss the advantages and disadvantages of each technique, and recommend circumstances when each technique would be most appropriate.

Line-Intercept Sampling

Becker (1991) used ground-based, line-intercept sampling (McDonald 1980; Kaiser 1983) of lynx (*Lynx canadensis*) tracks in the snow, along with movements of radio-collared lynx, to obtain a population estimate. Using aerially surveyed transects for wolverine tracks in the snow, Becker (1991) obtained an estimate of wolverine density. This line-intercept method, which we will refer to as transect intercept probability sampling (TIPS), relies on probability sampling (Horvitz and Thompson 1952) of tracks in the snow that intercept aerial transects. The probability of encountering a track is determined from the length and orientation of the track. We briefly describe the estimator below; Becker (1991) presented a more detailed description of the TIPS design.

Sample Design

Using a TIPS design for tracks in snow requires the following assumptions: (1) all animals move and leave tracks; (2) tracks intersecting the transect are seen, are readily recognizable, and can be followed to the animals' current locations and the locations where the tracks began (e.g., since the end of a snowstorm); (3) animal movements are independent of the sampling process; (4) pre- and post-snowstorm tracks can be distinguished; and (5) the size of the animal group that made the tracks can be correctly enumerated. Additional survey requirements include a rectangular-shaped study area with randomly selected transects that are oriented perpendicular to the long axis (x axis) of the study area, which should be oriented parallel to the general movement patterns of the animals. This orientation will result in a higher probability of encountering animals and a more precise population estimate. Readers should consult Kaiser (1983) for information on surveys with irregular-shaped study boundaries and transects of unequal length.

Transects of equal length are randomly selected from a systematic sample design. This process is repeated to obtain a variance estimate (Kaiser 1983; Becker 1991). To maximize precision, the number of transects per systematic sample should be sufficient to ensure that most of the variability is within the systematic sample, leaving a small amount between the

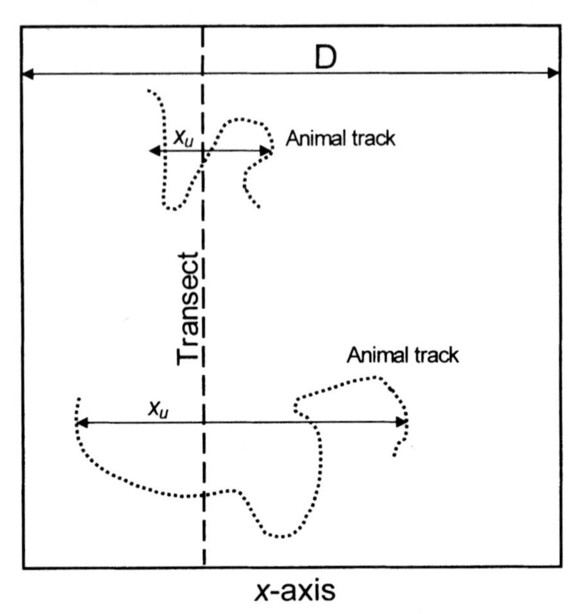

Figure 13.1. Example of distance calculations for the TIPS estimator, where D is the length of the x axis and x_u is the length of track movement relative to the x axis.

samples upon which the variance is calculated (Cochran 1977). During the survey of each transect, all fresh tracks of the species of interest are followed to determine the distance traversed parallel to the x axis (Figure 13.1). Becker (1991) used the following notation: S_i denotes the ith systematic sample; i indexes the systematic sample $i = 1, 2, \ldots r$; q is the number of transects per sample; T_y is the population total; D is the length of the x axis; x_u is the x-axis distance traveled by the uth group; y_u is the size of the uth group; and P_u is the probability of encountering tracks of the uth group.

An unbiased estimator of the population total of the ith systematic sample is

$$\hat{T}_{yi} = \sum_{u \in Si} \left({}^{y_u}\!/\!_{p_u} \right), \tag{13.1}$$

where

$$p_u = \begin{cases} {}^{x_u}\!/\!\left({}^{D}\!/\!_{q} \right) & for \quad x_u \leq \left({}^{D}\!/\!_{q} \right) \\ 1 & otherwise \end{cases} . \tag{13.2}$$

The population estimate and variance are calculated in the normal manner:

$$\hat{T}_y = \sum_i \left(\hat{T}_{yi} \big/ r \right) \tag{13.3}$$

and

$$\hat{\text{Var}}(\hat{T}_y) = \left[\sum_i \left(\hat{T}_{yi} - \hat{T}_y \right)^2 \right] \Big/ \left[r(r-1) \right]. \tag{13.4}$$

Survey Results

TIPS designs have been used to estimate densities of cougars (*Felis concolor*, Van Sickle and Lindzey 1991), wolves (Becker and Gardner 1992; Ballard et al. 1995), and wolverines (Becker 1991; Becker and Gardner 1992; Table 13.1). Becker (1991) also used a TIPS design to estimate lynx density, but we excluded those data from Table 13.1 because the coefficient of variation (CV) contained additional variability associated with using radio-collared lynx. As measured by the CV (Sarndal et al. 1992; Table 13.1), it is difficult to obtain precise results with these surveys. Van Sickle and Lindzey (1991) performed a simulation study of the TIPS design on cougar tracks and reported that the CV improved as the number of transects per systematic sample and/or cougar track density increased. However, their cougar estimate was based on a partial survey (Van Sickle and Lindzey 1991), so we excluded this data set from our analysis of CV. The median CV of the 9 completed surveys , excluding cougars, listed in Table 13.1 was 23.7%, which was exceptionally high. We performed a regression analysis to predict CV using transect density, number of tracks, track density, mean group size, and mean inclusion probability as explanatory variables. We found that the CV decreased as transect density (km transect/1,000 km^2 study area) increased (P = 0.039, R^2 = 0.480; Table 13.2). Based on regression coefficients in Table 13.2, we recommend using a transect density of 255 km/1,000 km^2 to obtain an expected CV of 10%.

The most successful application of the design, based on CV, was a 1991 aerial survey of wolverines in a 2,700-km^2 area of the Talkeetna Mountains, Alaska (Becker and Gardner 1992; Figure 13.2). The survey consisted of six systematic samples (A–F), each containing three 38.9-km transects, and was flown with three Supercubs. Nine sets of wolverine tracks made by 11 wolverines intersected the survey transects. Upon completion of the survey, the three planes did a search of the unsurveyed areas for additional

Table 13.1.

Cougar, wolf, and wolverine densities (no./1,000 km^2) estimated from surveys based on a TIPS design.

Species	Location	Date (mo/yr)	Study area (km^2)	Estimated density (SE)	CV (%)	df	No. groups	Mean inclusion probability	Transect density (km/1,000 km^2)
Cougar[a]	Southcentral UT	2/1989	360	39.44 (17.50)	44.4	NA[b]	5	0.46	5.56
Wolf[c]	Alphabet Hills, AK	2/1990	4,556	14.69 (5.71)	38.9	3	3	0.19	2.63
Wolf[c]	Lake Louise, AK	2/1990	5,201	7.03 (5.19)[d]	73.8	3	4	0.23	3.08
Wolf[c]	Alphabet Hills, AK	3/1990	5,335	23.28 (6.15)	26.4	6	8	0.35	6.56
Wolf[e]	Kobuk River, AK	5/1990	6,464	7.84 (1.86)	23.7	6	6	0.57	5.41
Wolf[e]	Minto Flats, AK	3/1991	5,011	6.66 (1.41)	21.2	6	4	0.33	6.98
Wolf[f]	Fog Lakes, AK	3/1992	4,701	3.78 (1.48)	39.1	6	5	0.14	7.45
Wolf[g]	Umiat, AK	4/1992	10,378	5.11 (0.80)	15.7	7	5	0.27	3.87
Wolverine[h]	Chugach Mtns., AK	3/1988	1,871	5.18 (1.05)	20.3	3	4	0.44	6.41
Wolverine[c]	Talkeetna Mtns., AK	2/1991	2,700	4.69 (0.61)	13.0	5	9	0.44	6.67

[a] Van Sickle and Lindzey (1991).
[b] Degrees of freedom were not available because a *t*-distribution was not used.
[c] Becker and Gardner (1992).
[d] Corrects arithmetic error in Becker and Gardner (1992).
[e] Ballard et al. (1995).
[f] E. Becker, unpublished data, Alaska Department of Fish and Game.
[g] G. Carroll and E. Becker, unpublished data, Alaska Department of Fish and Game.
[h] Becker (1991).

Table 13.2.

Regression coefficients used to obtain predicted coefficient of variation values for TIPS designs.

Variable	Regression coefficient	SE	Pr(> |t|)
Intercept	66.11	14.92	0.003
Transect density	–0.220	0.087	0.039

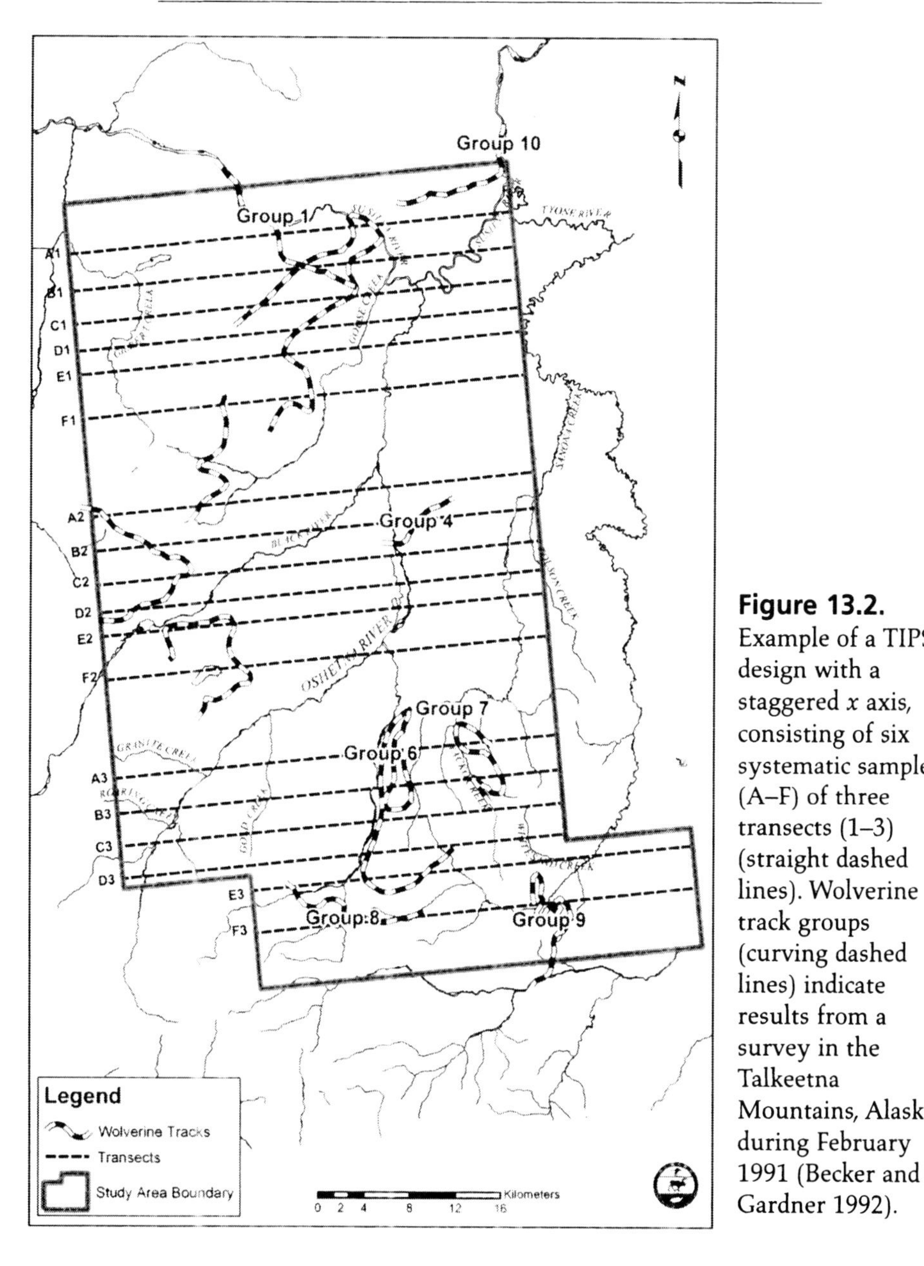

Figure 13.2. Example of a TIPS design with a staggered x axis, consisting of six systematic samples (A–F) of three transects (1–3) (straight dashed lines). Wolverine track groups (curving dashed lines) indicate results from a survey in the Talkeetna Mountains, Alaska, during February 1991 (Becker and Gardner 1992).

wolverine tracks. The goal was to test the TIPS assumptions by trying to find additional wolverine tracks that intersected completed transects. We located and recorded an additional lone wolverine track that did not intersect any of the transects. The survey estimate of 12.7 ($\widehat{SE} = 1.64$) wolverines was very close to the 12 known to be in the study area and resulted in an estimated CV of 13.0%. Because we were certain that at least 12 wolverines were present in the area at the time of the survey, 12 would be a valid lower limit for confidence intervals constructed from this survey (Becker and Gardner 1992).

Sample Design Alternatives

Here we describe alternatives to the previous sample design. These alternatives may provide a better opportunity to meet assumptions, which would produce more reliable estimates of abundance or density.

Use of Radio Telemetry

Some species are so difficult to track, especially from an airplane, that one cannot meet the assumption that all tracks intersecting a transect can be observed and followed. Under these conditions, it is still possible to obtain a population estimate using radio telemetry. This can be done by estimating the distance traveled along the x axis by the population ($\hat{T}_x$) and dividing that estimate by the average x-axis distance moved by a sample of radio-collared individuals. To estimate the x-axis distance moved by the population, no actual distance measurements are needed because the x_u term replaces y_u in equation (13.1) and is canceled by the x_u term in the denominator (by definition p_u (equation 13.2) contains an x_u term). Determining the number of distinct individuals whose tracks intersect transects often requires some backtracking on the ground. Readers requiring additional details should consult the lynx example in Becker (1991:734). Implementation of this design with GPS radio-collars, programmed with this application in mind, would be even more efficient.

Staggered and Stacked Designs

The normal rectangular shape used to define the study area can be modified to better fit the area under investigation. The easiest is a staggered x axis, which was used in the February 1991 wolverine survey of the Talkeetna Mountains, Alaska (Becker and Gardner 1992; Figure 13.2). In surveying large study areas, data quality is often enhanced by using transects of moderate length (e.g., 20–35 km).

To avoid using long (> 35 km) transects in large study areas, it may be necessary to stack the x axes (Figure 13.3). Calculation of x_u is straightforward unless the track is contained in more than one layer. Projecting the track location into the first panel of the systematic sample will ensure that the correct distance is obtained. For example, suppose the length of the x axis, D, equals 200 km, with 100 km in each of two layers, and there are seven systematic samples of four transects. If an observed track moves from 50 to 62 km on the x-axis coordinate of the first layer (0–100 km) then crosses into the second layer (100–200 km) and moves from coordinates 162 to 155 km, we might naively measure the x-axis distance moved at 19 km. However, when the coordinates are projected back into the first panel (0–50 km), we obtain the correct x_u of 12 km to apply to equation (13.1).

Some study areas may have travel routes and corridors that intersect at right angles to one another. To maximize the probability of encountering a track, we could run transects in both directions. Additionally, if the study area was large, we would want to use transects of moderate lengths. This situation can be accommodated by using a stacked x axis in two dimensions (e.g., x and y axes), where the study area is broken into squares with a length equal to that of the transect. Within the square, transects will be oriented in both directions. To calculate x_u, the track is projected down to both the x and y axes, and the coordinates are then projected into the first panel to obtain the correct x_u. Figure 13.4 is an example of such a design that was used for a wolf survey at Fog Lakes in southcentral Alaska (E. Becker, unpublished data).

Multiday Surveys

We designed TIPS surveys to be completed in 1 day. A multiday survey can be performed by breaking the study area into subareas, each of which can be surveyed in a single day. The additional assumption is that animals do not move from a nonsurveyed subsurvey area to a surveyed one between sample periods (e.g., at night before the next survey). Animals that would move between survey periods would have no probability of being observed in the survey, and this would violate probability-sampling assumptions.

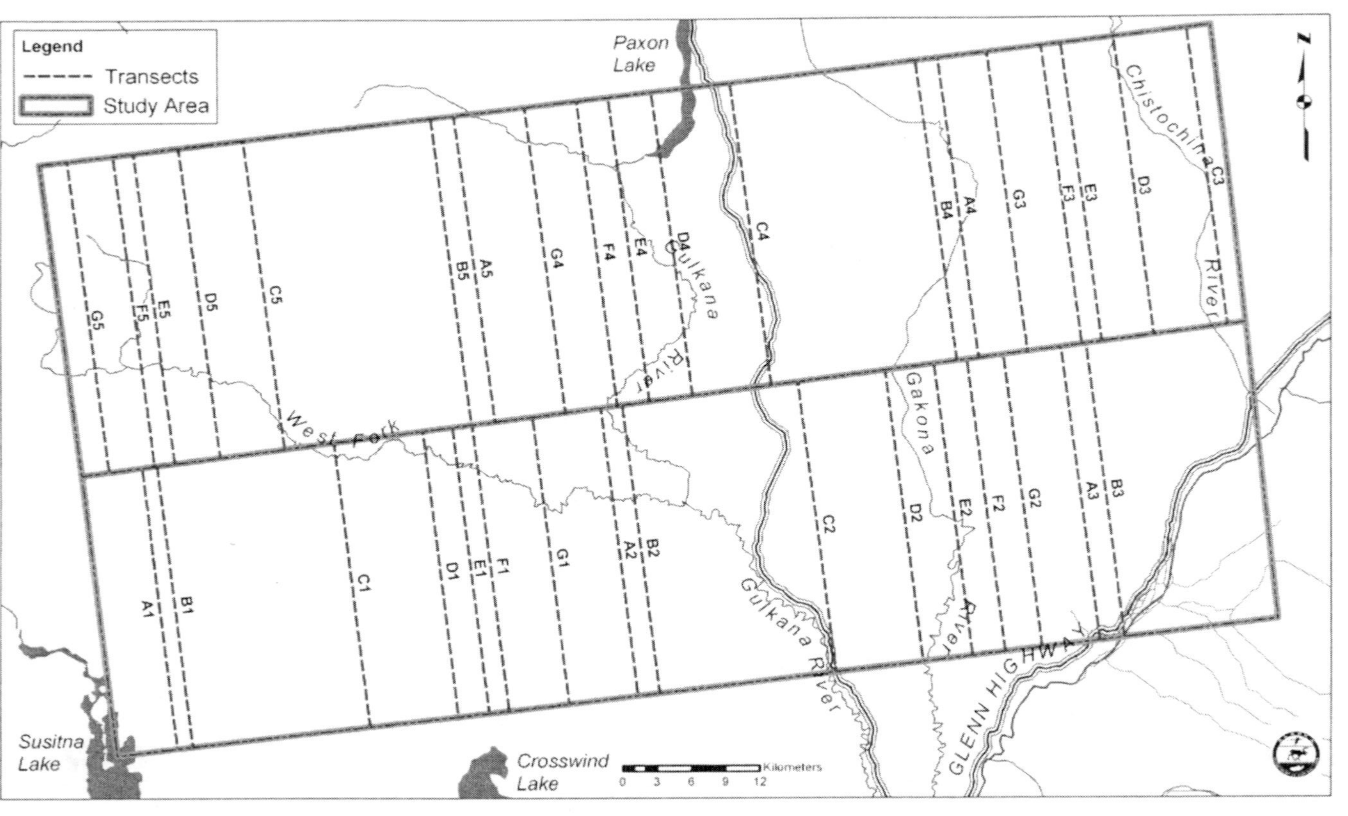

Figure 13.3. Example of a TIPS design with two stacked *x* axes, consisting of seven systematic samples (A–G) of five transects (1–5) (dashed lines).

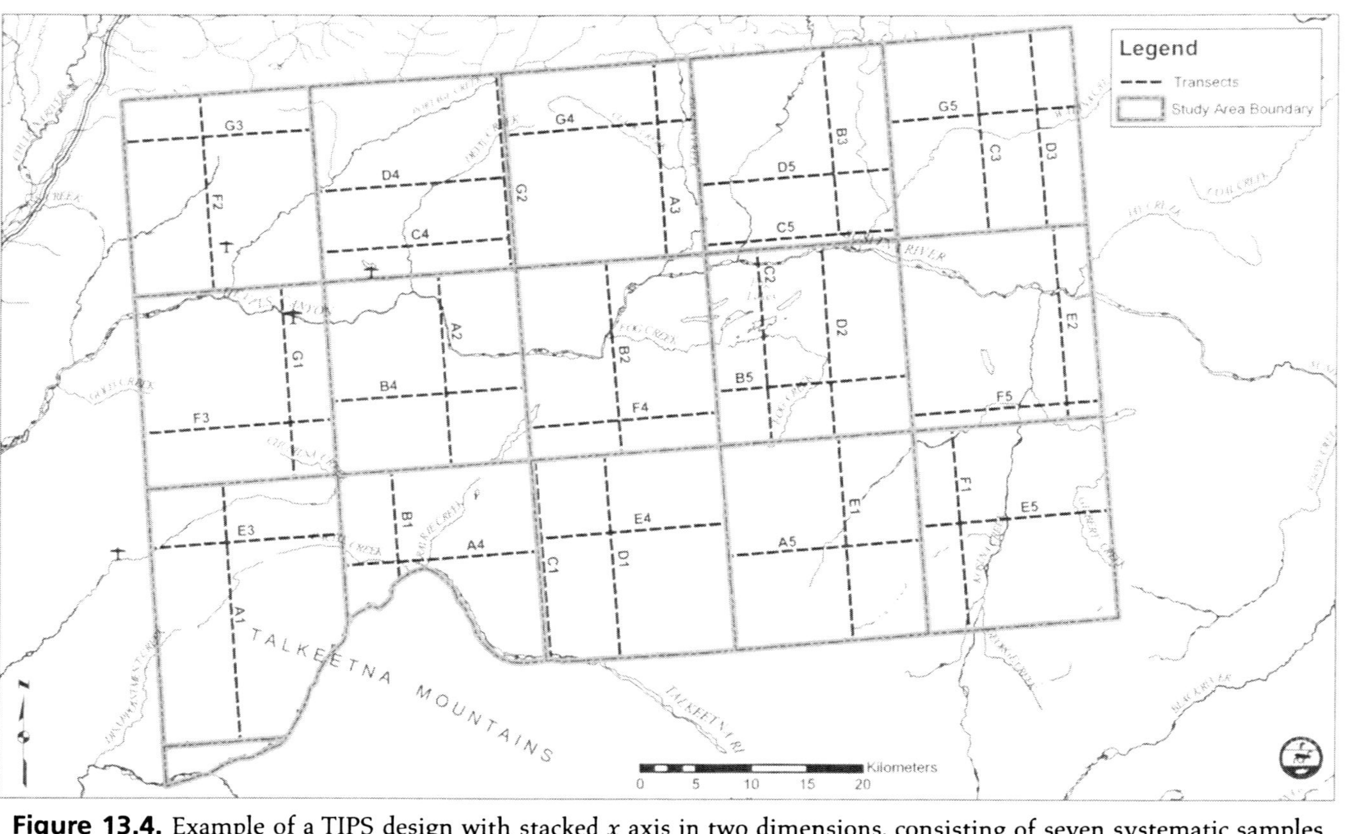

Figure 13.4. Example of a TIPS design with stacked *x* axis in two dimensions, consisting of seven systematic samples (A–G) of five transects (1–5) (dashed lines).

Stratified Network Sampling

Difficulties in obtaining precise results with the TIPS survey for wolves, as noted by large CVs in Table 13.1, highlighted the need for further technique development. Additional requirements for wolf surveys included the ability to deal with moderate to heavy track deposition of prey species that could reduce the sightability of target species' tracks, thick overstory canopy, incorporation of wolf distribution knowledge into the survey design, and the flexibility to survey large study areas over several days. Becker et al. (1998) developed the sample unit probability estimator (SUPE) to meet these requirements.

A SUPE design should be considered when the investigator has knowledge of the spatial distribution of a scarce or low-density population that makes tracks in the snow. A SUPE is a form of stratified network (snowball) sampling design (Thompson 1992; Becker et al. 1998). It requires a small, fixed-wing aircraft or helicopter for finding and following fresh tracks in snow and thereby obtaining data for calculating the probability of encountering those tracks during the survey. This information then is used in a probability estimator to derive a population estimate (Horvitz and Thompson 1952).

Becker et al. (1998:969) listed the sample design assumptions as:

> *(1) all animals of interest move during the course of the study; (2) their tracks are readily recognizable from a small, low-flying aircraft; (3) tracks are continuous; (4) movements are independent of the sampling process; (5) pre- and post-snowstorm tracks can be distinguished; (6) post-snowstorm tracks in searched sample units (SUs) are not missed; (7) post-snowstorm tracks found in selected SUs can be followed (forward and backward) to determine, without error, all SUs containing those tracks; and (8) group size is correctly enumerated.*

A study area may be so large that not all the selected sample units can be surveyed in 1 day. In this situation, we need two additional assumptions: (1) animals do not move from unsampled to sampled areas and they leave no fresh tracks in the unsampled areas; and (2) no animals are double counted by moving from sampled to unsampled areas. Difficult but localized snow conditions, such as patchy, drifting, or hard-packed snow, may obscure short sections of track. However, an unbiased estimator can be obtained if a one-to-one correspondence can be established between the track segments and the animals of interest that made them (Becker 1991).

Sample Design

Sample units first are grouped into unique strata denoting the relative likelihood of observing a fresh track of the target species. The stratification is based upon knowledge of harvest patterns, abundance, and distribution of the target species along with the location and abundance of its prey base (Becker et al. 1998). A simple random sample without replacement of SUs from each stratum is selected for survey from a small, low-flying airplane (Becker et al. 1998) or helicopter (Patterson et al., in press) with a pilot and biologist team experienced in tracking the target species.

The sampling procedure is initiated shortly after a snowstorm. We recommend commencing within 24–48 hours for wolves and 12–24 hours for wolverines to allow time for animals to make tracks. We have found it easier to enumerate wolverine tracks after waiting a shorter period of time. Although both species can quickly travel long distances, wolverines are more likely than wolves to circle back to a hole or den site, which can cause confusion in determining the number of animals making the track. For this reason, it is best to try to complete wolverine surveys within 2–3 days, whereas wolf surveys can easily extend to 4–5 days. Sampling involves searching selected SUs for fresh tracks (i.e., tracks made since the snowstorm) of the target species. These tracks are followed in both directions to determine which sample units contain the fresh tracks and the number of animals that made those tracks. Search intensity depends upon the overstory, lighting conditions, and amount of track deposition from other animals, but it should be sufficient to ensure that the model assumptions are met. Becker et al. (1998) provided a more detailed description of the sampling procedure and ways to increase sampling efficiency.

SUs are usually squares between 10 and 41 km^2, although any shape may be used and they do not have to be uniform in size. The size of the SU depends upon the relationship between the information available for stratification and the general movement patterns of the target species. If the survey is in mountainous terrain, one should consider a small square (10–25 km^2) or a rectangular sample unit with the short axis oriented to minimize time-consuming elevational movements. Becker et al. (1998:970) stated, "search efficiency generally increases with a larger SU because of less time spent per area determining SU boundaries. Study area size will be limited by the ability to obtain good survey conditions over the entire region and the ability to complete the sample design within a weather window that allows the sample design assumptions to be met." In some cases, study area boundaries may have to be adjusted to exclude areas with inadequate snow conditions for the technique, which can be done pre- or postsurvey.

Becker et al. (1998) used the rule that if more than half of the track was out of the study area that observation was not used in the estimate (population membership rule). An alternative would be to plot the track location with a GPS and determine the proportion of the track within the study area. This proportion would be multiplied by the group size to obtain a group size estimate for the "members of the study area," which is the group size number used to obtain the population estimate. The purpose of these membership rules is to deal with animals whose tracks bisect the study boundary. The purpose of the SUPE survey is to estimate the number of animals in the study area at the time of the survey (i.e., a snapshot of the number of animals in the area). Although this approach may exclude animals that use the study area but were outside of the area at the time of the survey or may include transitory individuals, it results in an unbiased calculation of animal density.

If more than one group of fresh tracks intersects an SU, the data should be recorded separately if the pilot-biologist team can separate the two groups along their entire set of tracks; otherwise, they should treat them as one group (Becker et al. 1998). Becker et al. (1998) should be consulted for details on how to implement a SUPE survey over large areas that require several days to complete.

Becker et al. (1998:970) used the following notation to estimate the population and its variance in a SUPE:

> *T_y is the population total, u and v index the animal group observations, y_u is the group size (1,2,3,...) for the uth group, r is the number of groups whose tracks were in selected SUs, p_u is the inclusion probability (i.e., the probability that fresh tracks from the uth group are observed with this sample design), and p_{uv} is the joint inclusion probability (i.e., the probability that both the uth and v^{th} animal groups are observed in this sample design).*

Using standard probability sampling results (Horvitz and Thompson 1952; Thompson 1992), the population estimate and variance are as follows:

$$\hat{T}_y = \sum_{u=1}^{r} \frac{y_u}{p_u} \quad (13.5)$$

and

$$\hat{\text{Var}}(\hat{T}_y) = \sum_{u=1}^{r} \frac{(1-p_u)}{p_u^2} y_u^2 + 2\sum_{u=1}^{r} \sum_{v=u+1}^{r} \left(\frac{1}{p_u p_v} - \frac{1}{p_{uv}} \right) y_u y_v . \quad (13.6)$$

In probability sampling, "unlikely" observations make a larger contribution to the population estimate than "likely" observations, since the inclusion probability (p_u) is in the denominator of the estimator (equation (13.5)). Examination of the variance formula in equation (13.6) highlights the importance of using a design that will maximize the size of the inclusion probabilities, especially for larger group sizes. Becker et al. (1998) presented formulas to assess the contribution of each observation to the variance.

Using network-sampling results (Thompson 1992), Becker et al. (1998) calculated the following inclusion probabilities:

$$p_u = 1 - \prod_{h=1}^{H} \binom{M_h - m^*_{hu}}{n_h} \Big/ \binom{M_h}{n_h} \qquad (13.7)$$

and

$$p_{uv} = p_u + p_v - 1 + \prod_{h=1}^{H} \binom{M_h - m^*_{hu} - m^*_{hv} + m^*_{huv}}{n_h} \Big/ \binom{M_h}{n_h}. \qquad (13.8)$$

The notation definitions follow Becker et al. (1998:970), where

Π denotes the multiplication operator, h indexes the number of strata (i.e., $h = 1,2,...H$), M_h is the number of SUs in the hth stratum, n_h is the number of SUs searched in the hth stratum, m^*_{hu} is the number of SUs in the hth stratum that contain tracks from the uth group of animals, m^*_{hv} is the number of SUs in the hth stratum that contain tracks from the vth group of animals, and m^*_{huv} is the number of SUs that contain tracks from the uth and v^{th} group of animals. The combinatorial notation $\binom{M_h}{n_h}$ denotes the number of ways to pick n_h things from M_h and is calculated as $\frac{M_h!}{(M_h - n_h)!n_h!}$, where $n_h! = n_h(n_h - 1)(n_h - 2)...(1)$ and, by definition, $0! = 1$.

Becker et al. (1998:971) presented a numerical example of these calculations. Confidence intervals can be constructed using a t-distribution with $r - 1$ degrees of freedom (Thompson 1992).

Survey Results

SUPE-estimated wolf densities ranged from 4.1 to 25.4 wolves/1,000 km^2, with the highest estimate occurring in Algonquin Provincial Park, Ontario, Canada (Patterson et al., in press; Table 13.3). The highest estimated wolf density in Alaska was 12.1, which occurred in a 1995 survey of the McGrath

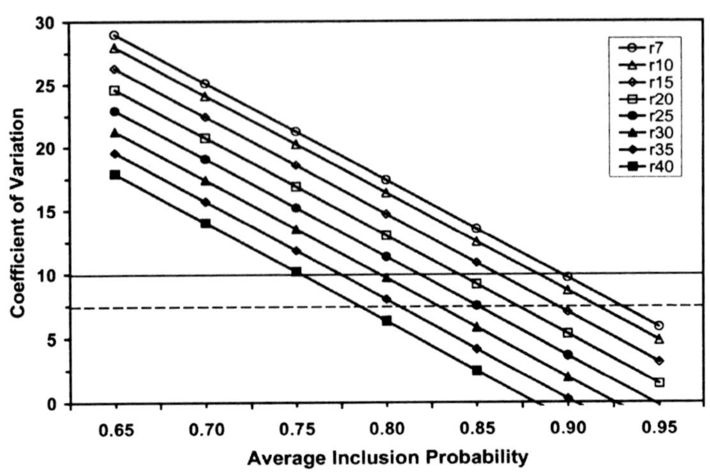

Figure 13.5. Predicted coefficients of variation (CV) for a SUPE design based on a regression analysis of mean inclusion probabilities and number of groups (r) observed as predictors. CV levels of 10% (solid line) and 7.5% (dashed line) are shown for reference.

area of interior Alaska. Survey efficiency was measured by CV and ranged between 6.7 and 25.0% (Table 13.3). A regression analysis of these data ($P < 0.0001$, $R^2 = 0.91$; Table 13.4) indicated that more efficient surveys have high mean inclusion probabilities and/or many group sightings (r).

For planning SUPE surveys, readers can calculate expected CV values from Table 13.4 or obtain them from Figure 13.5. Based on this analysis and our experience, we give recommendations on sampling fractions for three-strata SUPE surveys in Table 13.5 for different expected levels of groups sighted (r). The median of the mean inclusion probabilities listed for the 12 wolf surveys in Table 13.3 is 0.82, which implies that 25 group sightings will be needed to obtain a CV of 10% (Figure 13.5). If this is not feasible, more sampling effort will need to be expended to increase the mean inclusion probability. This recommendation is based on wolf data. Theoretically, one should expect more precise estimators for wolverines, cougars, and other non-pack species. We will continue to evaluate CVs for other species as results become available.

Sample Design Alternatives

Here we discuss two modifications to the previous design that may produce more reliable estimates of abundance or density. The usefulness of these alternatives will depend on the context of the survey.

Table 13.3.

Wolf density estimates (no./1,000 km²) from surveys based on SUPE designs, presented in order by CV

Location[a]	Date (mo/yr)	Study area (km²)	Estimated density (SE)	CV (%)	No. groups	Mean inclusion probability	Sampling fraction
GMU13A-West, AK[b]	2/2001	5,967	7.93 (0.53)	6.68	8	0.92	0.43
GMU13A, AK[c]	2/1996	10,360	6.37 (0.49)	7.69	13	0.90	0.43
GMU13A-West, AK[c]	2/1996	4,931	9.90 (0.87)	8.79	10	0.90	0.47
Umiat, AK[d]	4/1994	10,443	4.12 (0.45)	10.92	8	0.86	0.35
GMU21D, AK[e]	3/1994	31,373	8.16 (0.91)	11.15	37	0.74	0.31
Algonquin Park, ON[f]	2/2002	3,425	25.40 (3.29)	12.95	17	0.80	0.33
GMU21B, AK[g]	3/1996	12,616	5.42 (0.75)	13.84	14	0.83	0.34
GMU13A-West, AK[b]	3/2002	8,329	7.40 (1.10)	14.86	11	0.80	0.40
GMU19D, AK[h]	2/1995	13,468	12.14 (1.96)	15.90	17	0.80	0.32
GMU19D, AK[h]	3/1997	10,443	4.17 (0.69)	16.55	9	0.84	0.31
GMU13B, AK[c]	3/1996	7,128	4.65 (0.85)	18.06	5	0.81	0.38
GMU14C, AK[i]	2/1995	3,590	7.65 (1.91)	24.97	5	0.70	0.42

[a] GMU = Game Management Unit; Alaska is divided into 26 GMUs, most of which contain subunits.
[b] H. Golden, unpublished data, Alaska Department of Fish and Game.
[c] E. Becker, unpublished data, Alaska Department of Fish and Game.
[d] G. Carroll and E. Becker, unpublished data, Alaska Department of Fish and Game.
[e] Becker et al. 1998.
[f] Patterson et al., in press.
[g] Huntington and Becker 1998.
[h] J. Whitman and E. Becker, unpublished data, Alaska Department of Fish and Game.
[i] R. Sinnott and E. Becker, unpublished data, Alaska Department of Fish and Game.

Table 13.4.

Regression coefficients used to obtain predicted coefficient of variation values for SUPE designs.

Variable	Regression coefficient	SE	Pr(> |t|)
Intercept	81.50	7.00	<0.0001
No. of groups (*r*)	–0.34	0.06	0.0004
Mean inclusion probability	–77.19	8.20	<0.0001

Table 13.5.

Recommended sampling fraction, by stratum, for SUPE designs.

Expected no. observations (*r*)	Stratum		
	High	Medium	Low
7–10	0.67	0.40	0.20
11–25	0.63	0.38	0.18
> 25	0.60	0.35	0.16

Multiday Surveys

SUPE surveys can be conducted over many days if necessary. For example, Becker et al. (1998) needed 10 days to complete a wolf survey of a large (31,373 km^2) study area. During the course of a multiday survey, it is important to note environmental conditions, such as wind and the effect it is having on the tracks. After 2 days of surveying, the initial tracks will be 3–4 days old and may have noticeably deteriorated due to weather conditions, i.e., tracks lose their sharp, crisp edges and look more diffuse. Additionally, frost and debris may cover the bottom of the track. Starting about the third day of the survey, the definition of a fresh track should be modified to include tracks with the freshest appearance as well as those with some moderate signs of age. Each day after surveying, the crew should discuss their definition of a fresh track to ensure consistency among observers and to avoid crews following tracks that are much too old. Weathering will occur at different rates between habitats, with more sheltered habitats, such as conifer forests, having the slowest weathering. If a track in nonsheltered habitat has crisp edges, it is a fresh track. Tracks that look fresh in sheltered habitat but appear old in nonsheltered habitat are not considered fresh. Track condition can be assessed upon finding the animals; it may be possible to approximately age the tracks if beds were found while backtracking. If a fresh snowstorm is encountered during the survey, the survey should be suspended in that area for 24 hours to give the animals time to make new tracks.

Incomplete Surveys

Large study areas may not receive uniform snowfall, resulting in portions of those areas having insufficient fresh snow to conduct the survey. In such circumstances, it is possible to delete the portions that cannot be surveyed, redefine the study area, and proceed with the survey as normal. However, the redefined area must be surveyed as a complete block without missing any selected sample units. Upon completion of the survey, sampling fractions will have to be recomputed for the new survey area. Discovery of survey conditions usually occurs during the survey so that a decision must be made in flight and coordinated among all survey crews. We have found it helpful to have one crew designated to coordinate all survey activities by staying in radio contact with other crews and maintaining a master map of quadrats surveyed.

TIPS versus SUPE Designs

Selection of the appropriate survey design must be based on several factors, such as survey area size, overstory cover, track sightability, the ability of the survey team to stratify the area, observer experience, and the best flight pattern for finding tracks. We designed TIPS surveys to be completed in 1 day, but multiday surveys are possible. However, even though the basic, multiday assumptions are the same, a SUPE design is better suited for use in large areas where multiple survey days are required because it more effectively handles partial surveys with incomplete snowfall. It is easier to meet assumptions of the SUPE design if vegetation cover, track deposition by other species, or other factors make track sightability of target species problematic. A TIPS design does not take into account existing knowledge of the relative likelihood of encountering tracks in different locations, whereas a SUPE design uses this knowledge to increase the precision of the estimate. SUPE designs generally allow for greater scrutiny of tracks than the TIPS design, which would be particularly valuable to less experienced observers or for tracks that are difficult to identify from the air. Assuming a survey with the absence of the above problems, Becker et al. (1998) hypothesized that a TIPS design would be the best one for wolverines, based on the assumption that a linear search pattern is the most efficient way to find their tracks.

To test this hypothesis, we used data from the previously discussed 1991 wolverine survey of the Talkeetna Mountains, Alaska (Becker and Gardner 1992), to simulate both TIPS (Figure 13.2) and SUPE (Figure 13.6) surveys. For purposes of the simulation, we assumed that only 12 observed wolverines were present in the study area and that we correctly recorded track locations and group sizes. We stratified the area into medium-low and high strata, based on a SUPE design by Golden (1997) for that area, for a simulated SUPE survey using 112, 23.5-km^2 (4.86 × 4.86 km) sample units. The stratification scheme consisted of 61 high and 51 medium-low sample units. Based on our experience, observers in a Supercub can survey approximately 15 of these sample units per day. Using a 1-day sampling effort of three Supercubs would result in 45 sample units being searched. To meet these requirements, we partitioned the sampling effort for sample units to be surveyed as 57.4% (35) in the high stratum and 20% (10) in the medium-low stratum. Based on 10,000 simulations, we obtained a CV of 17.8%. To compare this statistic with that resulting from a TIPS design, we simulated the 1991 TIPS survey with 10,000 iterations of a TIPS survey consisting of six systematic samples with three transects per sample and obtained a CV of 15.1%. For this particular data set, the TIPS design was 18% more efficient than the specified SUPE design.

The reader should keep in mind that this gain in efficiency of the TIPS design is offset by greater difficulty in meeting design assumptions and by more restrictions on the study area size than in the SUPE design. Additionally, wolverines that emerge from a den and do not travel far are not problematic with the SUPE design, assuming that the track would be seen if that sample unit was searched. The worse case scenario would be for the sample unit to be in the low strata and the resulting inclusion probability to be in the 0.16–0.20 range if the recommended sampling guidelines were followed. Conversely, this type of movement would make the TIPS estimator very unstable. Encountering such an occurrence on a TIPS survey would result in an extremely large density estimate with very low precision due to an extremely small inclusion probability. We have yet to encounter this low-movement situation during surveys, although we have seen it on rare occasions during wolverine capture operations.

A SUPE design following the recommended sampling fractions listed in Table 13.5 would have resulted in 67% of the high and 30% of the

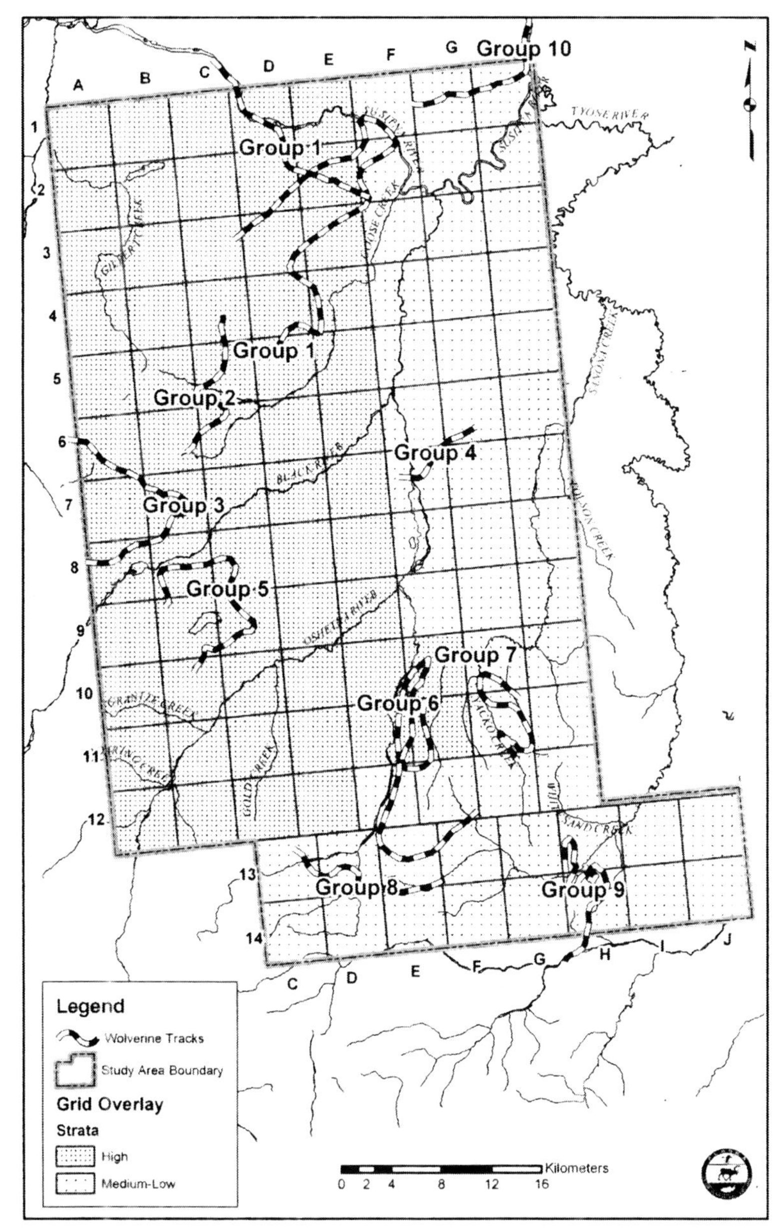

Figure 13.6. Example of a SUPE design with two strata. Wolverine track groups (curving dashed lines) used in this simulation indicate results from a survey in the Talkeetna Mountains, Alaska, February 1991 (Becker and Gardner 1992).

medium-low strata being sampled, and it would have required an additional plane to implement. This would have increased the cost from approximately $3,150 to $4,200 (calculated using three Super Cubs for 7 hours each at $150/hour in the Talkeetna Mountains). However, based on 10,000 simulations of this design, we obtained a CV of 12.2%, compared with 15.1% for the TIPS design. Therefore, this SUPE design was 24% more efficient than the TIPS design.

The TIPS design would have confidence intervals constructed using a *t*-statistic with degrees of freedom (df) based on the number of systematic samples minus 1 (5 in our simulations). The df of a SUPE design is the number of track sightings (r) encountered minus 1. Based on average values from the simulations, the SUPE design would use a *t*-statistic with the df equal to 5.8 and 6.6, respectively, for the two SUPE simulations. The advantages would be more pronounced in favor of the SUPE design as the number of track sightings increased. In many instances, the precision in the lower bound of the confidence intervals can be improved by using the known number of animals observed during the survey. For this to work, opportunistic observations have to be tracked to determine if they are indeed previously unobserved animals. Population membership rules also would apply to these observations.

Additional SUPE advantages are increased flexibility in dealing with weather problems (e.g., fog and poor light) and in handling logistic problems (e.g., delays in obtaining survey aircraft) that sometimes hinder the start of a survey. The SUPE design also is easier to use than the TIPS in steep or rugged terrain where it would be difficult to fly straight transects. Although more expensive than fixed-wing aircraft, helicopters can overcome sightability problems due to overstory cover (Patterson et al., in press) and enable the meeting of SUPE design assumptions. It is unlikely that the use of a helicopter in these situations would allow the TIPS design assumptions to be met.

A copy of the updated program SUPEPOP, which performs the calculations needed to obtain estimates with a SUPE design, can be found at the following FTP site: ftp://ftpr3.adfg.state.ak.us/MISC/PROGRAMS/SUPEPOP/. This FTP site also contains files with instructions for wolf and wolverine SUPE surveys, wolverine TIPS surveys, and appropriate data forms.

Acknowledgments

The Alaska Department of Fish and Game and Federal Aid in Wildlife Restoration provided funding for this project. We wish to thank all

ADF&G staff and those from other agencies who participated in conducting surveys reported here as unpublished data. We especially want to thank all pilots who participated in the surveys and acknowledge their skill and expertise in identifying and following tracks and for their safe handling of the aircraft. We greatly appreciate the assistance of Elizabeth Solomon, ADF&G, for preparing the figures in this chapter.

References

Ballard, W. B., M. E. McNay, C. L. Gardner, and D. J. Reed. 1995. Use of line-intercept track sampling for estimating wolf density. pp. 469–480 in L. N. Carbyn, S. H. Fritts, and D. R. Seip, eds., *Ecology and Conservation of Wolves in a Changing World*. Canadian Circumpolar Institute, Edmonton, Alberta.

Becker, E. F. 1991. A terrestrial furbearer estimator based on probability sampling. *Journal of Wildlife Management* 55:730–737.

Becker, E. F., and C. Gardner. 1992. *Wolf and Wolverine Density Estimation Techniques*. Alaska Department of Fish and Game and Federal Aid in Wildlife Restoration, Research Progress Report, Grant W-23-5, Juneau, Alaska.

Becker, E. F., M. A. Spindler, and T. O. Osborne. 1998. A population estimator based on network sampling of tracks in the snow. *Journal of Wildlife Management* 62:968–977.

Caughley, G. 1977. *Analysis of Vertebrate Populations*. Wiley, New York.

Cochran, W. G. 1977. *Sampling Techniques*, 3rd ed. Wiley, New York.

Golden, H. N. 1997. *Furbearer Management Technique Development*. Alaska Department of Fish and Game, Research Progress Report, Federal Aid in Wildlife Restoration Grant W-24-5, Juneau, Alaska.

Hayashi, C. 1978. A new statistical approach to estimating the size of an animal population: The case of a hare population. *Mathematical Scientist* 3:117–130.

_____. 1980. Some statistical methods of estimating the size of an animal population. pp. 85–91 in K. Matusita, ed., *Recent Developments in Statistical Inference and Data Analysis*. North-Holland, New York.

Hayashi, C., T. Komazawa, and F. Hayashi. 1979. A new statistical method to estimate the size of animal population—Estimation of population size of hare. *Annals of the Institute of Statistical Mathematics* 31:325–348.

Horvitz, D. G., and D. J. Thompson. 1952. A generalization of sampling without replacement from a finite universe. *Journal of the American Statistical Association* 47:663–685.

Huntington, O. H., and E. F. Becker. 1998. *1996 Nowitna NWR Wolf Census: Koyukuk/Nowitna National Wildlife Refuge Complex Alaska, Game Management Unit 21B*. U.S. Fish and Wildlife Service Progress Report FY97-05, U.S. Fish and Wildlife Service, Galena, Alaska.

Kaiser, L. 1983. Unbiased estimation in line-intercept sampling. *Biometrics* 39:965–976.

McDonald, L. L. 1980. Line-intercept sampling for attributes other than cover and density. *Journal of Wildlife Management* 44:530–533.

Otis, D. L. 1994. Introductory comments. *Transactions of the North American Wildlife and Natural Resources Conference* 59:147–148.

Patterson, B. D., N. W. S. Quinn, E. F. Becker, and D. B. Meier. In Press. Estimating wolf densities in forested areas using network sampling of tracks in the snow. *Wildlife Society Bulletin.*

Raphael, M. G. 1994. Techniques for monitoring populations of fishers and American martens. pp. 224–240 in S. W. Buskirk, A. Harestad, M. G. Raphael, and R. A. Powell, eds., *Martens, Sables, and Fishers: Biology and Conservation.* Cornell University, Ithaca, New York.

Reid, D. G., M. B. Bayer, T. E. Code, and B. McLean. 1987. A possible method for estimating river otter, *Lutra canadensis,* populations using snow tracks. *Canadian Field-Naturalist* 101:576–580.

Ross, S. 1976. *A First Course in Probability.* Macmillan, New York.

Sarndal, C.-E., B. Swensson, and J. Wretman. 1992. *Model Assisted Survey Sampling.* Springer-Verlag, New York.

Sudman, S., M. G. Sirken, and C. D. Cowan. 1988. Sampling rare and elusive populations. *Science* 240:991–996.

Thompson, I. D., I. J. Davidson, S. O'Donnell, and F. Brazeau. 1989. Use of track transects to measure the relative occurrence of some boreal mammals in uncut forest and regeneration stands. *Canadian Journal of Zoology* 67:1816–1823.

Thompson, S. K. 1992. *Sampling.* Wiley, New York.

Van Sickle, W. D., and F. G. Lindzey. 1991. Evaluation of a cougar population estimator based on probability sampling. *Journal of Wildlife Management* 55:738–743.

14

Sampling Rockfish Populations: Adaptive Sampling and Hydroacoustics

Dana H. Hanselman
and Terrance J. Quinn II

At least 96 species of rockfish (genus *Sebastes*) inhabit the northern Pacific Ocean (Love et al. 2002). Many of the more abundant rockfish that inhabit the continental slope share several important features related to sampling. These features include a prominent swim bladder, deep demersal existence, and patchy distributions. Capturing rockfish from great depth induces certain mortality from bursting of the swim bladder, which makes mark-recapture studies infeasible. The aggregated distribution of some of the major commercial slope rockfish makes them difficult to sample precisely under conventional designs. We illustrate some of these problems and some potential solutions with the Pacific ocean perch (*Sebastes alutus*).

S. alutus is the dominant commercial rockfish species in Alaska (Hanselman et al. 2003a). It is a very long-lived species (> 80 years), and its spatial distribution is highly aggregated, which makes the species particularly vulnerable to overfishing. An accurate biomass estimate of *S. alutus* is needed to properly assess the status of the stock. The multispecies groundfish trawl surveys conducted by the National Marine Fisheries Service (NMFS) have not adequately sampled *S. alutus* due to their highly aggregated distribution. This poor sampling occurs both because of the small area inhabited by *S. alutus* when compared to more uniformly distributed flatfish and because much of its primary bottom habitat is inaccessible to standard survey gear (Lunsford 1999). As a result, large fluctuations in biomass estimates occur with extremely wide

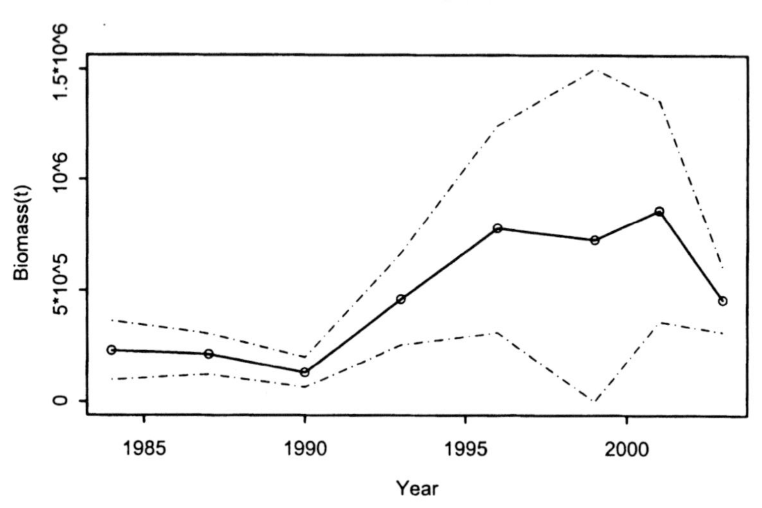

Figure 14.1. Survey biomass estimates of *S. alutus* in the Gulf of Alaska from NMFS groundfish trawl surveys. Dashed lines represent 95% confidence bands.

confidence intervals (Figure 14.1). *S. alutus* is a slow-growing, long-lived fish with moderate fecundity; therefore, these rapid upward and downward changes in biomass in the absence of heavy exploitation are unlikely. Adaptive cluster sampling (ACS; Thompson 1990) is one technique that has been examined to improve the precision of biomass estimates. Field studies of ACS showed some improvement in precision but perhaps not enough to justify the additional sampling effort involved (Hanselman et al. 2003b).

This study examines two related parts of this rockfish sampling problem. First, we briefly review the results of two ACS experiments for this species and use simulation to explore conditions under which ACS performs poorly. Second, we examine the usefulness of hydroacoustic data for improving biomass estimates. These data were recorded from vessels used in bottom-trawl surveys by the NMFS in the Gulf of Alaska during summer 2001 and in the Aleutian Islands and Bering Sea during summer 2002. We examine the quality of the direct relationship between echo-integrated hydroacoustic tracks and survey catches and then develop a model that can be used to predict catches for use in several sampling design experiments. These include an ACS design, a double sampling design, and the use of hydroacoustics for stratification.

ACS Simulation

ACS has recently been the focus of much attention (see Thompson and Seber 1996; *Environmental and Ecological Statistics*, Volume 10, 2003; Chapter 5, this volume). The main attraction of this sampling design is its ability to produce a more precise estimator for highly aggregated populations with less effort than surveying additional random stations. If a population is highly skewed, then the adaptive estimators are more likely to exhibit central tendency than conventional random sampling, in that the distribution of sample means is less skewed (Conners and Schwager 2002; Hanselman et al. 2003b).

A basic ACS survey for rockfish starts with any type of random sampling design, such as simple or stratified random sampling, until a random tow surpasses a criterion value (c). Next, samples are conducted in the neighborhood of that random tow until the catch drops below the criterion. The samples that do not exceed the criterion on the outside of the network are called edge units. These values are not used in the estimates. The resultant network can be a variety of different shapes, including the cross and linear patterns (Figure 14.2). In part one, we use the cross pat-

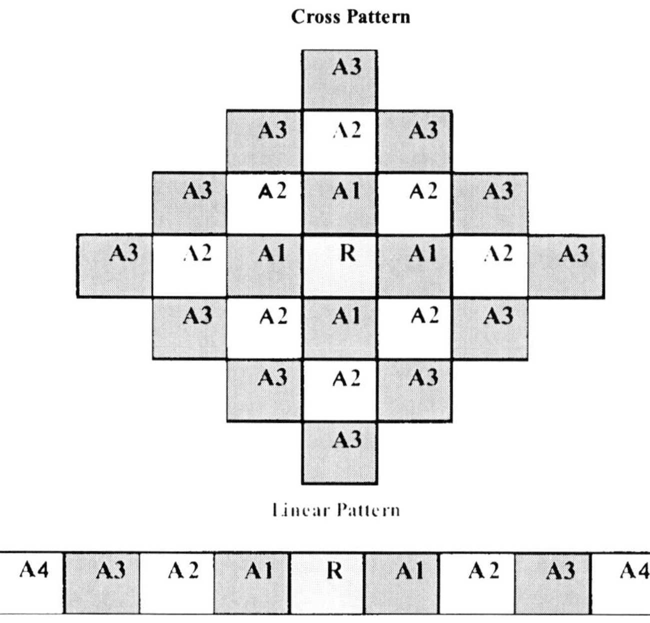

Figure 14.2. Two example network patterns for ACS, where R is the initial random station that exceeds the criterion value and $A1$ is the first level of adaptive sampling.

tern (left, right, top, bottom), and in part two we use the linear pattern (left and right only). The shapes used determine the efficiency of the ACS design relative to random sampling. Brown (2003) reported that the linear design had the smallest maximum efficiency gain over SRS but also was less likely to be much worse than SRS if the population sampled was not rare or clustered.

Most of the literature on ACS concerns the theory and simulation of the design. Few have tested it on real marine populations. We conducted two ACS field experiments on rockfish in the Gulf of Alaska. The first experiment (Hanselman et al. 2001) used a stratified random design with a criterion value determined by order statistics. Order statistics can be a useful way of setting the criterion value when little is known a priori about the population being sampled. Basically, an initial random sample is conducted, and then the catch-per-unit effort (CPUE, kg/km^2) values are ordered from highest to lowest. The scientist chooses how many of the top stations to adaptively sample and uses the next highest CPUE value as the criterion. In our study, we chose a relatively small area with four strata determined roughly by habitat type. We used a small initial sample size of $u \sim 15$ in each stratum, and then we set the criterion value by ordering the random tows by CPUE values. We adaptively sampled the top three stations using the fourth highest station's CPUE as the criterion value. An example of the results from one stratum is below:

	Adaptive stations			*Criterion*	*Remainder of random initial random sample*										
Order	*1*	*2*	*3*	4	5	6	7	8	9	10	11	12	13	14	15
CPUE	*5951*	*4681*	*3888*	464	332	311	194	125	108	100	83	54	51	28	27

The standard errors of the resulting abundance estimates from the 1998 study were no better than if we had simply added more random samples (Figure 14.3a).

The second experiment (Hanselman et al. 2003b) used a systematic random sample over a broad area in the Gulf of Alaska. In this experiment we used a fixed criterion that was determined from prior survey data. This technique allowed us to immediately begin adaptively sampling when the criterion was exceeded. This method increased logistical efficiency because we did not have to finish the initial random sample before adaptively sampling the station. Adaptively sampling immediately after the random tows also helped to meet the assumption of geostationarity because there was less chance of fish movement during the sampling time frame. When we

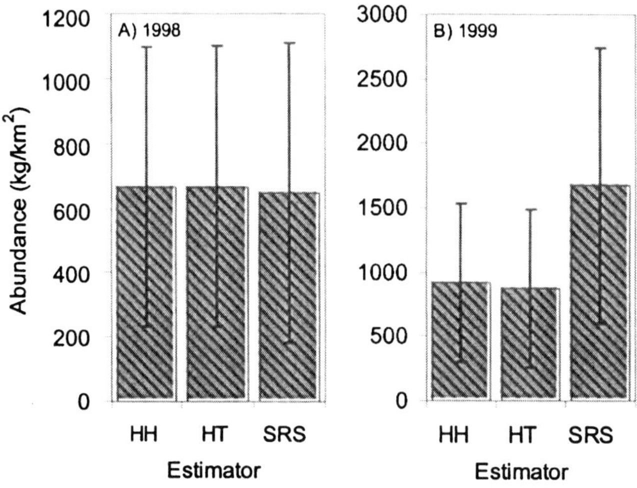

Figure 14.3. Results of 1998 (A) and 1999 (B) ACS experiments for *S. alutus*. HH = Hansen-Hurwitz estimator, HT = Horvitz-Thompson estimator, SRS = simple random sampling estimator. Sample size of v' (initial sample size plus adaptive sample minus edge units) is used for comparison at the same theoretical level.

used this type of criterion value and a larger unstratified area, we obtained slightly higher precision for the abundance estimates than if we had just added more random samples (Figure 14.3b).

The main result from the two rockfish studies was that most of the gains achieved by ACS were in terms of logistical efficiency, and the gains expected in statistical efficiency were minimal. This led to exploring the properties of ACS using simulated data from another author and another data set representing actual survey data. We use simulations to illustrate why ACS did not work well for *S. alutus* during our study and when it might work well.

Materials and Methods

We simulated two populations for analysis. Each population was modeled using a 40×40 grid (total number of sampling units, U, is 1,600). The first population type was modeled after the "highly aggregated" population used in Su and Quinn (2003), and the second type was modeled to reflect the characteristics of a population of *S. alutus*. We generated both using a Poisson cluster process (Diggle 2001). The characteristics of the two popu-

Table 14.1.

Summary statistics of two simulated populations. Populations were created in 40 × 40 grids (U = 1,600) with a Poisson cluster process (Diggle 2001). Note that μ is the population mean, CV is the coefficient of variation, and Pzr is the proportion of zero cells in the population.

Population	μ	CV	Pzr
Highly aggregated[a]	190.6	3.8	0.84
S. alutus[b]	191.1	4.3	0.16

[a] Based on "highly aggregated" population in Su and Quinn (2003).
[b] Based on CPUE data from 2001 biennial survey between 150 and 300 m, scaled to equal the mean of population 1.

lations are shown in Table 14.1. We compared the populations by their overall coefficient of variation (CV) and their proportion of zero (Pzr) cells in the grid. We set the means equal to the mean used in Su and Quinn (2003). The CV and Pzr for population 1 were set equal to Su and Quinn's (2003), and for population 2 they were set equal to survey data for *S. alutus* from the 2001 NMFS biennial survey.

We then sampled each population with six initial sample sizes (u_1) ranging between 40 and 240 and representing sampling fractions between 2.5 and 15%. We performed 1,000 replications for each initial sample size. We set the criterion value for network sampling at the population mean (μ) and one-half the population mean (0.5μ). Adaptive sampling was performed in the cross pattern. We then calculated summary statistics for SRS and ACS (the Hansen-Hurwitz (HH) and Horvitz-Thompson (HT) estimators). Details of these estimators were reviewed in Chapter 5 of this volume.

We then examined the efficiency of ACS by comparing the results from the two different populations. For comparison, the relative efficiency was computed as the variance of SRS divided by the variance of ACS; a value above one indicated a more efficient ACS design, whereas a value below one indicated a less efficient design. The designs were compared using two different final sample sizes: (1) v is the final sample size including edge units, which is the practical level when all units must be surveyed; and (2) v' is the final sample size without edge units that compares the designs at the same theoretical level (the sample size used in the estimators). We also investigated the effect of initial sample size on both efficiency and final sample sizes.

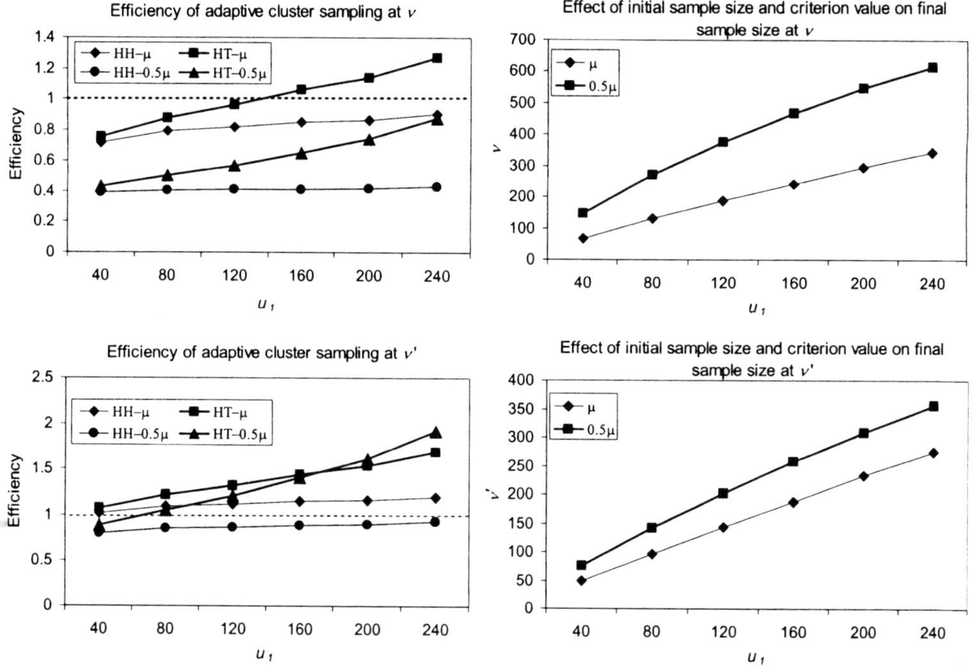

Figure 14.4. Efficiency of ACS versus SRS for a "highly aggregated" population ($U = 1{,}600$) from Su and Quinn (2003). HH-μ is the Hansen-Hurwitz estimator at the μ criterion level. HT-0.5μ is the Horvitz-Thompson estimator at the 0.5μ criterion level. Sample size u_1 is the initial random sample size, v is final sample size including edge units, and v' is final sample size excluding edge units. Efficiency is relative to SRS where the dashed line means equally efficient (1).

Results

Results of the adaptive sampling simulations were sensitive to both initial sample size and criterion value. The relative efficiency of ACS compared to SRS varies depending on which final sample size (v or v') is used. In population 1 (Figure 14.4) at v (final sample size, including edge units), the HT estimator was more efficient than SRS only at large initial sample sizes (> 160) and criterion μ. At v' (final sample size without edge units) for population 1, ACS was slightly more efficient than SRS regardless of initial sample size (Figure 14.4), except in the HH estimator at criterion μ. In population 2, the HT estimator became more efficient than SRS at an initial sample size between 120 and 160, depending on the criterion value used, whereas the HH estimator was never more efficient than SRS (Figure 14.5). The HT estimator was more efficient than the HH estimator, all fac-

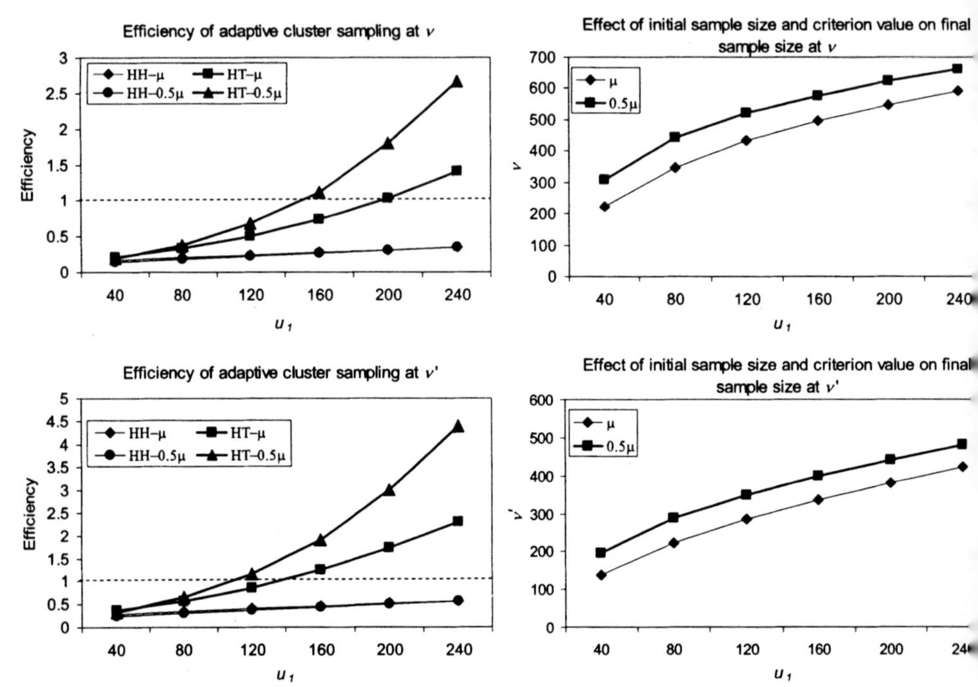

Figure 14.5. Adaptive sampling simulation results for an *S. alutus*-like simulated population (U = 1,600). HH-μ is the Hansen-Hurwitz estimator at the μ criterion level. HT-0.5μ is the Horvitz-Thompson estimator at the 0.5μ criterion level. Sample size u_1 is the initial random sample size, v is final sample size including edge units, and v' is final sample size excluding edge units. Efficiency is relative to SRS where the dashed line means equally efficient (1).

tors held equal. The HH estimator rarely attained efficiency greater than SRS (Figures 14.4, 14.5).

Results in the HT estimator were sensitive to the criterion value chosen in both populations. The choice of mean CPUE for the criterion value usually yielded a more efficient design at any of the initial sample sizes in population 1. The difference in final sample size between the two criterion values was much greater for population 1 (Figure 14.4). Generally, the criterion value of one-half mean CPUE was a better choice for population 2 in terms of maximum possible efficiency, but required a very large initial sample size. In both populations the criterion value of mean CPUE was as efficient as one-half CPUE when the final sample size was the same (Figures 14.4, 14.5). For example, in Figure 14.5, the initial sample size of 200 for criterion μ has approximately the same final sample size and efficiency as an initial sample size of 160 for criterion 0.5μ.

Hydroacoustics

Hydroacoustics is a technique in fisheries and fisheries research that uses sound pulsed through the water to detect organisms in the water column as well as bottom depth and structure. This is done by using an echo-sounder that measures the sound that is reflected back to a transducer, which is the physical unit attached to the vessel or a towed sled that sends and receives the sound pulses. The properties of sound transmitted through water are affected by water temperature, salinity, depth, and many other properties. The returned sound then is usually evaluated with software that adjusts for these different properties to obtain measures of organisms in the water column. These measures can vary from general statistics such as the average amount of backscatter per unit volume to the tracking of individual schools and fish.

The first successful experiment to detect fish acoustically was more than 70 years ago (Kimura 1929). The technology has improved significantly, but the concepts and techniques remain relatively similar (MacLennan and Simmonds 1992). One notable improvement was the introduction of echo-integration (Dragesund and Olsen 1965), which is still widely used today to estimate absolute abundance for a number of fish stocks. Echo-integration is basically summing the intensity of the backscatter for a given volume of water divided by the number of pings. Newer technology such as split-beam and dual-beam transducers (two pulses at the same frequency with a different size swath or two beams of different frequencies, respectively) has allowed easier identification of specific fish in the field without prior laboratory experiments and made calibration between surveys a simple task (MacLennan and Simmonds 1992).

Hydroacoustics have been used extensively by trawlers to locate rockfish (Major and Shippen 1970), but they have seldom been used for rockfish stock assessment. Earlier hydroacoustics research that has been applied to rockfish includes species differentiation (Richards et al. 1991), exploratory surveys for biomass estimation (Kieser et al. 1993), and quantifying above-bottom schools (Starr et al. 1996). More recently, Stanley et al. (1999a) used hydroacoustics to show that yellowtail rockfish have nocturnal dispersion, but diel behavior did not significantly affect biomass estimates. New rockfish research on the Canadian Pacific coast has concentrated on estimating biomass of areas of high widow rockfish abundance (Stanley et al. 1999b, 2000, 2002). In these studies, Stanley et al. were able to use hydroacoustics to estimate abundance and delineate the size of a high-density cluster for comparison with estimates from fishers. A key difference between Pacific ocean perch and widow rockfish is that *S. alutus* can be found in more trawlable topography and are usually closer to bottom than widow rockfish (Brodeur 2001; Stanley et al. 2000; Zimmerman 2003). For Pacific ocean perch in the Gulf of Alaska, Krieger et al. (2001) showed a relatively strong relationship between catch rates and raw acoustic signal in a small study area. However, none of these studies attempted to use hydroacoustics to improve large-scale sampling designs for rockfish. In this study we use hydroacoustic data collected from 2 years of NMFS bottom-trawl surveys to investigate how to improve precision of abundance estimators for Pacific ocean perch.

Materials and Methods

Hydroacoustic data were recorded on Simrad ES-60 (Simrad, http://www.simrad.com) echo-sounders on three vessels equipped with 38 KHz single-beam transducers. The vessels used in the Gulf of Alaska in 2001 were F/V *Vesteraalen* (38 m stern trawler) and F/V *Morning Star* (45 m stern trawler). The vessels used in the Aleutian Islands and Bering Sea in 2002 were F/V *Morning Star* and F/V *Sea Storm* (38 m stern trawler). Data during summer 2001 were recorded only during trawl hauls, whereas data recorded in summer 2002 were generally recorded continuously from before the first haul of the day until after the last one was completed. Nets were standard survey gear for the Gulf of Alaska, Aleutian Islands, and Bering Sea. They were towed an average of 15 min per haul (Martin and Clausen 1995). We used catch data from the NMFS' RACEBASE (Oracle database containing all historical survey data), which included the catch weight, composition, numbers, location, time, distance fished, depth, and net dimensions.

We analyzed the raw hydroacoustic data with SonarData's Echoview software (SonarData, http://www.sonardata.com), which we calibrated to the settings of each vessel. We analyzed all hauls from both years that contained any *S. alutus* and all hauls in the primary depth stratum for *S. alutus* (150–300 m). The hydroacoustics were matched with each bottom trawl with a combination of GPS readings and time-stamps. We approximated the distance of the net behind the boat by using basic geometric methods (Pythagorean Theorem, ignoring wire curvature). The track then was echo-integrated from 0.5 m off bottom to the average net height measured by the net sounder (Scanmar). By using this small offset from the bottom, we may have included some of the "acoustic dead zone" (Stanley et al. 2000), but we considered it a reasonable choice considering that the mean net height was all within the possible dead zone. This choice may be reasonable considering that the NMFS groundfish survey trawls on relatively smooth bottom topography and that the weather in the summer minimized pitch and roll of the vessel. This combination of factors should have minimized this dead zone. The echo-integration resulted in a number of different measures of acoustic backscatter, but the mean volume backscattering (*Sv*) proved to have the best relationship with *S. alutus* CPUE (kg/km^3). We then used this variable to form a predictive model to convert *Sv*s to predicted catches for the sampling designs described below.

Analytical Method and Sampling Designs

We computed 352 CPUE values (kg/km^3) from the three vessels and two years. The CPUE data were highly skewed and required transformation. We chose the flexible Box-Cox transformation, which was a power transformation that used maximum likelihood theory to estimate the optimum transformation. The distribution of these transformed values is shown in Figure 14.6. We compared these CPUE values to their respective *Sv*s from echo-integration. The data used are summarized in Table 14.2.

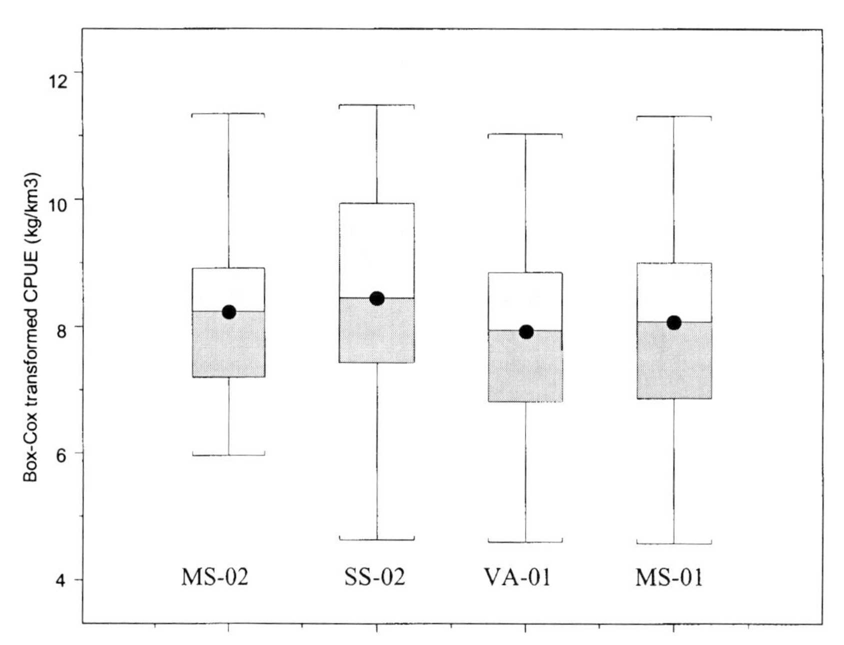

Figure 14.6. Boxplots of Box-Cox-transformed CPUE (kg/km^3) values for three vessels and two years of the NMFS groundfish surveys. MS02=F/V *Morning Star* 2002, SS02=F/V *Sea Storm* 2002, VA01=F/V *Vesteraalen* 2001, MS01=F/V *Morning Star* 2001.

Table 14.2.

Data collected from different vessels, years, and areas for hydroacoustic analysis of S. alutus catches.

Vessel	Year	Area	Continuous?[a]	Sample size[b]
Vesteraalen	2001	Gulf of AK	No	83
Morning Star	2001	Gulf of AK	Some	87
Morning Star	2002	Aleutian Islands/Bering Sea	Yes	85
Sea Storm	2002	Aleutian Islands	Yes	97

[a] Continuous means that hydroacoustics were recorded between tows.
[b] Sample size refers to the number of tows used in the analysis of *S. alutus* catch data.

We first regressed these transformed CPUE values against *Sv* (Sokal and Rohlf 1995). We then examined sets of nonlinear models that added variables including depth, longitude, and catch composition. Depth can be a strong predictor for groundfish because many species occupy a specific depth interval. We selected longitude as a possible predictor because of density changes from west to east in the Gulf of Alaska. Catch composition was chosen as a possible variable because it could be used to differentiate areas of hydroacoustic signal with high densities of other species from areas with high densities of *S. alutus*. None of these variables would have an effect on the hydroacoustic signal but would help predict when the signal may be representing *S. alutus*. We then selected the predictive model that yielded the best fit to the data in terms of R^2. We fitted each vessel's data to the model separately first, but confidence intervals for parameters overlapped across year and vessel; therefore, the final model was fitted with all the data pooled.

Because continuous hydroacoustic data were collected only in the 2002 surveys, these data were used for testing new sampling designs. For the 2002 data we examined a total of 182 hauls and their respective hydroacoustic tracks to be used in several sampling designs.

For the first design, we used the hydroacoustic model predictions to simulate a linear adaptive cluster sample around the tows that exceeded a predetermined criterion value. For the initial sample (u), we used data from the 84 trawl hauls that contained *S. alutus* in the Aleutian Islands/Bering Sea data collected on the F/V *Morning Star* in 2002. We used the mean CPUE of the F/V *Sea Storm* from the same year as the criterion value to invoke ACS. We have performed simulations that have shown that, in an aggregated population, mean CPUE tends to be a reasonable criterion value that provides an improvement in precision with a relatively small final sample size. Additionally, by taking this value from the other vessel, it is independent of the sample (Hanselman et al. 2003b).

We used the resulting adaptive networks to estimate mean abundance for the area sampled by the vessel. We used both the Hansen-Hurwitz-type (HH) and Horvitz-Thompson-type (HT) estimators (Thompson and Seber 1996) and compared them to standard SRS estimation. Smaller standard errors or coefficients of variation for the adaptive sampling estimators would indicate that adding the hydroacoustic model predictions resulted in an improvement in precision. We used nonparametric bootstrapping to estimate model variance and include this in the final variance. Equations are shown in Appendix 14.1.

We used double sampling (Thompson 2002) as the second design to incorporate hydroacoustics into abundance estimation. In double sampling, more precise estimates can be gained by using the relationship between an auxiliary variable that is easy to collect and the variable of interest that is more expensive or time-consuming. In this design we used the observed tows as our variable of interest and used a larger sample of hydroacoustic model predictions of CPUE as our auxiliary variable. If the two variables have a high correlation and an intercept at the origin, then a ratio estimator can lead to a dramatic improvement in precision with negligible bias (Thompson 2002).

For this study, we sampled the Aleutian Island data from the F/V *Sea Storm* in 2002. We took a subsample of 40 tows and their respective hydroacoustic tracks and took another random sample of 80 hydroacoustic tows in the vicinity of these 40, allocating two auxiliary samples per actual tow. The position of hydroacoustic tows was generated with uniform random numbers between –10 and 10 km away from each original station, which was then set to be the center of a hydroacoustic tow of equivalent length. The model predictions from these tows then were combined in a ratio estimate of mean abundance and compared with the SRS estimates. The form of the estimators can be found in Appendix 14.1 with a simple example showing the gain in precision from using a ratio estimator with auxiliary data.

Hydroacoustic data may be useful for optimizing stratification of a survey. We present an example of using raw acoustic backscatter (*Sv*) as a basis for pooling the strata from the 2001 biennial groundfish survey conducted by the NMFS. We used data from the western and central Gulf of Alaska (the eastern Gulf was not surveyed in 2001). The original design used 59 strata to attempt to minimize variance for all species. Twenty-eight of these strata contained *S. alutus* in the 2001 survey. We used the hydroacoustic model predictions from 2001 for this area to pool these strata into four larger strata assigned by the four quartiles of the hydroacoustic predictions. No data were used between tows. We used standard stratified random sampling estimators on combined strata to determine the overall mean and variance of the abundance estimates (Thompson 2002). Although this method could be biased because the sampling design was originally stratified under an optimal allocation design, it illustrates the possible usefulness of employing hydroacoustics as a means for stratification.

Results

When the transformed CPUE values were directly regressed on *Sv* (Figure 14.7), the coefficient and intercept were significant, but the fit was marginal ($R^2 = 0.12$). The best model chosen to predict CPUE with hydroacoustics related the transformed CPUE to the natural log of *Sv* and the localized catch composition. The model took the form:

$$CPUE^* = a \ln(-Sv)^b + (COMP)^c,$$

where $CPUE^*$ was the predicted Box-Cox-transformed CPUE value with $\lambda = 0.05$ and $L = 2500$ (Sokal and Rohlf 1995), *Sv* was the mean volume backscattering value, and *COMP* was the percentage of *S. alutus* in the closest tow. We used the pooled data ($u = 352$) from all three vessels for both years. We estimated parameters using nonlinear least squares and obtained estimates of $a = 0.0539$, b - 12.12, and $c = 0.075$ ($P < 0.0001$) with an $R^2 = 0.82$ (Figure 14.8). We used this model to generate predictions for the following sampling design results.

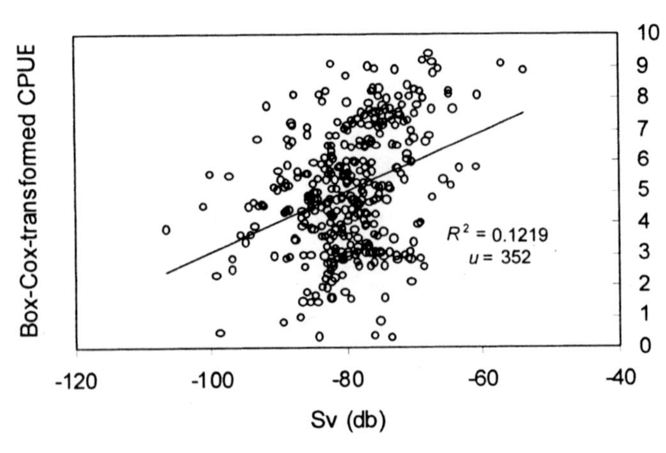

Figure 14.7. Fit of Box-Cox-transformed CPUE (kg/km^3) of Pacific ocean perch versus mean volume backscatter (*Sv*) for the pooled data for all vessels during all years.

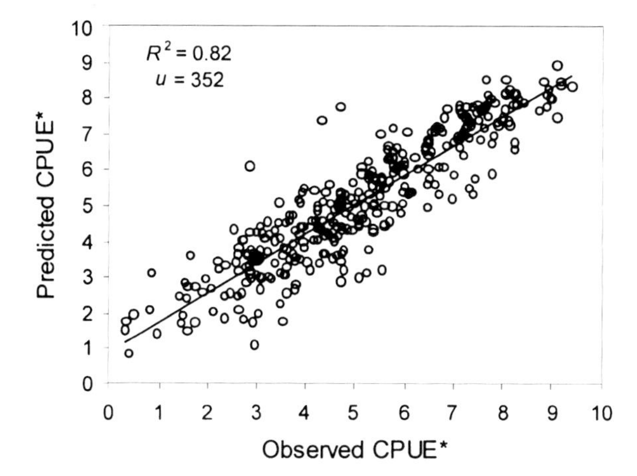

Figure 14.8. Plot of model predicted CPUE* versus the observed transformed CPUE* (Box-Cox power transformed) for pooled data for all vessels during all years.

Adaptive Cluster Sampling

The ACS design allows for any number of patterns for adding samples as long as they are symmetric. We conducted our survey in a nearly linear pattern because the vessel moved along a transect from one sampling location to the next (see Figure 14.9). We added normal measurement error to the *COMP* variable of adaptive model predictions with mean equal to the *COMP* value of the tow with a CV of 0.33. This additional error simulated within-network variability more appropriately than model error alone and allowed *COMP* to vary naturally as the survey moved away from the original tow. Previous work on adaptive sampling data for *S. alutus* showed that the sill of variograms (a measure of pairwise correlation of CPUEs with distance) produced for high abundance strata was approximately 10 km, roughly equating to average cluster size (Hanselman et al. 2001). Hence, we added adaptive samples linearly with model predictions until either they dropped below the criterion value or they were more than 10 km away from the original tow.

Six tows of 84 exceeded the criterion value, and the hydroacoustic model was used to add an additional 55 samples around these networks. Networks were bordered both by units not exceeding the criterion and by the 10 km distance limit. For this data set, there were lower estimates of mean abundance for ACS than for the SRS estimators and fairly large gains in precision for both adaptive estimators (Table 14.3). If the same number of random samples was taken as was included in the adaptive estimator, SRS yielded a CV closer to the adaptive estimators (36%). However, the hydroacoustic samples required no extra ship time.

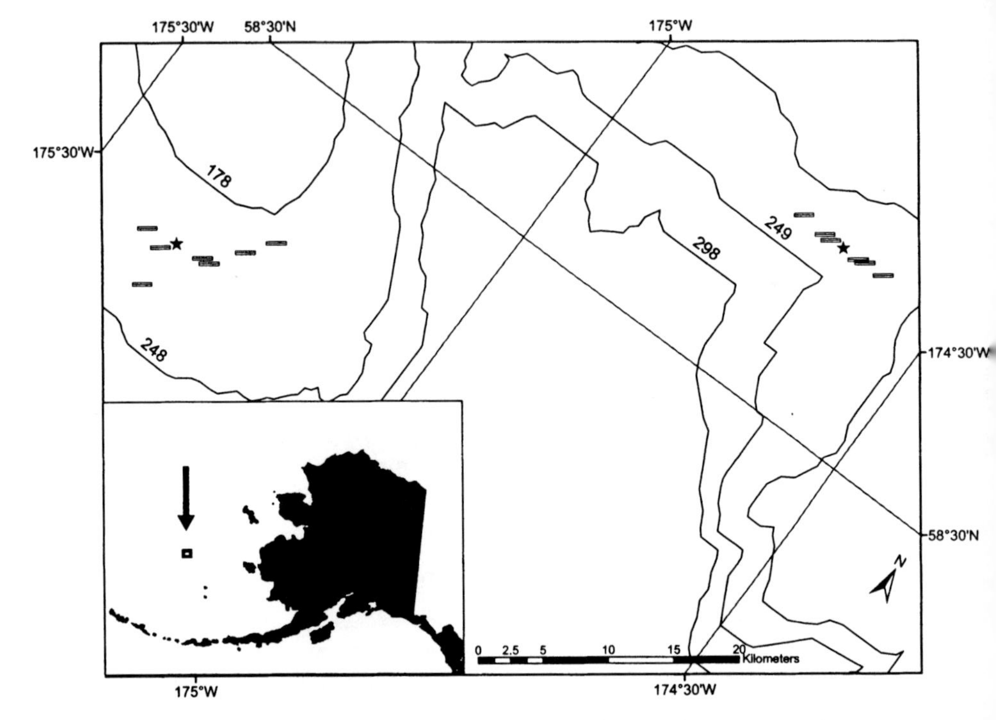

Figure 14.9. Two locations of simulated ACS for Pacific ocean perch. Data are from F/V *Morning Star* 2002 in the Bering Sea. Stars are survey tows; the thick lines are the predicted tow tracks from hydroacoustic observations. Thin contour lines are bathymetry (in m).

Table 14.3.

Results of a hydroacoustic adaptive sampling experiment, where $\hat{\mu}$ is the estimator of mean abundance (kg/km^2), SE is the standard error, and CV is the coefficient of variation. Simple random sampling-v' was conducted with the same sample size as adaptive sampling.

Estimator	$\hat{\mu}$	SE	CV (%)
Simple random sampling	10,773	5,042	47
Simple random sampling-v'	10,773	3,878	36
Hansen-Hurwitz	5,912	1,683	28
Horvitz-Thompson	6,086	1,681	28

Double Sampling

A requirement for good performance of the ratio estimator is that the two variables being used have a correlation near one and that the variables are linearly related with an intercept at the origin. The subsample of 40 tows (y_i) and their respective 40 hydroacoustic model predictions (x_i) was well correlated ($r = 0.9$) with an intercept near the origin. The estimate of the ratio between the two variables was 1.07. When we used the 80 additional hydroacoustic predictions (x_i) in the ratio estimator, it performed efficiently. However, if the model estimates are back-transformed instead, the relationship becomes nonlinear and the estimator performs poorly compared to SRS. Locations of some of the stations and their auxiliary hydroacoustic predictions are shown in Figure 14.10. The results from ratio estimation on the transformed scale are quite promising, with 95% confidence intervals roughly half as wide as SRS (Figure 14.11). However, when the estimation is calculated on the transformed scale and transformed back to

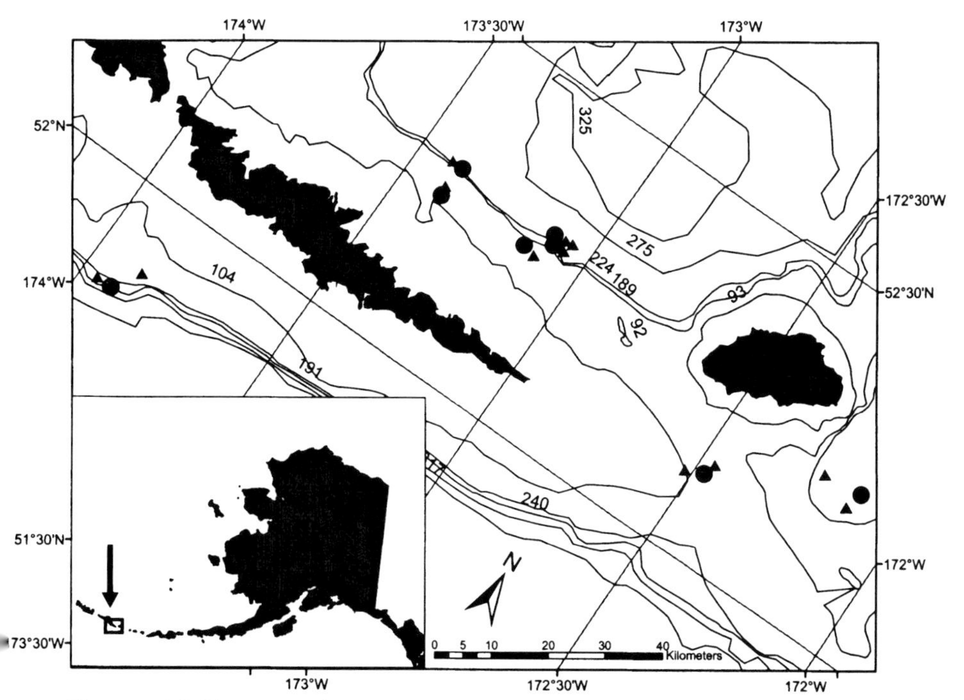

Figure 14.10. Map of some double-sampling locations from the F/V *Sea Storm* 2002 in the Aleutian Islands. Gray circles represent the center of the tow sample. Black triangles represent the random acoustic model predictions for double sampling. Numbered contour lines are bathymetry (in m).

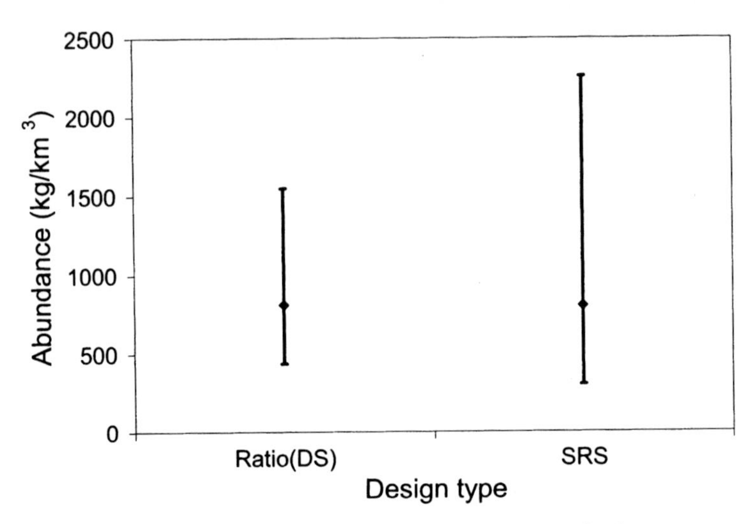

Figure 14.11. Ninety-five percent confidence intervals for a ratio estimate (double sampling) and an SRS estimate for mean abundance of one ship's catch in the Aleutian Islands 2002.

its original scale, the estimator is now the geometric mean rather than the arithmetic mean. Although this is a valid statistic, the geometric mean is more difficult to use in the standard stock assessment procedures employed by the NMFS for rockfish, but it could be useful as a biomass index.

Stratification

The pooled stratification assigned the 28 strata into four new strata (Table 14.4), with stratum 1 being the bottom quartile of model predictions and stratum 4 being the top quartile. These pooled strata were not contiguous and would have been difficult to show on a map, given the size of the Gulf of Alaska and the relatively small size of some of the strata on the continental slope.

The pooling of strata resulted in four strata containing eight of the old strata from the western Gulf and 20 from the central Gulf. The western Gulf appeared to contain more of the moderate strata, whereas the central Gulf contained the very high and low densities. The depth distribution showed that the strongest hydroacoustic signal corresponded with the inclusion of the 100–300 m depth stratum from the survey for strata 2–4. This depth range is where the density of *S. alutus* is highest (Hanselman et al. 2001). Stratum 1 looks much like stratum 4 in terms of areas and depths pooled because of the aggregation of the species. These areas may have been

Table 14.4.

Results of pooling strata for 2001 NMFS groundfish survey (eastern Gulf was not surveyed in 2001). Strata were ranked by quartile of hydroacoustic density, with four being the highest.

Stratum	No. of old strata	Combined area (km^2)	Average density (kg/km^2)	Western[a]	Central[b]	100–300 m depth[c] (%)
1	7	48,615	0.57	1	6	71
2	7	58,246	2.59	3	4	29
3	7	42,832	55.8	3	4	43
4	7	43,752	98.0	1	6	86
Total	28	193,445	35.4	8	20	39

[a] Number of strata from the western Gulf.
[b] Number of strata from the central Gulf.
[c] Refers to the percentage of strata that were in a depth between 100–300 m.

geographically proximate, but *S. alutus* simply were not encountered. This pooled stratification resulted in a substantial improvement in the precision of the estimate with only a minor change in the point estimate (Figure 14.12). Although this method may be biased because it is pooling strata of different sizes, the point estimate changes less than 4% between stratifica-

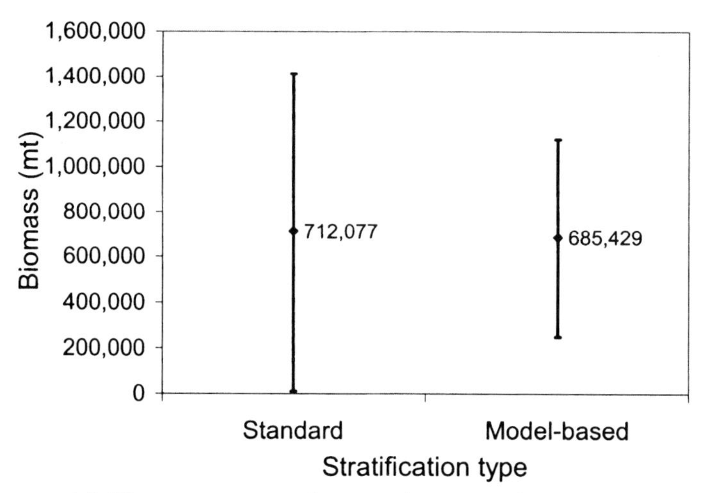

Figure 14.12. Comparison of using the original stratification of the 2001 NMFS biennial groundfish survey for *S. alutus* and using a hydroacoustic model-based approach. Error bars represent 95% confidence intervals. Data used were for the western and central Gulf of Alaska, and hydroacoustic data were taken from 2001 F/V *Morning Star* and the 2001 F/V *Vesteraalen.*

tions. This small change in the point estimate illustrates that the hydroacoustic data aided in gaining precision without much bias.

Discussion

Results from the first part of our study illustrated situations where ACS did not perform well on large marine populations. The results for population 1 showed that small improvements in precision could be gained at most initial sample sizes with either criterion value. This population was less variable than population 2 with more zero catches. ACS for the *S. alutus*-like population 2 was less efficient than SRS at lower sample sizes at either criterion value, but became much more efficient at very high initial sample sizes (> 10% of population). This result was consistent with the two field experiments performed on *S. alutus,* in which ACS provided relatively small gains in precision (Hanselman et al. 2001, 2003b) when small initial sample sizes were used. The simulations also confirmed results determined by Hanselman et al. (2003b) that a higher criterion value reduces sampling effort at a small cost to gains in precision. Su and Quinn (2003) also showed large possible gains with the HT estimator on a smaller population ($U = 400$) when greater than 10% of the population is included in the initial sample. Therefore, if we were to reap the potential large gains in precision with an adaptive design, a trawl survey would need to cover 5% or more of the sampling units for the population for the initial sample alone. Unfortunately, in an area as large as the Gulf of Alaska, the current trawl survey covers less than 0.5% of the possible sampling units. Because this survey is already large in terms of resources (> 204 vessel days, > 1,000 scientist days), a large increase in coverage is unlikely.

Our studies on the usefulness of ACS for large clustered marine populations have shown that the design may not be appropriate when a much larger initial sample size is unlikely to be obtained. It might be a useful design to obtain a good estimate of a small area where a large initial sample size is possible. We have also shown that ACS can capture fine-scale variability when insufficient information is available for stratification to capture large-scale variability (Hanselman et al. 2001). However, the collection of hydroacoustic data is simple and nearly cost-free. Determining how to use these data to improve rockfish estimation is the logical next step.

The addition of hydroacoustic data into sampling designs for *S. alutus* showed promising results in the second part of our study. The data were not collected randomly, but the data used in the analysis were subsampled randomly from the whole data set, allowing the assumptions of random sampling theory to apply. Considering that the additional data required practically no additional ship time, and only the time of one analyst, it may be an efficient way to gain precision.

We needed a model to use raw backscatter to predict CPUE values from the untrawled sections of the survey. The model we used had an excellent fit to the data but may have been hyperstable (little variability in predicted catch) because it used a localized catch composition to fit the data well. This type of model is more appropriate for use in double sampling or stratification than it is for the adaptive sampling design, which requires "true" samples to be unbiased. Double sampling and stratification do not have this restriction, and auxiliary samples can simply be an "eyeball" estimate or any correlated variable that permits many samples to be collected easily.

Using the hydroacoustic data as a way to adaptively sample around units without actually performing tows is appealing. However, the unbiased estimator intended by the original design likely has a bias of unknown size using the model-based predictions. ACS is most efficient when the within-network variability is high. In this example, the measurement error added into the *COMP* variable resulted in relatively high within-network variability. Therefore, adding hydroacoustic samples linearly around tows yielded substantial gains in precision, but the estimate was lower than random sampling. This might be because of bias, or it could be similar to our applications of ACS to rockfish (Hanselman et al. 2001; Hanselman et al. 2003b), in which adaptive sampling yielded smaller estimates of mean abundance than did SRS. However, this bias cannot be determined because true abundance is unknown. The validity of a model that uses catch composition becomes questionable as one moves farther away from the original tow.

In double sampling, it is not important if the hydroacoustic model is completely accurate, just that it is highly correlated with the variable of interest. In our example, this was true and the gains were substantial, with a reduction of about one-half in the width of the 95% confidence intervals. Although a ratio estimate is not design-unbiased (its expected value does not equal the mean), a small bias is usually worth the tradeoff for a large gain in precision. Double sampling with the untransformed data resulted in a less precise estimate than SRS (a 20% greater SE). Transformation of the data was necessary for the data to work well with the ratio estimator. Therefore, the problem with the use of a geometric mean in stock assessment remains. Taylor (1986) suggested several estimators for obtaining an arithmetic mean from the geometric mean of a Box-Cox transformation, but these estimators did not work well because λ is near zero in our transformation (i.e., highly skewed). It would work fine as an index, but for an index to be useful, we would need a time series of measurements where any bias was relatively constant. We have collected data for two surveys (2001, 2003) in the Gulf of Alaska and should continue to collect this data because they may become more useful as we accumulate a longer time series.

Using the hydroacoustic model for stratification showed some potential usefulness. Our heuristic approach of pooling strata of different sizes was not statistically unbiased, but changed the abundance estimate very little while reducing the confidence intervals substantially. This indicates that hydroacoustics may be useful for stratification of future designs for rockfish or for real-time stratification. One application of this type is to stratify on the fly by allocating more tows to patches of higher density hydroacoustic signal and fewer tows to areas of low-density signal as the survey progresses (Everson et al. 1996). This study was done with icefish, and experiments are in progress to test the methods on rockfish (Paul Spencer, pers. comm.).

Conclusions and Future Work

Future work for sampling *S. alutus* should concentrate on obtaining more samples within their habitat range, perhaps with a vessel equipped with a more rugged net to assess areas that have been previously sampled only via hydroacoustics. This would allow a more random and representative sample of *S. alutus* habitat and their distribution than the current gear allows. The idea of using the low cost and readily available hydroacoustic information that is currently being collected on survey vessels is tempting. Collecting hydroacoustic data from commercial vessels throughout the season is also a possibility. Further comparison of what the net is catching versus what the echo-sounder is recording should be done by submersibles or towed sleds to validate the use of hydroacoustic signal so close to the bottom. The most promising use of the data thus far is in a double sampling design. Future work with these data should concentrate on how to use them as an index in the current stock assessment model and on using hydroacoustic data for stratification of future surveys.

Acknowledgments

We would like to thank Jeff Fujioka at the Auke Bay Laboratory, who originally started looking at hydroacoustic data for rockfish. We also would like to thank the group at the Alaska Fisheries Science Center for their help in collecting the data, specifically Paul Spencer, Bill Flerx, Jerry Hoff, and the other field party chiefs in the 2001 and 2002 survey seasons. Finally, we thank the captains of the F/Vs *Morning Star*, *Vesteraalen*, and *Sea Storm* for allowing us to tinker with their echo-sounders. This study was funded by the NMFS, the National NMFS/Sea Grant Population Dynamics Fellowship, and the Alaska Fisheries Science Center Population Dynamics fellowship.

References

Brodeur, R. D. 2001. Habitat-specific distribution of Pacific ocean perch *(Sebastes alutus)* in Pribilof Canyon, Bering Sea. *Continental Shelf Research* 21:207–224.

Brown, J. A. 2003. Designing an efficient adaptive cluster sample. *Environmental and Ecological Statistics* 10:95–105.

Conners, M. E., and S. J. Schwager. 2002. The use of adaptive cluster sampling for hydroacoustic surveys. *ICES Journal of Marine Science* 59:1314–1325.

Diggle, P. J. 2001. *Statistical Analysis of Spatial Point Patterns*, 2nd ed. Oxford University Press, London.

Dragesund, O., and S. Olsen. 1965. On the possibility of estimating year-class strength by measuring echo-abundance of 0-group fish. *Fisk Dir. Skr. Ser. Havunders.* 13:47–75. (Cited in MacLennan and Simmonds 1992).

Everson, I., M. Bravington and C. Goss. 1996. A combined acoustic and trawl survey for efficiently estimating fish abundance. *Fisheries Research* 26:75–91.

Hanselman, D. H., J. Heifetz, J. Fujioka, and J. Ianelli. 2003a. *Gulf of Alaska Pacific Ocean Perch.* Stock Assessment and Fishery Evaluation Report for the Groundfish Resources of the Gulf of Alaska, North Pacific Fisheries Management Council, Anchorage.

Hanselman, D. H., T. J. Quinn II, C. Lunsford, J. Heifetz, and D. M. Clausen. 2001. Spatial implications of adaptive cluster sampling on Gulf of Alaska rockfish. pp. 303–325 in *Proceedings of the 17th Lowell-Wakefield Symposium: Spatial Processes and Management of Marine Populations.* University of Alaska Sea Grant Program, Fairbanks.

_____. 2003b. Applications in adaptive cluster sampling of Gulf of Alaska rockfish. *Fishery Bulletin* 101:501–512.

Kieser, R. B., K. Cooke, R. D. Stanley, and G. E. Gillespie. 1993. *Experimental Hydroacoustic Estimation of Rockfish (*Sebastes *spp.) Biomass off Vancouver Island, November 13–25 1991.* Canadian Manuscript Report of Fisheries and Aquatic Sciences 2185.

Kimura, K. 1929. On the detection of fish groups by an acoustic method. *Journal of the Imperial Fisheries Institute* 24:41–45.

Krieger, K., J. Heifetz, and D. Ito. 2001. Rockfish assessed acoustically and compared to bottom-trawl catch rates. *Alaska Fishery Research Bulletin* 8(1):71–77.

Love, M. S., M. Yoklavich, and L. Thorsteinson. 2002. *The Rockfishes of the Northeast Pacific.* University of California Press, Berkeley.

Lunsford, C. 1999. *Distribution Patterns and Reproductive Aspects of Pacific Ocean Perch* (Sebastes alutus) *in the Gulf of Alaska.* M.S. thesis, University of Alaska-Fairbanks, Juneau Center.

MacLennan, D. N., and J. E. Simmonds. 1992. *Fisheries Acoustics*. Chapman and Hall, New York.

Major, R. L., and H. H. Shippen. 1970. *Synopsis of Biological Data on Pacific Ocean Perch,* Sebastodes alutus. FAO Fisheries Synopsis 79, and U.S. Dept. of Commerce, National Marine Fisheries Service NMFS/S 79, Circular 347.

Martin, M. H., and D. M. Clausen. 1995. *Data Report: 1993 Gulf of Alaska Bottom Trawl Survey*. NOAA Technical Memorandum NMFS-AFSC-59.

Quinn, T. J. II, and R. B. Deriso. 1999. *Quantitative Fish Dynamics*. Oxford University Press, New York.

Richards, L. J., R. Keiser, T. J. Mulligan, and J. R. Candy. 1991. Classification of fish assemblages based on echo integration surveys. *Canadian Journal of Fisheries and Aquatic Sciences* 48:1264–1272.

Sokal, R. R., and F. J. Rohlf. 1995. *Biometry: The Principles and Practice of Statistics in Biological Research,* 3rd ed. Freeman, New York.

Stanley, R. D., A. M. Cornthwaite, R. Kieser, K. Cooke, G. Workman, and B. Mose. 1999a. *An Acoustic Biomass Survey of the Triangle Island Widow Rockfish* (Sebastes entomelas) *Aggregation by Fisheries and Oceans, Canada and Canadian Groundfish Research and Conservation Society, January 16–February 7, 1998*. Canadian Technical Report of Fisheries and Aquatic Sciences 2262.

Stanley, R. D., R. Kieser, K. Cooke, A. M. Surry, and B. Mose. 2000. Estimation of a widow rockfish *(Sebastes entomelas)* shoal off British Columbia, Canada, as a joint exercise between stock assessment staff and the fishing industry. *ICES Journal of Marine Science* 57:1035–1049.

Stanley, R. D., R. Kieser, and M. Hajirakar. 2002. Three-dimensional visualization of a widow rockfish *(Sebastes entomelas)* shoal over interpolated bathymetry. *ICES Journal of Marine Science* 59:151–155.

Stanley, R. D., R. Kieser, B. M. Leaman and K. Cooke. 1999b. Diel vertical migration by yellowtail rockfish *(Sebastes flavidus)* and its impact on acoustic biomass estimation. *Fishery Bulletin* 97: 320–331.

Starr, R. M., D. S. Fox, M. A. Hixon, B. N. Tissot, G. E. Johnson, and W. H. Barss. 1996. Comparison of submersible-survey and hydroacoustic-survey estimates of fish density on a rocky bank. *Fishery Bulletin* 94:113–123.

Su, Z., and T. J. Quinn. 2003. Estimator bias and efficiency for adaptive cluster sampling with order statistics and a stopping rule. *Environmental and Ecological Statistics* 10:17–41.

Taylor, J. M. G. 1986. The retransformed mean after a fitted power transformation. *Journal of the American Statistical Association* 81:114–118.

Thompson, S. K. 1990. Adaptive cluster sampling. *Journal of the American Statistical Association* 412:1050–1059.

_____. 2002. *Sampling,* 2nd ed. Wiley, New York.

Thompson, S. K., and G. A. F. Seber. 1996. *Adaptive Sampling*. Wiley, New York.

Zimmerman, M. 2003. Calculation of untrawlable areas within the boundaries of a bottom trawl survey. *Canadian Journal of Fisheries and Aquatic Sciences* 60:657–669.

Appendix 14.1

Equations for Estimation of Abundance in Adaptive Sampling and Double Sampling

a. The Hansen-Hurwitz Estimator in Adaptive Cluster Sampling with Hydroacoustic Adaptive Units.

From Thompson (2002:294):

$$\hat{\mu}_{HH} = \frac{1}{u}\sum_{i=1}^{u}\frac{\hat{y}_i^*}{x_i} = \frac{1}{u}\sum_{i=1}^{u}\hat{w}_i^* \text{ and } s_w^2 = \sum_{i=1}^{u}\frac{(\hat{w}_i^* - \hat{\mu}_{HH})^2}{u-1},$$

where $\hat{\mu}_{HH}$ is the Hansen-Hurwitz estimator of the population mean, u is the sample size, $\hat{y}_i^*$ is the estimator of the network total from the actual tow y_{i1} that exceeds a criterion value and its corresponding hydroacoustic samples $\hat{y}_{i2}, y_{i3}, \ldots \hat{y}_{iu,\,y>c}$, and $\hat{w}_i^*$ is the estimator of the network mean. The best-fit model from the hydroacoustic data set is $\hat{y}_{ij} = a(Sv) + b(COMP)^c$ with $u = 352$ and $R^2 = 0.82$, where Sv is the acoustic backscatter and $COMP$ is the catch composition for the species. The variance estimator of y_{ij} is $\hat{\text{Var}}(\hat{y}_{ij}) = Sv^2\text{Var}(a)_{\text{boot}} + \text{Var}[b(COMP)^c]_{\text{boot}}$, which is generated from nonparametric residual resampling (Quinn and Deriso 1999).

Note $\hat{\text{Var}}(\hat{w}_i^*) = \sum_{2}^{x|y>c}\text{Var}(\hat{y}_{ij}) + \sum_{j=1}^{x}\text{Cov}(\hat{y}_{ij}, \hat{y}_{ij+1})$, where $\sum_{j=1}^{x}\text{Cov}(\hat{y}_{ij}, y_{ij+1})$ is assumed to be zero (Sokal and Rohlf 1995:567). Thus,

$\hat{\text{Var}}(\hat{\mu}_{HH} = \left(\frac{1-u}{U}\right)\frac{s_w^2}{u} + \frac{1}{Uu}\sum_{i=1}^{u}\frac{\hat{\text{Var}}(\hat{w}_i^*)}{x_i}$, where U is the total number of sampling units (two-stage estimator from Thompson 2002).

b. The Ratio Estimator with Hydroacoustic Model Predictions as an Auxiliary Variable.

From Thompson (2002:158), the ratio estimator is: $\hat{r} = \frac{\sum_{i=1}^{u} y_i}{\sum_{i=1}^{u}\hat{x}_i}$, where $\hat{r}$ is the ratio of the sum of observed catches y_i to the sum of corresponding catches $\hat{x}_i$ predicted by the hydroacoustic model. Further, $\hat{\mu}_x = \frac{1}{u}\sum_{i=1}^{u'}\hat{x}_i$ where u' is the number of auxiliary predictions $\hat{x}_i$. The estimator of the population mean is $\hat{\mu}_r = \hat{r}\hat{\mu}_x$ and its variance estimator is

$$\hat{\text{Var}}(\hat{\mu}_r) = (1 \frac{u'}{U})\frac{s_y^2}{u'} + \left[\frac{u' - u}{u'u(u-1)}\right]\sum_{i=1}^{u}(y_i - \hat{r}\hat{x}_i)^2.$$

It does not matter that $\hat{x}_i$ is estimated because it is auxiliary and could be compared to an "eyeball" estimate. The best-fit model from the hydroacoustic data set is $\hat{x}_i = a(Sv) + b(COMP)^c$ with $u = 352$ and $R^2 = 0.82$, where Sv is the acoustic backscatter and $COMP$ is the catch composition for the species.

Here we present a numerical example of using the ratio estimator with auxiliary data. Given $U = 100$, $u = 3$, and $u' = 11$, we collect auxiliary data x_i for u (2, 6, 8) and for u' (4, 7, 2, 5, 7, 8, 10, 12). The observed data y_i are 3, 7, and 9, and the sample variance s_y^2 is 9.33. The estimated ratio of the subsample y_i and x_i is

$$\hat{r} = \frac{\sum_{i=1}^{u} y_i}{\sum_{i=1}^{u} x_i} = \frac{(3+7+9)}{(2+6+8)} = \frac{19}{16} = 1.19.$$

The estimated mean of the auxiliary variable is

$$\hat{\mu}_x = \sum_{i=1}^{u'}\frac{x_i}{u'} = \frac{(2+6+8+4+7+2+5+7+8+10+12)}{11} = 6.45.$$

The ratio estimate of the mean is

$$\hat{\mu}_r = \hat{r}\hat{\mu}_x = 1.19 \times 6.45 = 7.7.$$

The variance of ratio estimate is

$$\hat{\text{Var}}(\hat{\mu}_r) = (1 - \frac{u'}{U})\frac{s_y^2}{u'} + \left[\frac{u' - u}{u'u(u-1)}\right]\sum_{i=1}^{u}(y_i - \hat{r}\hat{x}_i)^2 = \left(1 - \frac{11}{100}\right)\left(\frac{9.33}{11}\right) + \left[\frac{11-3}{11 \times 3 \times (3-1)}\right](0.66) = 0$$

and the variance of the SRS estimate is

$$\hat{\text{Var}}(\hat{\mu}_{\text{SRS}}) = \left(1 - \frac{u}{U}\right)\frac{s_y^2}{u} = \left(1 - \frac{3}{100}\right)\left(\frac{9.33}{3}\right) = 3.01.$$

15

Survival Estimation in Bats: Historical Overview, Critical Appraisal, and Suggestions for New Approaches

Thomas J. O'Shea, Laura E. Ellison, and Thomas R. Stanley

There are more than 1,100 species of bats, representing nearly 20% of all species of mammals (Simmons, in press). Bats occur on every continent except Antarctica, and although they occur in greatest diversity in the tropics, their distribution extends at least to tree line at high latitudes. Given this wide distribution and high species richness, it is not surprising that bats display a large number of feeding specializations and play key functional roles in ecosystems. Insectivorous bats are best known to temperate zone biologists, but at lower latitudes other species are frugivorous, nectarivorous, piscivorous, carnivorous, or even sanguivorous (blood-feeders). Bats play important ecological roles as pollinators and dispersers of fruit seeds in warm climates and as major consumers of insects in most terrestrial ecosystems. These attributes are of significant economic and ecological benefit, but some bats can also be detrimental to human affairs. In the neotropics vampire bats negatively affect livestock health, and bats may play an as yet poorly understood role in disease transmission cycles.

In light of their importance, bats are also of global conservation concern. Many species of bats are endangered or are considered "sensitive" species in the United States, and several nations provide legal protection to bats. Negative impacts on bat populations have stemmed from a variety of factors. These include habitat loss, declines in roost availability, disturbance to roosts (especially at winter hibernacula in caves), intentional killing of bats in vulnerable aggregations, hunting for food (particularly Old World fruit bats), and mortality due to environmental contaminants.

Considering the high diversity, wide distribution, ecological significance, economic importance, and conservation issues of bats, surprisingly little attention has been given to the application of modern approaches toward understanding their population biology. This is particularly true for estimation of survival. Some of this inattention has been a result of logistic difficulties. Most species of bats are difficult to observe because they are active at night and roost during the day in secretive or inaccessible places. All bats fly and thus can be highly mobile. However, cryptic behavior, mobility, and logistic difficulties are common to many of the other problematic groups of animals that are discussed in this book. The lack of detailed information on survival in bats may also be in part due to past lack of widespread recognition of their importance, particularly in comparison with other groups of mammals such as large herbivores, marine mammals, and economically important predators, furbearers, and crop pests. In addition to logistic challenges, thorough study of the population biology of bats can require a long-term commitment. Bats are long-lived for mammals of their size and reproduce slowly. Maximum longevities of several species of bats of temperate zones have been documented at more than 25 years (e.g., Ransome 1995; Keen and Hitchcock 1980) and in some cases at more than 30 years (reviewed by Barclay and Harder 2003). Bats typically give birth to single young (or much less frequently twins) just once annually (Tuttle and Stevenson 1982).

Our objectives in this chapter are to provide an overview of past sampling aimed at estimation of survival in bats, a parameter critical to understanding their population biology. We then point out some of the deficiencies of past work in comparison with the potentially more reliable inferences that may be possible through the application of new and powerful tools that are rapidly developing for the analysis of survival in wildlife populations. In highlighting deficiencies we do not cast aspersions on past efforts. In most cases these investigations provided very hard-earned results; they were as thorough and allowed as much inference as was possible given the limitations of the methods available. Instead we seek to emphasize to researchers the potential benefits of taking new directions using techniques that have only recently become practical for application to bats. We provide examples of such approaches from the recent literature and from our own nascent work on big brown bats (*Eptesicus fuscus*) in Colorado. We hope this review will serve not only to stimulate bat biologists to consider new approaches in their studies of survival, but also to stimulate thought toward designing new applications for sampling this intriguing group of mammals.

Bat Banding

Past research on bat survival relied heavily on banding for marking of individuals. We therefore provide an overview of the history and range of findings from bat-banding studies, emphasizing those in the United States.

History and Overview of Banding Studies in Bats

Allen (1921) first applied bands to bats. He placed numbered aluminum bird bands on the legs of four female eastern pipistrelles (*Pipistrellus subflavus*) from a roost at a residence in Ithaca, New York in 1916. He recovered three of the banded bats at the same location in 1919. A second study was completed soon thereafter, documenting returns of banded *Eptesicus fuscus* in California (Howell and Little 1924). Allen (1921:55) noted, "The valuable results that are now being obtained by banding birds could no doubt be duplicated with bats if only enough persons would cooperate in the project of banding."

Banding of bats did indeed grow over the next half-century. Managing and coordinating bat banding was an official function within the U.S. Fish and Wildlife Service (USFWS) beginning in 1932, but it had no legislative mandate and the work never achieved a scale comparable to that of banding birds. From the 1930s through the mid-1960s, the USFWS and its predecessors issued about 1.5 million numbered metal bands for placement on bats (Greenhall and Paradiso 1968; estimated at 1.9 million by 1982 in Hill and Smith 1984). Banding of bats in the United States grew slowly in the 1930s (Mohr 1952) but intensified during the 1950s and 1960s. Most banding efforts were aimed at local or regional studies carried out independently by field biologists, particularly those interested in bats that use caves (Mohr 1952; Greenhall and Paradiso 1968). Modifications to design and techniques were made during this period (Barclay and Bell 1988; Greenhall and Paradiso 1968). Banding also took place in at least 14 European nations, Mexico, Japan, and Australia, involving additional hundreds of thousands of bats (Greenhall and Paradiso 1968; Hill and Smith 1984; Kuramoto et al. 1985).

Banding provided previously unattainable information on the natural history of bats. In addition to investigations of survival, recaptures of banded individuals revealed patterns in reproduction (Pearson et al. 1952), migration (see below), longevity (Ransome 1995; Keen and Hitchcock 1980), growth (see review by Tuttle and Stevenson 1982), energetics and physiology (Kurta et al. 1989, 1990), and aspects of behavior such as site fidelity (Rice 1957; Tinkle and Patterson 1965), social organization (Dwyer 1970; Bradbury and Vehrencamp 1976), and homing (Davis 1966a; Wilson and Findley 1972). A prime example of discoveries made about bat migrations stems from banding of Mexican free-tailed bats (*Tadarida brasiliensis mexicana*). These bats form huge colonies in the southwestern United States that are the largest aggregations of mammals on Earth. Banding allowed discovery of seasonal migrations between these locations and Mexico, with documentation of some stopover points and movements among caves (Constantine 1967; Cockrum 1969; Glass 1982). Recovery of banded Mexican free-tailed bats was low, but outlines of migratory patterns emerged from compilations of anecdotal case records. Cockrum (1969) reported banding 162,892 free-tailed bats at 71 localities in Arizona and adjacent regions during 1952–1967. Only 539 (0.3%) were recovered at locations other than the point of banding (just 3,240 [2%] were recovered at the original banding sites), and nearly all of these were recaptured just once. Nonetheless, one-way migratory journeys of 1,600–1,800 km were documented (Villa-R. and Cockrum 1962; Constantine 1967; Cockrum 1969; Glass 1982). Serendipitous recoveries of a few bats in migration revealed travel rates of at least 64 km per night. Banding records of this species also showed complex geographic patterns in migratory tendencies in the western United States (Cockrum 1969) that prompted speculation about genetic distinctiveness in these areas. Such distinctiveness was not born out by subsequent genetic studies (McCracken et al. 1994; McCracken and Gassel 1997; Svoboda et al. 1985).

Findings from banding studies also documented other aspects of natural history, the annual cycle of reproduction and movements, and regional migrations to hibernacula of bats in the eastern United States, particularly in New England and New York (Griffin 1945). Davis and Hitchcock (1965) banded about 71,000 little brown bats (*Myotis lucifugus*) in this region. They noted movements between hibernacula and nurseries of up to 320 km and rates of travel of at least 43 km per night. Banding studies also showed that in certain locations, some species of bats make only local seasonal movements (e.g., *Corynorhinus townsendii*, Pearson et al. 1952,

Humphrey and Kunz 1976; *Macrotus waterhousii*, Cockrum et al. 1996). Banding studies also provided information on seasonal migrations of European bats over various distances, ranging from local movements (e.g., 1–64 km; Hooper 1983) to longer distances (e.g., 250–300 km in *Myotis dasycneme*; Sluiter et al. 1971). Distances of movements away from banding localities ranging up to 120 km were recorded for several species of bats in Japan (Kuramoto et al. 1985), and Dwyer (1966) reported complex seasonal movement patterns of nearly 600 km for banded *Miniopterus schreibersii* in Australia.

Critical Appraisal of Bat Banding

Although much was learned about the natural history of bats from banding studies, the technique poses problems. Chief among these are injuries and infections from banding, and obliteration of numbers on bands by chewing or by overgrowth of tissues on the forearm (bat biologists switched from banding legs to using wing bands in the late 1930s [Trapido and Crowe 1946]). These problems varied by species, with some more sensitive than others. Among U.S. bats, *Corynorhinus townsendii*, for example, was reported to be very sensitive to injury from bands (Humphrey and Kunz 1976), whereas *Pipistrellus subflavus* was not thought to be sensitive (Davis 1966b). Partial rectification was sought by shifting from bird bands to a lipped design made specifically for bats based on modifications developed by a few European bat banders (Hitchcock 1957), but injuries continued to be reported. About 42% of the Mexican free-tailed bats tagged with bird bands had visible irritation, wing tears, embedding in bone, or fleshy outgrowths with evidence of infection by about 300 days post-banding; 73% of those tagged with lipped bat bands had injuries, although these were described as less severe than those caused by bird bands (Herreid et al. 1960). Skeletal damage also occurred (Perry and Beckett 1966). In contrast, Happold and Happold (1998) reported less injury in *Pipistrellus nanus* marked with lipped bands in Malawi and concluded that the effects were not serious. Injuries to wings due to banding were also apparent from studies of some European bats (e.g., Sluiter et al. 1971).

In Australia, Baker et al. (2001) provided perhaps the most thorough documentation of injurious effects of bands on bats. Their records and analyses showed that injury rates and severity varied with species (17 species of insectivorous bats were banded with different types of bands and recaptured over multiple years), person applying bands, band type (bird bands were less injurious to some species than bat bands), band material, and size of band (Baker et al. 2001). Bats also may suffer marked premature tooth wear from chewing at bands (Young 2001), which could result in shortened longevity.

In addition to direct injury, much of the banding of bats in the United States and Europe took place at winter hibernacula in caves or mines (Mohr 1952). The energetic cost of arousal from hibernation due to disturbance (that is, banding activities) can be extreme. The amount of energy mobilized from fat for a single arousal can cost as much as 68 days of the supply a bat needs for hibernation (Thomas et al. 1990). Thus, bats disturbed during hibernation for banding have a reduced capability of surviving over winter and lower prospects of countering starvation if faced with food shortages during uncertain spring conditions. Banding of bats at hibernacula very likely drove population declines of bats at some sites in the United States (e.g., Tinkle and Patterson 1965; Keen and Hitchcock 1980). Daan (1980) suggested that marked declines in counts of six of eight species of bats in 16 limestone caves in The Netherlands during 1940–1979 also were likely a result of banding, perhaps due to very high adult mortality. Daan (1980:101) noted that banding as a source of mortality "is certainly serious enough to affect the conclusions obtained with the technique."

In the United States, attitudes moved from a formal resolution of the American Society of Mammalogists in favor of bat banding in 1964 (Greenhall and Paradiso 1968) to an official policy to cease issuing of bat bands by the USFWS in 1973, primarily because of disturbance (Jones 1976). Similarly, Australia began an official bat banding program in 1960 (Young 2001). Subsequently, a permanent moratorium on banding eastern horseshoe bats (*Rhinolophus megaphyllus*) was instituted in Australia in the 1970s following recognition that up to 88% of banded bats had moderate to severe injuries (Young 2001). In 1996 Australia extended the moratorium on banding to all microchiropteran bats in three families (Baker et al. 2001). Recommendations to curtail banding also were made in Europe (e.g., Rybar 1973; Hooper 1983).

Suggestions for Future Studies

Biologists continue to band bats for various purposes. The use of bat bands in studies of survival should be eschewed unless investigators can empirically determine that resulting injuries and biases are very minor. Further research is needed on alternative methods of permanently marking bats.

Barclay and Bell (1988) list a number of possible alternatives. Necklaces and ear tags have been used on bats (Appendix 15.1; Mohr 1952). However, injuries can result from necklaces (Barclay and Bell 1988) and ears have very critical functions in orientation and prey detection. Tag loss also may be a problem with these methods. Necklaces were lost in five of 73 (6.8%) doubly marked (bands and necklaces) Jamaican fruit bats, *Artibeus jamaicensis* (Gardner et al. 1991) and in 6.5% of short-tailed fruit bats, *Carollia perspicillata* (Fleming 1988). Loss of ear tags also has been reported (Hoyle et al. 2001). Freeze-branding is another possible method of permanent marking that has only recently been explored in bats (Sherwin et al. 2002).

Participants at a 1999 workshop on monitoring bat populations recommended exploring various potential new marking techniques such as microtaggants and passive integrated transponders (PIT tags) (Working Group A report in O'Shea and Bogan 2003). PIT tags are small (ca. 0.1 g, 11 mm length), long-lasting implant devices that have been extensively used in fisheries and to identify pets. They are passive, and instantaneously emit a unique binary ultrasonic code only when activated and simultaneously received by reader devices at very close range. PIT tags have been used to mark big brown bats in captivity (Barnard 1989) and in field studies of activity patterns, social structure, and genetics of a few species of bats (Brooke 1997; Horn 1998; Kerth et al. 2002). They have not yet been applied to estimation of survival in bats, although we provide a case example below that suggests some promise. There also is a need for a clearinghouse for information about such techniques and to coordinate contacts for ongoing studies of marked bats (e.g., when remote recoveries are made), including banding studies (Working Group C report in O'Shea and Bogan 2003).

Estimation of Survival

There have been two previous reviews of the study of survival in bats. Keen (1988) provided a critical appraisal of the literature through 1984, and Tuttle and Stevenson (1982) reviewed those studies appearing through 1982. Here we provide an updated chronological review, critique, and suggestions for future approaches.

History and Overview of Survival Studies in Bats

We surveyed the literature on survival estimation in bats published since 1950 (Appendix 15.1). We found 42 studies with information relevant to estimating survival in 36 species of bats in the Americas, Europe, Asia, and Australia. Earlier efforts in the 1930s and 1940s (summarized by Mohr 1952) simply expressed the percentage of banded bats recaptured at sites where originally banded. To our knowledge, the first published attempts at calculating survival estimates for bats were those of Beer (1955) for hibernating big brown bats in Minnesota (Appendix 15.1). As pointed out by Keen (1988), these estimates were based on an incorrect application of Hickey's (1952) ad hoc (Williams et al. 2002) procedures for estimating survival from recoveries of bands from dead birds.

Soon afterward, a series of papers appeared based on banding of thousands of bats of several species at winter hibernacula in The Netherlands over many years. This work provided survival estimates based on the "method of Bezem" (Sluiter et al. 1971:7; Sluiter et al. 1956; Bezem et al. 1960). This was a life table approach that pooled recaptures of unaged cohorts. Bezem's technique involved calculating the proportion of bats caught in any one year out of those bats caught the previous year, and plotting the logarithm of the proportion against the number of years since banding. Under unlikely assumptions of constant rates of survival and constant capture probabilities, the coefficient of regression of a straight line fitted to these points was taken as the annual survival estimate for the study period and the y-intercept was taken as the capture probability. Sluiter et al. (1956) gave a detailed mathematical rationale for this approach and provided justification via simulation. They acknowledged drawbacks to their methods and assumptions but noted that without using the regression technique, "the computational labour [for estimating annual survival and capture probabilities] rapidly becomes enormous as the number of years increases" and "therefore we have turned to a method which, though it may lead to slightly erroneous results, has the advantage that the necessary calculations are very easy" (Sluiter et al. 1956:74). This work was done a half-century ago when this justification was perhaps understandable.

Bezem's approach (with modifications), coupled with life table analyses derived from the slopes of the regression lines under an assumption of constant survival, prevailed in survival estimation in bats for the next 30 years (Appendix 15.1). This was despite the appearance of Cormack's (1964) paper and others on maximum-likelihood-based methods, the increasing availability of electronic computing techniques, and the specific rejection by Caughley (1966) of the assumptions (particularly constant rate of survival) made by Bezem. Maximum-likelihood methods, introduced by R. A. Fisher (1922), provide the means for estimating parameters from sample data given an appropriate model and are "the backbone of statistical inference used in nearly every area of science" (White 2001:37). For descriptions of properties and methods of maximum-likelihood estimation, particularly as they relate to capture-recapture models, see White et al. (1982), Burnham and Anderson (2002), and Williams et al. (2002).

Studies that relied on regression varied in their methods of computations of the proportions that formed the regression points, which in some studies are not clearly specified. Bezem and coworkers (1960) included all observations of any banded bat 1 year apart as survival 1 year after marking; thus, each banded bat could be counted several times in the estimate of each regression point (Keen 1988). The "intuitive regression" (Keen 1988:167) method employed by most workers used a minimum-number-known-alive (MNA) approach in calculating the proportions in the regression points (a bat not seen between successive captures was calculated as present or alive at the intervening periods when it was not captured). Some studies made study-specific modifications to computations; for example, Davis (1966a) used an ad hoc computation to correct percentage recaptured by the estimated number alive but not recaptured. However, justification for the form of the correction was not provided and properties such as bias were unknown. Data used for both Bezem's and the intuitive regression methods were not appropriate for regression because they were not statistically independent (Keen 1988).

Of the 42 studies of bats we reviewed (Appendix 15.1), 7 (17%) used the maximum-likelihood-based Cormack-Jolly-Seber (CJS) model to estimate survival of bats, whereas 35 (83%) used ad hoc methods. The CJS model is a preferred, maximum-likelihood method based on open populations that allows estimation of survival and capture probabilities, and conditions on the capture, marking, and release of individuals at each of several times during a study (Williams et al. 2002). The ad hoc methods included simple reporting of recapture percentages by time period (about 20% of all studies); some form of "intuitive regression," often on pooled unaged cohorts with life tables derived by assuming constant rates of survival (about 50%); life tables of known age cohorts (2%); and other ad hoc methods (about 13%). Accordingly, estimates of variance for survival rates were made only in a small number of cases. A few studies used combinations of analytical approaches (e.g., Boyd and Stebbings 1989; Fleming 1988; Keen and Hitchcock 1980). Most studies of survival in bats were primarily descriptive reporting of estimates (about 76%); a few studies involved statistical null hypothesis testing, usually regarding sex (Appendix 15.1). These generally involved tests of differences in slopes of the regression lines or χ^2 tests of proportions recaptured at various time periods in age and sex categories.

Application of more useful modern approaches to estimating survival in bats occurred in two phases: (1) the first CJS-based estimates of survival appeared during 1980–1990 (Keen and Hitchcock 1980; Hitchcock et al. 1984; Boyd and Stebbings 1989; Gerell and Lundberg 1990); and (2) three very recent studies that used CJS estimators to evaluate specific hypotheses about survival in bats in Australia and Germany. Baker et al. (2001) reported that survival estimates of banded bats varied by type of band in *Nyctophilus geoffroyi* but not in two other species of Australian bats. Hoyle et al. (2001) adopted the information-theoretic approach (Burnham and Anderson 2002) to model selection and evaluated biological hypotheses about survival in the large Australian ghost bat, *Macroderma gigas*. The preferred model included sex, season, and rainfall as factors or covariates influencing survival in this species (Hoyle et al. 2001). Sendor and Simon (2003) also used the information-theoretic approach to select models for the effects of age, time, sex, season, and winter severity on annual survival in pipistrelles (*Pipistrellus pipistrellus*) captured with mist nets and banded at a site in Germany that was used both as a hibernaculum and in summer. Based on the most parsimonious models, age had the strongest effect on survival in the pipistrelles: survival was lower in bats in their first year than in adults. No differences in survival of the sexes were apparent, and there was no effect of winter severity, time, or season on survival.

Although methods to estimate survival from most past studies had flaws, some generalizations about the life history of bats emerged from these efforts. Survival and longevity are higher in bats than in most other mammals of similar size, and survival through the first year of life (or first year of banding in unaged cohorts) is usually lower than in subsequent years. Survival estimates for juvenile male bats often seem lower than in juvenile females, but this may in part be due to emigration and differential dispersal from natal roosts. It is difficult to draw many other reliable conclusions about survival in bats; accuracy and precision of most estimates are unknown, despite the tremendous efforts that have been involved in acquiring past recapture data.

Critical Appraisal of Survival Studies in Bats

Most previous studies of survival in bats suffer from shortcomings in at least one of three areas. The first of these was failure to specify the population of interest, or target population (defined in sampling theory as a collection of elements about which we wish to make inferences; Cochran 1977; Scheaffer et al. 1986), and to employ a sampling design that allows valid inferences to be made from the sample population (a collection of sample units drawn from a sample frame; Cochran 1977, Scheaffer et al. 1986) to the target population. For example, some study sites were likely chosen on the basis of convenience sampling (Anderson 2001:1294), where locations were based on prior knowledge that certain sites had bats and perhaps were closer to the researchers or less difficult to work in than others. This was not problematic insofar as the investigator limited inferences to the study sites actually sampled (assuming appropriate sampling at the sites). However, in past studies survival estimates were sometimes taken to apply to some larger population or even the entire species. A related problem was that much of the past estimation of survival was based on studies in which banding may have been carried out with multiple objectives (such as determining ranges of movements) and marking was not designed specifically for survival estimation. Consequently, the sample population was not representative of the target population, and this prevented valid inferences about survival.

A second shortcoming was the use of field methods that reduced survival in the sample population so it was no longer representative of the target population or that introduced biases of unknown magnitude into the sample data. For example, potentially injurious banding was used in 34 of the 36 studies that involved marked bats (Appendix 15.1), and 17 (40%) of the 42 studies estimated survival but involved sampling bats in hibernacula where disturbance-induced arousals can be a major source of overwinter mortality (11 of the 16 [68%] studies of marked bats in the United States and Canada involved sampling at hibernacula). Thus, the majority of survival studies based on recaptures of bats used field methods known or suspected to cause injury and reduced survival in marked individuals. Consequently, survival estimates for marked individuals in these studies could not represent survival rates for the unmarked portion of the population. Other sources of bias in the sample data included effects due to tag loss, inconsistent sampling efforts, and heterogeneity in responses to past capture and handling. Tag loss problems include effective tag loss, in cases where numbers on bands are obliterated from chewing or overgrowth of tissues on the forearm, or physical tag loss. Physical tag loss has also been reported for the two alternative marking methods (necklaces and ear tags) employed thus far (Gardner et al. 1991; Fleming 1988; Hoyle et al. 2001). The end result of tag loss was that, on average, fewer recoveries were likely recorded than actually occurred, and this likely biased survival estimates negatively. Stevenson and Tuttle (1981) pointed out other sources of bias from field methods in studies of survival in bats. These included unequal effort at recapturing bats among years and locations; widely varying characteristics of roosts and locations of bats within roosts that affected accessibility and capture probabilities; and differential segregation and accessibility of age and sex classes at capture sites. Individuals of some species of bats also have been shown to avoid recapture within hibernacula based on their experience from captures in previous years (Stevenson and Tuttle 1981).

The third shortcoming of many of the bat studies we surveyed was the use of ad hoc procedures for estimating survival, many of which invoked biologically unrealistic assumptions (such as constant rates of survival, despite year-to-year environmental stochasticity) or failed to state their assumptions. This included the use of life table approaches (Appendix 15.1), which are now considered ad hoc and circular (Anderson et al. 1981; Williams et al. 2002). Ad hoc estimators of survival are undesirable for the simple reason that they have no underlying theoretical justification or probabilistic framework in which to derive variance estimators or to evaluate important properties such as unbiasedness, efficiency, or consistency. Consequently, it is unknown if the calculated quantity is actually estimating the parameter of interest or if it is better in some sense than other ad hoc procedures. This leads to the problem described in Tuttle and Stevenson (1982:122) that "The student of chiropteran survival analyses will be considerably frustrated in attempting to compare the results of various studies." Obstacles to comparison occur because different estimators may be estimating different quantities and have no associated measures of precision.

Suggestions for Future Studies of Survival in Bats

Future research should avoid the shortcomings so prevalent in past studies. As a starting point, investigators must strive to clearly define their population of interest and then make every effort to construct an appropriate sample frame (a complete list of sample units; Thompson et al.

1998). In some cases it may not be possible to construct a sample frame that exactly matches the target population. For example, if an investigator wishes to make inferences about the population of bats that roosts in a particular geographic area (the target population), then he or she must first locate all the roosts in that area so as to choose a probability sample from that population. However, in constructing the sample frame, some roosts may be missed because they contain few bats, whereas roosts containing large numbers of bats will be less likely to be missed. In such cases, the sample frame will be a size-biased representation of the target population and inferences drawn from a sample of roosts will be applicable only to roosts in the sample frame, not to the population of roosts in the geographic area of interest.

Once an appropriate sample frame has been developed, well-established sampling procedures such as simple random sampling, stratified random sampling, or perhaps cluster sampling should be used to select the sample population. In the example above, one might first stratify roosts on the basis of size and then select a simple random sample of roosts from each stratum. Of course, once a roost has been selected for sampling, it is desirable to obtain a representative sample of individuals from each roost. This can sometimes be difficult to accomplish for bats, though certain measures can be taken to minimize potential biases. For instance, one might ensure that capture efforts are distributed equally among all parts of a roost (e.g., a cave) rather than just the most accessible parts. If there are multiple exits from a roost, one might attempt to capture individuals at all of these exits instead of just a single exit.

Methods used to acquire and mark individuals must be improved. Using bands to mark bats for survival studies is unacceptable without empirical determination of the effects of bands on the species or situation of interest, nor is it acceptable to disturb bats on a large scale in hibernacula to mark individuals or acquire recapture information. Alternative marking techniques need to be developed that allow permanent identification of individuals without causing injury and potentially biasing estimates (see above section on banding; preliminary results from application of PIT tags as a marking alternative are provided in the case study described below). In addition, marking efforts should be applied away from hibernacula to avoid the effects of disturbance-induced arousals on overwinter survival and should instead focus at alternative locations where bats occur in summer or at the mouths of caves during swarming in autumn prior to hibernation. Improved methods of acquisition and marking that do not reduce survival of the sample population or introduce biases into the sample data will yield more informative and reliable estimates of survival that can be more confidently applied to the unmarked portion of the population.

Our final suggestion is to abandon the ad hoc procedures used in past studies to estimate survival. Inferior approaches should not be followed simply as a legacy of precedent. Instead, we recommend that survival estimators be based on well-founded probabilistic models and a sound theoretical foundation, such as maximum likelihood theory, and that assumptions upon which models are built be stated explicitly. In particular, we believe the CJS open population capture-recapture model shows great promise for estimating survival in bats, as previously stressed by Keen (1988). This likelihood-based model conditions on first captures, allows inferences to be made to the marked population, and is based on the following assumptions (from Williams et al. 2002:422):

> *(1) every marked animal present in the population at sampling period i has the same probability p_i of being recaptured or resighted; (2) every marked animal present in the population immediately following sampling in period i has the same probability ϕ_i of survival until sampling period i+1; (3) marks are neither lost nor overlooked and are recorded correctly; (4) sampling periods are instantaneous (or in reality they are very short periods), and recaptured animals are released immediately; (5) all emigration from the sampled area is permanent; and (6) the fate of each marked animal with respect to capture and survival probability is independent of the fate of any other animal (see Williams et al. 2002:422–423 for a discussion of these assumptions and consequences of their violation).*

Generalizations to the CJS model allow survival, capture, and transition-state probabilities to be modeled as functions of individual or group covariates, and powerful software exists for estimating parameters under user-defined models (White and Burnham 1999). Transition-state probabilities are computed based on generalized CJS models that allow individuals to be distributed across multiple sites or phenotypic states, under the assumption that transitions among states follow a stochastic Markovian process (Williams et al. 2002). Furthermore, because the CJS model and its generalizations are likelihood based, the weight of evidence in favor of or against a particular candidate model can easily be evaluated using information-theoretic model selection procedures (Burnham and Anderson 2002). We anticipate that application of the CJS model in future bat studies will allow for more rapid advances in understanding bat population biology and in application of findings for management.

Big Brown Bats in Fort Collins: A Case Study Combining New Approaches

In 2001 we began a study of the ecology of rabies transmission in big brown bats that roost in human-occupied buildings in Fort Collins, Colorado. One facet of the study was to estimate survival and transition-state probabilities of marked bats as they related to exposure to rabies. Rabies is transmitted through biting. In addition to determining the influence of rabies exposure and other factors on survival of bats, understanding rabies transmission dynamics in our study population also required an understanding of various transition-state probabilities that may influence possible contact among bats. Transition states may include movements of bats among roosts, shifts in exposure histories and immune status of bats based on serology, or transitions from noninfective to infective (virus shedding) phases.

Our study differed from most previous efforts to estimate survival, capture, and transition-state probabilities by combining several different marking, recapture, and analytical approaches. We employed a PIT tag method of marking bats rather than bands, we studied bats at maternity colonies rather than hibernacula, and we "resighted" bats using passive PIT tag readers placed at roost entrances rather than repeatedly capturing and handling bats to read tags. Our objective was to evaluate a number of hypotheses about survival and transition-state probabilities using a generalized CJS model (after Lebreton et al. 1992). Specifically, we developed competing models for survival, capture, or transition-state probabilities and then obtained the maximum-likelihood estimates for these parameters. Models were constructed a priori to evaluate the influence of select categorical variables and individual covariates of likely biological importance, such as age, sex, time, body condition, ectoparasite burdens, and ecological aspects of principal roosting sites. We used the information-theoretic approach (Burnham and Anderson 2002) to rank and select among competing models (hypotheses). In this section we provide results from three examples of our preliminary studies that involved estimation of short-term survival, capture, and movement probabilities of bats. Results cover only 2 years of marking and recapture (2001–2002), so findings regarding longer term annual estimates based on these approaches are not yet available.

Methods Employed in the Fort Collins Study

The study of big brown bats in Fort Collins involved selection of roosts for sampling without prior knowledge of roost locations, uniquely marking individuals using technology not previously applied to studies of survival in bats, and using maximum-likelihood-based analytical methods. This work attempted to address some (but not all) of the deficiencies identified in many of the earlier studies of survival in bats.

Selection of Roosts for Sampling

Nearly all records of public exposure to rabid bats in Colorado are from encounters in and around buildings and parks in towns during summer (Pape et al. 1999). Therefore, we focused our sampling on summer maternity colonies (adult females and juveniles, adult males are widely dispersed) in buildings in Fort Collins (Figure 15.1). Selection of colony sites was not truly random. We located roosts for sampling by first capturing bats in mist nets at night as they flew over water for foraging or drinking in parks and open spaces. Thus individuals caught for tagging were probably encountered haphazardly. We then radio-tagged adult females and released them on site and drove the streets during the day to find colonies by tracking radio signals. For a more representative sample of bats, we also intentionally selected initial capture sites so that we included sections of the city spanning a range of ages of buildings. We radio-tagged 122 bats during summers 2001–2002 and located 54 buildings with colonies of at least 20 bats (some were occupied by bats for only a few days). Within this sample frame, we chose roosts for PIT tagging and biological sampling based on logistic considerations (equipment limitations, accessibility, potential cooperation). We also included a few colony sites in this total of 54 based on citizen knowledge. However, we did not select colony sites based on our own prior knowledge, and this differed from most previous studies of survival of bats.

Capturing, Marking, and Resighting Bats at Roosts

We captured bats as they exited from roosts at evening emergence, and thus captured only adults and volant juveniles (juveniles born that summer but probably at least 30 days old). In our examples below we did not

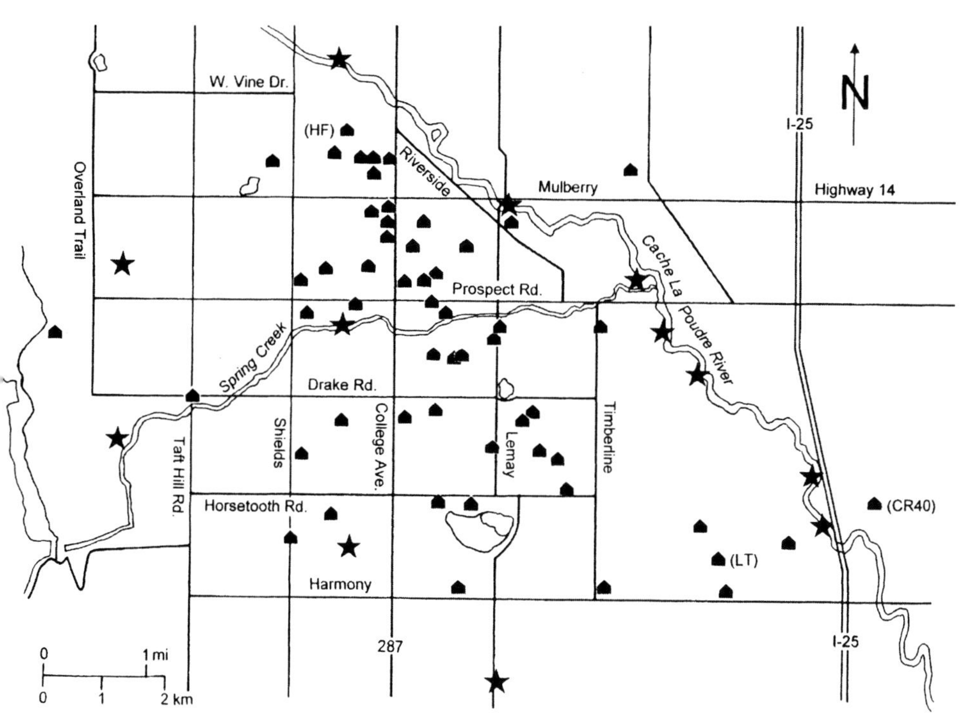

Figure 15.1. Fort Collins, Colorado, where big brown bats were sampled as part of a study of the ecology of rabies transmission in urban bats. Each square represents a one-square-mile section (2.6 km^2) of developed land in the city (population 120,000 people) bounded by major thoroughfares. Building symbols denote locations of structures used as roosts by big brown bats in 2001 and 2002, as discovered by radio-tracking adult females captured while foraging in parks and open spaces (capture locations denoted by stars). Abbreviations (LT, HF, CR40) correspond to specific roosts mentioned in tables and text of this chapter.

disturb bats by attempting to capture them within roosts. We used standard techniques and equipment for capturing flying or emerging bats, as described in detail in other sources (Greenhall and Paradiso 1968; Kunz and Kurta 1988; Kunz et al. 1996). Depending on locations of the exit points, we caught bats in harp traps (approximately 2 × 2 m frames with taut, closely spaced fine monofilament lines that flying bats collided with, then fell to collecting receptacles below) or in mist nets set on poles if exits were up to 5 m above ground. We captured exiting bats in mist nets suspended by pulleys from heights of 5–10 m in cases where exit points were high on the sides of buildings. We sometimes supplemented these with handheld nets and with funnel traps (custom-made, enclosed nylon tubes up to 10 m in length extending to the ground, see Kunz and Kurta 1988) placed over roost entrances.

We transported bats to the laboratory (isolated from each other in individual cotton specimen bags placed in disposable drink cups with lids) for biological sampling and for marking by subdermal injection of PIT tags. We followed transport, marking, and biological sample collection procedures described by Wimsatt et al. (in press) and then released the marked bats near the roost within 6 hours of capture. We PIT tagged 2,073 big brown bats at multiple sites in Fort Collins during 2001–2002. In 2002 we also marked newly PIT-tagged bats with small (3 mm diameter) circular freeze-brands over the left scapula. These spots were expected to grow out as small tufts of white pelage in the late summer moult (Sherwin et al. 2002). By August 2003 we reexamined 185 bats recaptured with white hair tufts, and just three lacked functional PIT tags (1.6%), suggesting that PIT tag loss was not a major problem.

In 2002 we installed 19 PIT readers at roosts in 14 different buildings. PIT readers are 15 or 30 cm diameter activating and receiving hoops that are each connected to a separate data logger and powered by 12 V automotive batteries (we used readers and PIT tags manufactured by AVID, Inc., Norco, California). PIT readers were installed over crevice openings where bats typically crawled over them upon emerging or entering. Data loggers stored the date, time, and identification number for every detection, and these were downloaded to a laptop computer three times each week throughout the summer. Tens of thousands of records were reduced to daily presence/absence for short-term survival studies.

Data Analysis

We used Program MARK (White and Burnham 1999) to generate estimates for parameters of interest. Encounter history files were created that included a listing of each individual bat, coded for every time period as either a "1" (captured), or "0" (not captured). We modeled data sets as "live recaptures," in which animals are reencountered only when recaptured alive on succeeding occasions. Two of our parameters of interest were capture probability (p, the conditional probability that a marked animal will be seen over a specified time interval given that it is alive and in the study population; Cormack 1964; Williams et al. 2002) and apparent survival (ϕ). Apparent survival cannot distinguish emigration from mortality, but is the probability that the animal is alive, remains on the study area, and is available for recapture (White and Burnham 1999). The identification of models followed the notation of Lebreton et al. (1992). For example, a model incorporating a group effect (such as age or reproductive status) on apparent survival and capture probabilities was coded as {ϕ (g), p (g)}. A model including time effects on survival and capture probabilities was coded {ϕ (t), p (t)}. We also calculated a 1-year return rate ($\hat{r}$ = [number recaptured at time t +1]/[number released at time t]) and confidence intervals based on simple binomial probability.

Example 1: Comparisons of Estimates from Conventional Captures vs. PIT Readers

We hypothesized that daily probabilities of capture would be higher and daily survival would be more precisely estimated using passive PIT readers placed at roost entrances than by using conventional methods of capturing bats to read marks by hand. Individual bats probably vary in their susceptibility to capture in nets and traps and may learn to avoid them. We predicted that survival estimates would be comparable using the two methods because CJS-based survival estimators are typically robust to heterogeneity in capture probabilities (Nichols and Pollock 1983). However, PIT readers can sample continuously whereas conventional captures must be conducted sparingly because of effort requirements, possible abandonment of roosts, and animal welfare considerations. Thus the larger number of samples obtainable by PIT readers should result in more precise estimates. We also predicted that estimates of overwinter return rates from 2001 to 2002 would be higher using reader records than estimates obtained by hand captures.

We created separate encounter history files for conventional captures and for PIT readers at roosts in three buildings. Conventional captures were conducted at least five times during summer 2002; PIT readers were operational over the intervals between first and last conventional captures at each of these three buildings. We analyzed samples separately by building because each differed in degree of difficulty in capturing bats. We did not use samples that occurred after mid-August because of postbreeding dispersal. We used Program MARK to adjust differing intervals between occasions and then converted these to daily rates. Capture events were typically 2 weeks apart (Table 15.1). We generated estimates of daily survival and capture probabilities for the two different techniques using a model incorporating age as a group effect (adult females and juveniles of both

Table 15.1.

Daily capture probability ($\hat{p}$) and apparent survival estimates ($\hat{\phi}$) for big brown bats at three different roosts in Fort Collins, Colorado (see Figure 15.1), during the summer of 2002. Estimates were compared using conventional capture techniques (mist nets and traps, with tag reading during handling) and PIT reader records collected passively between the first and last conventional capture events.

Roost	Method (n occasions)	$\hat{p}$(95% CI) Adults	$\hat{p}$(95% CI) Juveniles	$\hat{\phi}$(95% CI) Adults	$\hat{\phi}$(95% CI) Juveniles
HF[a]	Conventional captures (5)	0.091 (0.034–0.219)	0.064 (0.035–0.115)	0.974 (0.917–0.992)	Inestimable[b]
	PIT readers (30)	0.862 (0.851–0.872)	0.678 (0.637–0.716)	0.989 (0.987–0.991)	0.967 (0.955–0.976)
CR40[c]	Conventional captures (5)	0.494 (0.365–0.623)	0.447 (0.221–0.697)	0.987 (0.975–0.993)	0.987 (0.975–0.993)
	PIT readers (30)	0.743 (0.717–0.768)	0.938 (0.876–0.970)	0.986 (0.980–0.990)	0.976 (0.948–0.989)
LT[d]	Conventional captures (5)	0.204 (0.115–0.335)	0.334 (0.158–0.572)	0.984 (0.954–0.994)	0.985 (0.874–0.998)
	PIT readers (21)	0.858 (0.842–0.872)	0.860 (0.806–0.900)	0.988 (0.984–0.991)	0.973 (0.954–0.985)

[a] Conventional captures occurred on 14, 28 June, 12, 26 July, and 12 August 2002.
[b] Incalculable due to low capture probability of juveniles during conventional captures.
[c] Conventional captures occurred on 10, 24 June, 8, 22 July, and 8 August 2002.
[d] Conventional captures occurred on 11, 26 June, 10, 24 July, and 6 August 2002.

sexes). (These daily survival rates should not be converted to estimate annual rates because sampling may have included early dispersal; differences in severity of seasonal mortality were also likely and these estimates were made only in summer.) Several PIT reader input files were truncated from daily encounters to encounters for every 48-hour period to allow the model to reach numerical convergence in Program MARK.

Capture probabilities at all three roosts (Table 15.1) were up to 10-fold higher when estimated from PIT readers than when estimated from conventional techniques. Capture probabilities also varied more widely among roosts for conventional captures than for PIT readers. The probability of capturing bats was highest at roost CR40 and lowest at HF. The CR40 roost opening was within 5 m of the ground and most bats that emerged to forage were easily captured, whereas emergence points at HF were less predictable and were located 10 m above ground. Estimates of apparent survival were similar but confidence intervals were more precise when estimates were derived from PIT readers in comparison with conventional captures (Table 15.1).

Return rates of females from 2001 to 2002 were high when bats were resighted by PIT readers, and most of these individuals were never detected using conventional methods. We PIT tagged 404 adult females at seven roosts during 2001 that were monitored by readers in 2002; 81.9% [95% CI = (78–86%), $n = 331$] returned in 2002. This is higher than most rates reported in the literature for this species based on other methods (Appendix 15.1). Return rates of juvenile females (103 of 163, or 63%; 95% CI = 56–70%) also were higher. Two hundred fifty-two of 434 (58%) bats of both age groups combined that were detected by PIT readers during 2002 were never recaptured by hand at roosts, despite multiple conventional captures to obtain biological samples and compare capture techniques.

Example 2: Effects of Biological Sampling Techniques on Survival

We used techniques developed and described by Wimsatt et al. (in press) to sample blood of anesthetized big brown bats to obtain serum for rabies antibody determination. Obtaining blood samples from large numbers of bats under anesthesia and releasing them in the field was not common in most studies of bats, and this was the most invasive aspect of our biological sampling. Therefore, in this example we summarize findings of Wimsatt et al. (in press), who sought to determine if these procedures affected short-term survival of big brown bats.

Wimsatt et al. (in press) used a random number table to randomly assign individuals captured by hand during summer 2002 to a treatment (handled and sampled for several biological attributes, anesthetized, and blood sampled through intravenous puncture) or a reference group (handled and sampled, but not anesthetized and no blood taken). Bats were held until all were processed and released simultaneously near their roosts each night when sampling was completed. We obtained daily PIT reader records at roosts where the readers were operational for 14 consecutive days after capture. Separate encounter history files were created for adults and juveniles. We sampled adult females from six roosts ($n = 179$ treatment, 86 reference group) and juveniles from five roosts ($n = 87$ treatment, 92 reference group). We estimated daily survival over the 2 weeks following the

Table 15.2.

Maximum likelihood estimates of short-term apparent survival ($\hat{\phi}$) and capture probabilities ($\hat{p}$) for adult female and juvenile big brown bats by treatment (anesthetized and bled) and reference (not anesthetized, not bled) groups. Estimates were based on passive detection by PIT tag readers placed over roost entrances (see also Wimsatt et al., in press; reproduced with permission of the Journal of Wildlife Diseases).

	Adults		Juveniles	
Treatment	$\hat{\phi}$ (95 % CI)	$\hat{p}$ (95 % CI)	$\hat{\phi}$ (95 % CI)	$\hat{p}$ (95 % CI)
Anesthetized and bled	0.984 (0.977–0.989)	0.757 (0.738–0.775)	0.968 (0.954–0.978)	0.636 (0.604–0.668)
Not anesthetized, not bled	0.958 (0.942–0.978)	0.757 (0.725–0.785)	0.956 (0.940–0.968)	0.660 (0.626–0.692)

first capture, handling, and bleeding of the season (some individuals were bled more than once during the summer). We used Program MARK to estimate survival and capture probabilities for the two groups of bats and goodness-of-fit Test 1 in Program RELEASE (Burnham et al. 1987) to test for differences between the groups. We also attempted to assess possible longer term impacts of multiple sampling sessions by comparing return rates in 2002 of female bats (most males do not use maternity roosts) that had been captured, handled, and bled once or twice in 2001.

We did not detect a difference in 14-day daily survival and capture probabilities (Table 15.2; Wimsatt et al., in press) of bats that were sampled for blood under anesthesia and those that were not ($\chi^2 = 22.2, df = 27$, $P = 0.73$ for juveniles; $\chi^2 = 9.7$, $df = 18$, $P = 0.94$ for adults). Return rates in 2002 for bats bled once or twice during the summer of 2001 were high and confidence intervals overlapped widely with estimates of return rates for bats not bled in 2001 (Table 15.3). These results provide some assurance that the most invasive aspects of the biological sampling did not strongly bias our estimates of apparent survival.

Example 3: Feasibility of Using PIT Marking to Estimate Transition-State Probabilities

This third example was a preliminary study in which we had two objectives. Our first objective was to estimate daily probabilities of movement

Table 15.3.

First year (2001–2002) return rates ($\hat{r}$) for female big brown bats handled, but not anesthetized and bled, or sampled for blood under anesthesia once or twice during 2001 (after Wimsatt et al., in press; reproduced with permission of the Journal of Wildlife Diseases).

	Treatment		
Age	Not sampled	Sampled once	Sampled twice
Adults	58/72 $\hat{r} = 0.806$ $\hat{SE}(\hat{r}) = 0.0466$ 95% CI (0.698–0.881)	227/276 $\hat{r} = 0.822$ $\hat{SE}(\hat{r}) = 0.0230$ 95% CI (0.773–0.863)	42/50 $\hat{r} = 0.840$ $\hat{SE}(\hat{r}) = 0.0518$ 95% CI (0.711–0.918)
Juveniles	16/29 $\hat{r} = 0.552$ $\hat{SE}(\hat{r}) = 0.0924$ 95% CI (0.372–0.719)	75/113 $\hat{r} = 0.664$ $\hat{SE}(\hat{r}) = 0.0444$ 95% CI (0.572–0.745)	12/17 $\hat{r} = 0.706$ $\hat{SE}(\hat{r}) = 0.1105$ 95% CI (0.458–0.872)

based on PIT reader records. Our second objective was to use model selection procedures (Burnham and Anderson 2002) to evaluate a priori hypotheses about factors that may influence daily movements of bats among roosts.

Many studies using radio-tagging and other observational methods have shown that individuals of many species of bats (including big brown bats in maternity colonies) switch roosting sites from day to day (see Lewis 1995 for a review). However, probabilities and variances associated with such movements have never been estimated. Hypotheses in the literature suggest that movements of bats among roosts may be explained by a number of factors. Two prominent factors proposed to influence the probability of movement among roosts are day-to-day microclimatic shifts and ectoparasite infestations. To our knowledge, Lewis (1996) conducted the only field study that investigated the relative importance of these two possible explanations. Based on daily radio-tracking, she reported that individual pallid bats (*Antrozous pallidus*) with higher ectoparasite loads switched roosts more often (based on an ANOVA approach), but that changes in roost use did not appear to be in response to daily changes in temperature.

We used the model of Brownie et al. (1993) and Hestbeck et al. (1991), which generalized the CJS model to allow animals to move among multiple strata by modeling transition probabilities (ψ). To meet the first objective of this example we calculated ψ for members of a maternity colony of big brown bats that used four locations within one large building (HF); each location was monitored with a PIT tag reader at its opening. The locations were designated as northeast (NE), northwest (NW), southeast (SE), and southwest (SW). We assumed that bats roosted within walls near those locations. To meet the second objective we hypothesized that movement probabilities among these locations would be higher with more pronounced shifts in ambient temperature ranges. We also hypothesized that higher numbers of ectoparasites would increase movement probabilities.

We used multi-strata "live recaptures only" models in Program MARK to examine daily transition probabilities in relation to change in temperature the previous day (delta temperature [Δt], which we defined as the maximum temperature the previous day minus the minimum temperature that evening/morning) and total number of ectoparasites (mites of the genus *Steatonyssus*) counted on standard portions of the wings and body of each bat. We limited the analysis to those adult females captured in hand during 15–28 June 2002 at one stage of reproduction (lactation, $n = 43$; juveniles were not yet volant). We modeled number of ectoparasites as an individual covariate and Δt as an environmental covariate. The Δt spanned 11–23° C units over the 2-week period. Counts of ectoparasites ranged from 1 to 562 *Steatonyssus* sp. per bat [$\bar{X} = 130.6$ (SD = 156.5)]. We considered only models in which apparent survival was constant over the 2-week period. We estimated overdispersion (where sampling variance exceeds the theoretical, model-based variance; Burnham and Anderson 2002) by dividing the deviance of the most general model by its number of parameters. Deviance is the difference in –2 log (likelihood) of the current model and –2 log (likelihood) of the saturated model (the saturated model is the model with the number of parameters equal to the sample size; McCullagh and Nelder 1989; Lebreton et al. 1992). The most parsimonious models were selected using $QAIC_c$ (Akaike's Information Criterion corrected for overdispersed data) (Burnham and Anderson 2002).

Our results show that resightings from PIT readers in nearby alternative roosting locations can be used in multi-strata models to estimate movement probabilities and to model associated environmental and individual covariates (Tables 15.4 and 15.5). Estimated movement rates from the north to the south (that is, NE to SE and NW to SE) were generally higher over the period of analysis than movements from the south to the north (that is, SE to NE and SE to NW; Table 15.4). Estimated movement rates also were generally higher from the north to the south than between points on the same side of the building (that is, NE to NW and NW to NE). Movement rates

Table 15.4.

Estimates of transition probabilities ($\hat{\psi}$) and corresponding 95% confidence intervals for female, lactating big brown bats during 15–28 June 2002 at the HF building, Fort Collins, Colorado. NE, NW, SE, SW were the four strata (roost locations) used. Model {p All (.), ψ All (.)} was used to estimate probability of transitioning (see Table 15.5). Read across rows for probabilities of movement between strata (e.g., daily probability of moving from NE to SE was 0.438).

	Stratum			
Stratum	NE	NW	SE	SW
NE	–	0.047 (0.012–0.169)	0.438 (0.297–0.590)	0.058 (0.014–0.204)
NW	0.101 (0.042–0.225)	–	0.308 (0.195–0.450)	0.068 (0.023–0.186)
SE	0.005 (0.001–0.036)	0.030 (0.014–0.063)	–	0.219 (0.169–0.279)
SW	0.022 (0.004–0.102)	0.092 (0.052–0.157)	0.255 (0.165–0.371)	–

Table 15.5.

Results from Program MARK for multi-strata modeling of roost-switching by lactating female big brown bats within HF building, Fort Collins, Colorado. For each model we list the model name (p is capture probability and ψ is transition probability), the Akaike Information Criterion corrected for overdispersion ($QAIC_c$), the Δ $QAIC_c$, $QAIC_c$ weight, and number of parameters (K). Overdispersion ($\hat{c}$) was estimated to be 2.5. Strata were NE, NW, SE, and SW locations of building. Apparent survival ($\hat{\phi}$) was assumed to be constant for all models.

Model	Δ $QAIC_c$	$QAIC_c$	$QAIC_c$ Weight	*K*
{*p* All (.), *ψ* All (Δt)}[a,b]	481.49	0.00	0.38	21
{*p* All (.), *ψ* All (.)}	481.60	0.11	0.36	20
{*p* All (.), *ψ* All (Steat)}[c]	483.70	2.21	0.13	21
{*p* All (.), *ψ* All (Steat + Δt)}	483.71	2.22	0.13	22
{*p* All (time), *ψ* All (.)}	521.48	39.99	0.00	68
{*p* All (time), *ψ* All (Δt)}	523.00	41.51	0.00	69
{*p* All (time), *ψ* All (Steat)}	523.72	42.23	0.00	69
{*p* All (time), *ψ* All (Steat + Δt)}	525.30	43.81	0.00	70
{*p* All (.), *ψ* All (time)}	839.14	357.66	0.00	164
{*p* All (time), *ψ* All (time)}	1086.25	604.77	0.00	212

[a] "All" in model name means each of the four strata were modeled the same.

[b] Δt is high temperature minus low temperature the day before transition.

[c] Steat is the count of individual mites (*Steatonyssus* sp.) on each bat.

between SE and SW roosting sites also were generally higher than those between NE and NW (Table 15.4). These changes may have reflected the differential exposure of the roosts to the sun on north and south sides and a preference of bats to be on warmer south-facing sides as lactation progresses.

The best model in our example (Table 15.5) included constant capture probabilities and an effect of Δt on the transition probabilities. That is, probability of movement of bats varied on a daily basis and co-varied with the scope of temperature change the previous day. Based on the top four models, our results also showed some support for the hypothesis that the number of ectoparasites may have influenced whether a bat moved to another roost site. Models incorporating a time effect on the transition probabilities were less parsimonious than models incorporating the individual covariate (ectoparasites) and the environmental variable (Δt). Future applications in our study will include larger samples of bats and roosts and will consider additional biological hypotheses, locations, time frames, and alternative covariates (e.g., direct temperature measurements within roosts, counts of other ectoparasites) that may better represent the hypotheses under consideration. Although results from this example should be considered preliminary, they illustrate the potential uses of combining new marking approaches in a more rigorous analytical framework than those previously used in most studies of bat survival and movements.

Conclusions

Most studies of survival in bats used ad hoc procedures for estimation of survival rates. However, recent innovations in modeling of survival using maximum-likelihood-based analytical methods and new software applications provide better opportunities to obtain more reliable estimates. A few recent studies of bats have taken such a probabilistic approach to survival estimation (e.g., Baker et al. 2001; Boyd and Stebbings 1989; Hoyle et al. 2001; Sendor and Simon 2003). Future researchers should follow these recent examples, rather than repeat the ad hoc intuitive regression, life table, and other approaches more widely used in the past. Additionally, most previous studies of survival in bats had drawbacks in that methods of marking and recapturing individual bats had the potential to cause injury, decrease survival, or involved other possible biases. We recommend that alternative marking methods continue to be explored for studies of survival in bats. Recent case examples combining an alternative marking technology (PIT tags) with modern analytical approaches from our own work suggest that advances in estimating survival in bats are indeed feasible. We also recommend that future research on survival and transition rates in bats strive to construct biological hypotheses about these critical traits and that hypotheses be evaluated using freely available modern software such as Program MARK (available at http://www.cnr.colostate.edu/~gwhite/mark/mark.htm) and its embedded model selection procedures that are particularly well suited to the study of these probabilities.

Acknowledgments

We thank D. Anderson for key guidance to the Fort Collins study, G. White for tutelage in Program MARK, and R. Bowen for overarching support and database development. We thank the above as well as R. Reich, C. Rupprecht, V. Shankar, and J. Wimsatt for input to study planning; and M. Andre, T. Barnes, M. Carson, K. Castle, L. Galvin, D. Grosblat, B. Iannone, J. LaPlante, G. Nance, D. Neubaum, R. Pearce, V. Price, S. Smith, and T. Torcoletti for assistance in fieldwork in 2001 and 2002. J. Horn and T. Kunz gave advice in use of PIT tags. R. Pearce provided ectoparasite data. The Fort Collins bats and rabies study is supported by a grant from the joint NSF/NIH Ecology of Infectious Diseases Program to Colorado State University and by the U.S. Geological Survey. M. Dodson and L. Lucke helped obtain references. Manuscript drafts were improved by comments from W. Thompson, P. Cryan, and an anonymous reviewer.

References

Allen, A. A. 1921. Banding bats. *Journal of Mammalogy* 2:53–57.

Anderson, D. R. 2001. The need to get the basics right in wildlife field studies. *Wildlife Society Bulletin* 29:1294–1297.

Anderson, D. R., A. P. Wywialowski, and K. P. Burnham. 1981. Tests of the assumptions underlying life table methods for estimating parameters from cohort data. *Ecology* 62:1121–1124.

Baker, G. B., L. F. Lumsden, E. B. Dettmann, N. K. Schedvin, M. Schulz, D. Watkins, and L. Jansen. 2001. The effect of forearm bands on insectivorous bats (Microchiroptera) in Australia. *Wildlife Research* 28:229–237.

Barclay, R. M. R., and G. P. Bell. 1988. Marking and observational techniques. pp. 59–76 in T. H. Kunz, ed., *Ecological and Behavioral Methods for the Study of Bats*. Smithsonian Institution Press, Washington, D.C.

Barclay, R. M. R., and L. D. Harder. 2003. Life histories of bats: Life in the slow lane. pp. 209–253 in T. H. Kunz and M. B. Fenton, eds., *Bat Ecology*. University of Chicago Press, Chicago.

Barnard, S. M. 1989. The use of microchip implants for identifying big brown bats (*Eptesicus fuscus*). *Animal Keepers Forum* 16:50–52.

Beer, J. R. 1955. Survival and movements of banded big brown bats. *Journal of Mammalogy* 36:242–247.

Bezem, J. J., J. W. Sluiter, and P. F. Van Heerdt. 1960. Population statistics of five species of the bat genus *Myotis* and one of the genus *Rhinolophus*, hibernating in the caves of S. Limburg. *Archives Neerlandaises de Zoologie* 13:511–539.

Boyd, I. L., and R. E. Stebbings. 1989. Population changes of brown long-eared bats (*Plecotus auritus*) in bat boxes at Thetford Forest. *Journal of Applied Ecology* 26:101–112.

Bradbury, J. W., and S. L. Vehrencamp. 1976. Social organization and foraging in emballonurid bats. I. Field studies. *Behavioral Ecology and Sociobiology* 1:337–381.

_____. 1977. Social organization and foraging in emballonurid bats. IV. Parental investment patterns. *Behavioral Ecology and Sociobiology* 2:19–29.

Brenner, F. J. 1968. A three-year study of two breeding colonies of the big brown bat, *Eptesicus fuscus*. *Journal of Mammalogy* 49:775–778.

_____. 1974. A five-year study of a hibernating colony of *Myotis lucifugus*. *The Ohio Journal of Science* 74:239–244.

Brooke, A. P. 1997. Social organization and foraging behaviour of the fishing bat, *Noctilio leporinus* (Chiroptera:Noctilionidae). *Ethology* 103:421–436.

Brownie, C., J. E. Hines, J. D. Nichols, K. H. Pollock, and J. B. Hestbeck. 1993. Capture-recapture studies for multiple strata including non-Markovian transitions. *Biometrics* 49:1173–1187.

Burnham, K. P., and D. R. Anderson. 2002. *Model Selection and Multimodel Inference: A Practical Information-Theoretic Approach*, 2nd ed. Springer-Verlag, New York.

Burnham, K. P., D. R. Anderson, G. C. White, C. Brownie, and K. P. Pollock. 1987. Design and analysis of methods for fish survival experiments based on release-recapture. *American Fisheries Society Monograph* 5:1–437.

Caughley, G. 1966. Mortality patterns in mammals. *Ecology* 47:906–918.

Cochran, W. G. 1977. *Sampling Techniques*, 3rd ed. Wiley, New York.

Cockrum, E. L. 1969. Migration in the guano bat, *Tadarida brasiliensis*. pp. 303–336 in J. K. Jones, Jr., ed., *Contributions in Mammalogy*. Museum of Natural History, University of Kansas Miscellaneous Publications 51.

Cockrum, E. L., B. Musgrove, and Y. Petryszyn. 1996. Bats of Mohave County, Arizona: Populations and movements. *Occasional Papers The Museum Texas Tech University* 157:1–72.

Constantine, D. G. 1967. Activity patterns of the Mexican free-tailed bat. *University of New Mexico Publications in Biology* 7:1–79.

Cormack, R. M. 1964. Estimates of survival from the sighting of marked animals. *Biometrika* 51:429–438.

Daan, S. 1980. Long term changes in bat populations in The Netherlands: A summary. *Lutra* 22:95–103.

Davis, R. 1966a. Homing performance and homing ability in bats. *Ecological Monographs* 36:201–237.

Davis, R. B., C. F. Herreid II, and H. L. Short. 1962. Mexican free-tailed bats in Texas. *Ecological Monographs* 32:311-346.

Davis, W. H. 1966b. Population dynamics of the bat *Pipistrellus subflavus*. *Journal of Mammalogy* 47:383–396.

Davis, W. H., and H. B. Hitchcock. 1965. Biology and migration of the bat, *Myotis lucifugus*, in New England. *Journal of Mammalogy* 46:296–313.

Dwyer, P. D. 1966. The population pattern of *Miniopterus schreibersii* (Chiroptera) in north-eastern New South Wales. *Australian Journal of Zoology* 14:1073–1137.

———. 1970. Social organization in the bat *Myotis adversus*. *Science* 168:106–108.

Elder, W. H., and W. J. Gunier. 1981. Dynamics of a grey bat population (*Myotis grisescens*) in Missouri. *American Midland Naturalist* 105:193–195.
Fisher, R. A. 1922. On the mathematical foundation of theoretical statistics. *Philosophical Transactions of the Royal Society of London, Series A* 222:309–368.
Fleming, T. H. 1988. *The Short-Tailed Fruit Bat.* University of Chicago Press, Chicago.
Funakoshi, K., and T. A. Uchida. 1982. Age composition of summer colonies in the Japanese house-dwelling bat, *Pipistrellus abramus. Journal of the Faculty of Agriculture, Kyushu University* 27:55–64.
Gardner, A. L., C. O. Handley Jr., and D. E. Wilson. 1991. Survival and relative abundance. pp. 53–75 in C. O. Handley Jr., D. E. Wilson, and A. L. Gardner, eds., *Demography and Natural History of the Common Fruit Bat,* Artibeus jamaicensis, *on Barro Colorado Island, Panamá.* Smithsonian Contributions to Zoology 511, Washington, D.C.
Gerell, R., and K. Lundberg. 1990. Sexual differences in survival rates of adult pipistrelle bats (*Pipistrellus pipistrellus*) in south Sweden. *Oecologia* 83:401–404.
Glass, B. P. 1982. Seasonal movements of Mexican free-tailed bats *Tadarida brasiliensis mexicana* banded in the Great Plains. *Southwestern Naturalist* 27:127–133.
Goehring, H. H. 1972. Twenty-year study of *Eptescus fuscus* in Minnesota. *Journal of Mammalogy* 53:201–207.
Greenhall, A. M., and J. L. Paradiso. 1968. Bats and bat banding. *U.S. Fish and Wildlife Service Resource Publication* 72:1–47.
Griffin, D. R. 1945. Travels of banded cave bats. *Journal of Mammalogy* 26:15–23.
Happold, D. C. D., and M. Happold. 1998. Effects of bat bands and banding on a population of *Pipistrellus nanus* (Chiroptera: Vespertilionidae) in Malawi. *Zeitschrift fur Saugetierkunde* 63:65–78.
Herreid, C. F., R. B. Davis, and H. L. Short. 1960. Injuries due to bat banding. *Journal of Mammalogy* 41:398–400.
Hestbeck, J. B., J. D. Nichols, and R. A. Malecki. 1991. Estimates of movement and site fidelity using mark-resight data of wintering Canada geese. *Ecology* 72:523–533.
Hickey, J. J. 1952. Survival studies of banded birds. *U.S. Fish and Wildlife Service Special Scientific Report* 15:1–177.
Hill, J. E., and J. D. Smith. 1984. *Bats: A Natural History.* University of Texas Press, Austin.
Hitchcock, H. B. 1957. The use of bird bands on bats. *Journal of Mammalogy* 38:402–405.
Hitchcock, H. B., R. Keen, and A. Kurta. 1984. Survival rates of *Myotis leibii* and *Eptesicus fuscus* in southeastern Ontario. *Journal of Mammalogy* 65:126–130.
Hooper, J. H. D. 1983. The study of horseshoe bats in Devon caves: A review of progress, 1947–1982. *Studies in Speleology* 4:59–70.
Horn, J. W. 1998. *Individual Variation in Nightly Time Budgets of the Little Brown Bat,* Myotis lucifugus. M.A. thesis, Boston University, Boston.
Howell, A. B., and L. Little. 1924. Additional notes on California bats; with observations upon the young of *Eumops. Journal of Mammalogy* 5:261–263.
Hoyle, S. D., A. R. Pople, and G. J. Toop. 2001. Mark-recapture may reveal more about ecology than about population trends: Demography of a threatened ghost bat (*Macroderma gigas*) population. *Austral Ecology* 26:80–92.
Humphrey, S. R., and J. B. Cope. 1970. Population samples of the evening bat, *Nycticeus humeralis. Journal of Mammalogy* 51:399–401.
_____. 1976. Population ecology of the little brown bat, *Myotis lucifugus,* in Indiana and north-central Kentucky. *American Society of Mammalogists Special Publication* 4:1–81.

_____. 1977. Survival rates of the endangered Indiana bat, *Myotis sodalis. Journal of Mammalogy* 58:32–36.

Humphrey, S. R., and T. H. Kunz. 1976. Ecology of a Pleistocene relict, the western big-eared bat (*Plecotus townsendii*), in the southern Great Plains. *Journal of Mammalogy* 57:470–494.

Jolly, S. 1990. The biology of the common sheath-tail bat, *Taphozous georgianus* (Chiroptera: Emballonuridae), in central Queensland. *Australian Journal of Zoology* 38:65–77.

Jones, C. 1976. Economics and conservation. pp. 133–145 in R. J. Baker, J. K. Jones, and D. C. Carter, eds., *Biology of Bats of the New World Family Phyllostomatidae, Part 1.* Texas Tech University Press, Lubbock.

Keen, R. 1988. Mark-recapture estimates of bat survival. pp. 157–170 in T. H. Kunz, ed., *Ecological and Behavioral Methods for the Study of Bats.* Smithsonian Institution Press, Washington, D.C.

Keen, R., and H. B. Hitchcock. 1980. Survival and longevity of the little brown bat (*Myotis lucifugus*) in southeastern Ontario. *Journal of Mammalogy* 61:1–7.

Kerth, G., K. Safi, and B. Koenig. 2002. Mean colony relatedness is a poor predictor of colony structure and female philopatry in the community breeding Bechstein's bat (*Myotis bechsteinii*). *Behavioral Ecology and Sociobiology* 52:203–210.

Kunz, T. H., and A. Kurta. 1988. Capture methods and holding devices. pp. 1–30 in T. H. Kunz, ed., *Ecological and Behavioral Methods for the Study of Bats.* Smithsonian Institution Press, Washington, D.C.

Kunz, T. H., C. R. Tidemann, and G. C. Richards. 1996. Capturing small volant mammals. pp. 122–146 in D. E. Wilson, F. R. Cole, J. D. Nichols, R. Rudran, and M. S. Foster, eds., *Measuring and Monitoring Biological Diversity: Standard Methods for Mammals.* Smithsonian Institution Press, Washington, D.C.

Kuramoto, T., H. Nakamura, and T. A. Uchida. 1985. A survey of bat-banding on the Akiyoshi-dai plateau. IV. Results from April 1975 to March 1983. *Bulletin of the Akiyoshi-dai Science Museum* 20:25–44.

Kurta, A., G. P. Bell, K. A. Nagy, and T. H. Kunz. 1989. Energetics of pregnancy and lactation in free-ranging little brown bats (*Myotis lucifugus*). *Physiological Zoology* 62:804–818.

Kurta, A., T. H. Kunz, and K. A. Nagy. 1990. Energetics and water flux of free-ranging big brown bats (*Eptesicus fuscus*) during pregnancy and lactation. *Journal of Mammalogy* 71:59–65.

Lebreton, J.-D., K. P. Burnham, J. Clobert, and D. R. Anderson. 1992. Modeling survival and testing biological hypotheses using marked animals: A unified approach with case studies. *Ecological Monographs* 62:67–118.

Leigh, E. G. Jr., and C. O. Handley Jr. 1991. Population estimates. pp. 77–87 in C. O. Handley Jr., D. E. Wilson, and A. L. Gardner, eds., *Demography and Natural History of the Common Fruit Bat,* Artibeus jamaicensis, *on Barro Colorado Island, Panamá.* Smithsonian Contributions to Zoology 511, Washington, D.C.

Lewis, S. E. 1995. Roost fidelity in bats: A review. *Journal of Mammalogy* 76:481–496.

_____. 1996. Low roost-site fidelity in pallid bats: Associated factors and effect on group stability. *Behavioral Ecology and Sociobiology* 39:335–344.

Linhart, S. B. 1973. Age determination and occurrence of incremental growth lines in the dental cementum of the common vampire bat (*Desmodus rotundus*). *Journal of Mammalogy* 54:493–496.

Lord, R. D., F. Muradali, and L. Lazare. 1976. Age composition of vampire bats (*Desmodus rotundus*) in northern Argentina and southern Brazil. *Journal of Mammalogy* 57:573–575.

McCracken, G. F., and M. F. Gassel. 1997. Genetic structure of migratory and nonmigratory populations of Brazilian free-tailed bats. *Journal of Mammalogy* 78:348–357.

McCracken, G. F., M. K. McCracken, and A. T. Vawter. 1994. Genetic structure in migratory populations of the bat *Tadarida brasiliensis mexicana*. *Journal of Mammalogy* 75:500–514.

McCullagh, P., and J. A. Nelder. 1989. *Generalized Linear Models*, 2nd ed. Chapman & Hall, London.

Mills, R. S., G. W. Barrett, and M. P. Farrell. 1975. Population dynamics of the big brown bat (*Eptesicus fuscus*) in southwestern Ohio. *Journal of Mammalogy* 56:591–604.

Mohr, C. E. 1952. A survey of bat banding in North America, 1932–1951. *Bulletin of the National Speleological Society* 14:3–13.

Nichols, J. D., and K. H. Pollock. 1983. Estimation methodology in contemporary small mammal capture-recapture studies. *Journal of Mammalogy* 64:253–260.

O'Shea, T. J., and M. A. Bogan, eds. 2003. *Monitoring Trends in Bat Populations in the United States and Territories: Problems and Prospects.* U.S. Geological Survey Information and Technology Report ITR 2003-0003: 1–274

Pape, J. W., T. D. Fitzsimmons, and R. E. Hoffman. 1999. Risk for rabies transmission from encounters with bats, Colorado, 1977–1996. *Emerging Infectious Diseases* 5:433–437.

Pearson, O. P., M. R. Koford, and A. K. Pearson. 1952. Reproduction of the lump-nosed bat (*Corynorhinus rafinesquei*) in California. *Journal of Mammalogy* 33:273–320.

Perry, A. E., and G. Beckett. 1966. Skeletal damage as a result of band injury in bats. *Journal of Mammalogy* 47:131–132.

Ransome, R. D. 1995. Earlier breeding shortens life in female greater horseshoe bats. *Philosophical Transactions of the Royal Society of London B* 350:153–161.

Rice, D. 1957. Life history and ecology of *Myotis austroriparius* in Florida. *Journal of Mammalogy* 38:15–31.

Rybar, P. 1973. Remarks on banding and protection of bats. *Periodicum Biologorum* 75:177–179.

Scheaffer, R. L., W. Mendenhall, and L. Ott. 1986. *Elementary Survey Sampling*, 3rd ed. Duxbury Press, Boston.

Schowalter, D. B., L. D. Harder, and B. H. Treichel. 1978. Age composition of some vespertilionid bats as determined by dental annuli. *Canadian Journal of Zoology* 56:355–358.

Sendor, T., and M. Simon. 2003. Population dynamics of the pipistrelle bat: Effects of sex, age and winter weather on seasonal survival. *Journal of Animal Ecology* 72:308–320.

Sherwin, R. E., S. Haymond, D. Stricklan, and R. Olsen. 2002. Freeze-branding to permanently mark bats. *Wildlife Society Bulletin* 30:97–100.

Sidner, R. M. 1997. *Studies of Bats in Southeastern Arizona with Emphasis on Aspects of Life History of* Antrozous pallidus *and* Eptesicus fuscus. Ph.D. dissertation, University of Arizona, Tucson.

Simmons, N. B. In press. Chiroptera. in D.E. Wilson and D.M. Reeder, eds., *Mammal Species of the World*, 2nd ed. Smithsonian Institution Press, Washington, D.C.

Sluiter, J. W., P. F. Van Heerdt, and J. J. Bezem. 1956. Population statistics of the bat *Myotis mystacinus*, based on the marking-recapture method. *Archives Naeerlandaises de Zool* 12:63–88.

Sluiter, J. W., P. F. Van Heerdt, and A. M. Voute. 1971. Contribution to the population biology of the pond bat, *Myotis dasycneme*, (Boie, 1825). *Decheniana-Beihefte* 18:1–44.

Stebbings, R. E. 1966. A population study of bats of the Genus *Plecotus*. *Journal of Zoology* 150:53–75.

_____. 1970. A comparative study of *Plecotus auritus* and *P. austriacus* (Chiroptera, Vespertilionidae) inhabiting one roost. *Bijdragen Tot De Dierkunde* 40:91–94.

Stebbings, R. E., and H. R. Arnold. 1987. Assessment of trends in size and structure of a colony of the greater horseshoe bat. *Symposium of the Zoological Society of London* 58:7–24.

Stevenson, D. E., and M. D. Tuttle. 1981. Survivorship in the endangered gray bat (*Myotis grisescens*). *Journal of Mammalogy* 62:244–257.

Svoboda, P. L., J. R. Choate, and R. K. Chesser. 1985. Genetic relationships among southwestern populations of the Brazilian free-tailed bat. *Journal of Mammalogy* 66:444–450.

Thomas, D. W., M. Dorais, and J.-M. Bergeron. 1990. Winter energy budgets and cost of arousals for hibernating little brown bats, *Myotis lucifugus*. *Journal of Mammalogy* 71:475–479.

Thompson, M. J. A. 1987. Longevity and survival of female pipistrelle bats (*Pipistrellus pipistrellus*) on the Vale of York, England. *Journal of Zoology* 211:209–214.

Thompson, W. L., G. C. White, and C. Gowan. 1998. *Monitoring Vertebrate Populations*. Academic Press, San Diego.

Tinkle, D. W., and I. G. Patterson. 1965. A study of hibernating populations of *Myotis velifer* in northwestern Texas. *Journal of Mammalogy* 46:612–633.

Trapido, H., and P. E. Crowe. 1946. The wing banding method in the study of the travels of bats. *Journal of Mammalogy* 27:224–226.

Tuttle, M. D., and D. Stevenson. 1982. Growth and survival of bats. pp. 105–150 in T. H. Kunz, ed., *Ecology of Bats*. Plenum Press, New York.

Villa R., B., and E. L. Cockrum. 1962. Migration in the guano bat, *Tadarida brasiliensis mexicana* (Saussure). *Journal of Mammalogy* 43:43–64.

White, G. C. 2001. Statistical models: Keys to understanding the natural world. pp. 35–56 in T. M. Shenk and A. B. Franklin, eds., *Modeling in Natural Resource Management: Development, Interpretation, and Application*. Island Press, Washington, D.C.

White, G. C., D. R. Anderson, K. P. Burnham, and D. L. Otis. 1982. *Capture-recapture and Removal Methods for Sampling Closed Populations*. USDOE Report Number LA-8787-NERP, Los Alamos National Laboratory, Los Alamos, New Mexico.

White, G. C., and K. P. Burnham. 1999. Program MARK: Survival estimation from populations of marked animals. *Bird Study* 46:S120–S138.

Williams, B. K., J. D. Nichols, and M. J. Conroy. 2002. *Analysis and Management of Animal Populations*. Academic Press, San Diego.

Wilson, D. E., and J. S. Findley. 1972. Randomness in bat homing. *American Naturalist* 106:418–424.

Wimsatt, J., T. J. O'Shea, L. E. Ellison, R. D. Pearce, and V. R. Price. In press. Anesthesia and blood sampling for health screening of big brown bats, *Eptesicus fuscus*, with an assessment of impacts on survival. *Journal of Wildlife Diseases*.

Young, R. A. 2001. The eastern horseshoe bat, *Rhinolophus megaphyllus*, in south-east Queensland, Australia: Colony demography and dynamics, activity levels, seasonal weight changes, and capture-recapture analyses. *Wildlife Research* 28:425–434.

Zahn, A. 1999. Reproductive success, colony size and roost temperature in attic-dwelling bat *Myotis myotis*. *Journal of Zoology* 247:275–280.

Appendix 15.1.

Summary of studies of survival rates in bats See Mohr (1952) for summaries of studies reporting recapture rates before 1950. Abbreviations: A= adult, AF = adult female; AIC = Akaike's Information Criteria; AM = adult male; BM = Bezem method; CJS = Cormack-Jolly-Seber; F = female; GLG = growth layer groups in tooth sections; GOF = goodness-of-fit; IR = intuitive regression approach; J = juvenile; KAC = known-age cohorts; LT = life table; M = male; SC = survivorship curve; SE = standard error; subA = subadult; UAC = unaged cohorts.

pecies, ocation, Years	Methods and Sample	Hypotheses Tested	Summary of Findings, Comments	Reference
ntrozous pallidus, ptesicus fuscus, rizona, 1980–1995	Bands, 1,702 *A. pallidus*, 2,231 *E. fuscus*, 2 nurseries each; LT from KAC.	Descriptive; age, species contrasts; correspondence with environmental factors.	1st-year F survival lowest (means 0.42–0.49 *A. pallidus*, 0.53–0.64 *E. fuscus*). Annual mean AF survival 0.29–0.76 *A. pallidus*, 0.71–0.90 *E. fuscus*.	Sidner (1997)
rtibeus jamaicensis, nama, 1976–1980	Bead necklaces, mist-netting forest, roosts; 8,907. % recap or % known alive by half-year intervals, some KAC (J, subA, A). Sexes, ages, marking periods treated separately. Chi-square tests on % recap among groups. Leigh and Handley (1991) assumed constant, identical survival process as an earlier study; implies no individual heterogeneity in survival and capture probabilities; also used IR.	Descriptive; differential survival sexes, age classes, birth season of J.	J survival < A, J born spring > J born autumn. Suggested AF survival > AM. Tag loss. Leigh and Handley (1991) calculated annual survival of 0.58. Survival appears low, but survival and capture probabilities confounded.	Gardner et al. (1991); Leigh and Handley (1991)

Species, Location, Years	Methods and Sample	Hypotheses Tested	Summary of Findings, Comments	Reference
Balionopteryx plicata, Rhynchonycteris naso, Saccopteryx bilineata, S. leptura, Costa Rica, 1973	Ad hoc surrogate, females only, (% of females captured with used teats, parturition rates) *n* unspecified.	Descriptive, differences based on habitat stability.	*B. plicata* 54%, *R. naso* >78%, *S. bilineata* 78%, *S. leptura* >78%.	Bradbury anc Vehrencamp (1976, 1977)
Carollia perspicillata, Costa Rica, 1974–1984	Bands, beaded necklaces, 3,253, UAC IR/LT approach; annual changes in tooth wear classes; % in cave recaptured after 1 yr.	Descriptive; differential survival of AF, AM, J.	Young recap < A; young F recap < young M (likely differential dispersal);AM, AF similar % recap. Tag loss.	Fleming (198
Chalinolobus morio, Nyctophilus geoffroyi, Vespadelus vulturnus, Australia, years unspecified	Bands, 10–12 yr periods, 1,870–3,936 (varied with species), colony type unspecified; CJS estimate ϕ Program JOLLY, time-varying p_i and ϕ.	Band type, degree of injury affected ϕ. Differences in ϕ by band type within species tested by Behrens-Fisher t-tests.	Differences by band type in *N. geoffroyi* (0.19 ± 0.05 SE Monel bat bands, 0.42 ± 0.12 Mg-Al bird bands). Age, sex unspecified. Highest annual ϕ for *V. vulturnus* 0.59 ± 0.10; *C. morio* 0.72 ± 0.04.	Baker et al. (2001)
Corynorhinus townsendii, California, 1947–1951	Bands, AF, J, 2 nurseries, 85 and 60 AF, also hibernacula. % recaptured.	Descriptive	40–54 % J return after 1 yr; 70–80% AF return annually.	Pearson et a (1952)
Desmodus rotundus, Mexico and Argentina, 1969	GLG teeth, 217, sexes combined. SC (age vs. % in sample). GLG not validated with known-age specimens.	Descriptive	SC. Assumed constant reproduction, stationary age distribution.	Lord et al. (1976)
Desmodus rotundus, Argentina and Brazil, years unspecified	GLG teeth, 87, sexes combined. SC (number GLG vs. number of bats). GLG not validated with known-age specimens.	Descriptive	SC	Linhart (19
Eptesicus fuscus, Minnesota, 1940–1954	Bands, 3,871, hibernacula; UAC, incorrect ad hoc band recovery methods.	Descriptive	"over-all mortality rate of about 40 percent per year" (Beer 1955:245); survival lowest 1st yr.	Beer (1955

Species, Location, Years	Methods and Sample	Hypotheses Tested	Summary of Findings, Comments	Reference
Eptesicus fuscus, Minnesota, ca. 1951–1971	Bands, 960, hibernaculum; UAC, % alive vs. yrs after banding by sex.	Descriptive	Survival 60% 1st yr after banding, higher thereafter.	Goehring (1972)
Eptesicus fuscus, Minnesota, ca. 1951–1971	Bands, IR	Descriptive	Annual survival AM 0.82, AF 0.74.	Hitchcock et al. (1984), based on Goehring (1972)
Eptesicus fuscus, Ohio, 1969–1972	Bands, 110 and 142, 2 nurseries, KAC, IR/LT at 1–3 yrs after banding.	Descriptive	% recap colony 1, colony 2: yr 1 32%, 10%; yr 2 71%, 70%; yr 3 28%, 57%. Declines (88% and 60%) thought due to study.	Mills et al. (1975)
Eptesicus fuscus, Ohio and Pennsylvania, 1965–1967	Bands, ~39–72, 2 nurseries (Ohio, Pennsylvania); % of F recaptured after 1 yr.	Descriptive	Ohio: AF return 53% in 1966, 24% 1967; J 32% 1966, 17% 1967. Penn: AF 21% 1966, 10% 1967; J 14% 1966, 12% 1967.	Brenner (1968)
Macroderma gigas, Queensland Australia, 1976–1980	Ear tags, 268, mist nets, caves. Tropical. Recaps in 3-mo seasons. CJS model in MARK, GOF tests RELEASE; used MARK to test for trap dependency, bootstrap estimate overdispersion parameter. A treated separately from an age-structured model based on 163 bats first caught as J.	AIC model selection for A ϕ. Sex, time, season, year, rainfall (flooding roosts), temperature as group factors or covariates.	Best model for A ϕ included sex, season, rainfall. AF annual ϕ (0.57–0.77) > AM (0.43–0.66). AIC support for rainfall, sex, temperature effects for J. 1st yr ϕ JF 0.35–0.46, JM 0.29–0.42. Tag loss.	Hoyle et al. (2001)
Miniopterus schreibersi, NSW Australia, 1960–1963	Bands, toe clips, 8,775, nurseries and nonbreeding sites; estimated % as yearlings, adults, arithmetically computed survival that would yield this age distribution.	Descriptive	A survival ~0.75.	Dwyer (1966)

Species, Location, Years	Methods and Sample	Hypotheses Tested	Summary of Findings, Comments	Reference
Miniopterus schreibersi, Myotis macrodactylus, Rhinolophus ferrum-equinum, Japan, ca. 1966–1980	Bands, 2,795 *M. schreibersi*, 1,924 *M. macrodactylus*, 3,707 *R. ferrum-equinum*; nursery, hibernacula; % recaptured 1–14 yrs after banding.	Descriptive	AF % recaptured > AM in *M. schreibersi*, sexes similar in *Myotis*.	Kuramoto et al. (1985)
Myotis austroriparius, Florida, 1952–1955	Bands, 1,998, nurseries; % recap by age at 1, 2 or ≥ 3 yrs.	Descriptive	Recap ~47% yr 1, 18% yr 2, 35 % ≥ yr 3.	Rice (1957)
Myotis dasycneme, Netherlands, 1955–1962	Bands, 179 AF, nursery; BM.	Descriptive; compared estimates at winter vs. summer banding sites.	Annual AF survival similar summer (0.70) and winter (0.67) colonies; > % injury bats banded summer.	Sluiter et al. (1971)
Myotis dasycneme, 1940–57; *M. daubentoni*, 1945–1957; *M. emarginatus*, 1945–1955; *M. myotis*,1940–1955; *M. mystacinus*, 1945–1957; *Rhinolophus hipposideros*, 1942–1956; Netherlands	Bands, hibernacula 1,191 *M. dasycneme*; 920 *M. daubentoni*; 1,608 *M. emarginatus*; 650 *M. myotis*; 1,828 *M. mystacinus*; 1,717 *R. hipposideros*. BM.	Descriptive; differences by 1 cave groups in 3 species *Myotis*, sex in all (ANOVA on slopes and intercepts).	AF, AM survival similar all species; F = M capture rates in all, except *M. daubentonii*. No diffs in survival but diffs in capture rates between caves for 3 species. Annual survival for 5 species of *Myotis* range 0.64–0.80, 0.57 in *R. hipposideros*.	Bezem et al. (1960)
Myotis grisescens, Missouri, 1968–1978	Bands, 18,632 banded 2 yrs, 1 hibernaculum, UAC; IR by sex.	Descriptive	M annual survival 70%, F 73%; lower rates 1st yr.	Elder and Gunier (1981
Myotis grisescens, Alabama, Tennessee, ca. 1968–1976	Bands, ca. 3,946 at 3 nurseries, recap at 2 hibernacula. J, yearling and A cohorts. IR/LT; nonparametric trend tests on % recap.	Differential survival at declining and stable maternity colonies, sex and age class diffs.	Survival similar by sex, age. 1st yr estimates very variable, 0.06–0.73; 0.57–0.85 thereafter. Survival reflects changes in counts.	Stevenson an Tuttle (1981)

Species, Location, Years	Methods and Sample	Hypotheses Tested	Summary of Findings, Comments	Reference
Myotis leibii, Ontario, 1941–1948	Bands, 258, hibernaculum, CJS model.	Differential survival of sexes.	AM ϕ 0.76 ± 0.11 SE (range 0.60–0.89, 5 estimates). AF ϕ 0.42 ± 0.07 (range 0.17–1.0, 5 estimates). $P = 0.01$, z-test.	Hitchcock et al. (1984)
Myotis lucifugus, Ontario, ca. 1947–1962	Bands, 1,936 bats at 1 hibernaculum; CJS model, tested rates using z-test; compared results BM, IR/LT.	Differences by sex.	Annual survival of AM (0.82 ± 0.01) > AF (0.71 ± 0.02).	Keen and Hitchcock (1980)
Myotis lucifugus, Indiana and Kentucky 1953–1969	Bands, > 24,000; 386 cohorts from multiple nurseries, 278 cohorts from multiple hibernacula; IR/LT, estimates across multiple colonies, cohorts from nurseries defined as adult or immature at banding, UAC at hibernacula.	Descriptive	Not all roosts sampled annually, selected subsets with highest numbers alive as typical for semi-log plots. Survival 1st yr after banding (13–49%) < subsequent yrs (54–86%).	Humphrey and Cope (1976)
Myotis lucifugus, Pennsylvania, ca. 1965–1969	Bands, 2,914, hibernaculum; number recap annually, LT.	Descriptive	1st yr survival lowest.	Brenner (1974)
Myotis lucifugus, Alberta, ca. 1975	GLG, 90 AF collected at maternity colony, SC from % each age class.	Descriptive		Schowalter et al. (1978)
Myotis myotis, Germany, 1991–1994	Bands, 232 JF at 4 nurseries, reappearance rates 1 yr later; J carcasses as % of J present.	Descriptive	F reappearance after 1st winter varied by yr and colony (0–40%).	Zahn (1999)
Myotis mystacinus, Netherlands, 1942–1956	Bands, *n* not clearly specified; hibernacula, BM.	Descriptive, differences by cave, sex.	AM, AF survival similar at 0.79.	Sluiter et al. (1956)

Species, Location, Years	Methods and Sample	Hypotheses Tested	Summary of Findings, Comments	Reference
Myotis sodalis, Indiana, ca. 1956–1973	Bands, 4 M and 3 F cohorts at 2 hibernacula; UAC, IR/LT.	Descriptive	AM, AF SC similar. Lowest survival 1st yr after marking, last few yrs. A survival highest and constant at yrs 1–6 (F 0.76, M 0.70), lower and constant at yrs 6–10 (F 0.66, M 0.36).	Humphrey and Cope (1977)
Nycticeus humeralis, Indiana, 1958–1964	Bands, 526, 2 nurseries, % recap for UAC, J, A, 1–4 yrs after banding.	Descriptive	1st yr % recap 0.23–0.32; AF ca. 0.60.	Humphrey and Cope (1970)
Pipistrellus abramus, Japan, 1971–1975	Bands, 315 in 2 nurseries; SC, % known alive, LT; GLG in killed sample of 106 at 1 colony.	Descriptive	"Disappearance rates" (Funakoshi and Uchida 1982: 60) of F to 1 yr 0.18–0.29, 0.39 in 2nd yr; M 0.85–0.96 in 1st yr. Based on LT of GLG classes, survival in bats age 3–5 < bats age 1–2.	Funakoshi and Uchida (1982)
Pipistrellus pipistrellus, Sweden, 1981–1988	Bands, 1,253, bat boxes in summer, CJS model.	Differential survival of sexes due to male territoriality; seasonal climatic influences on annual variation.	Annual survival AM = 0.44 (ranges of variances 0.010–0.026), AF = 0.54 (0.005–0.097). $P < 0.05$, likelihood test.	Gerell and Lundberg (1990)
Pipistrellus pipistrellus, England, 1977–1984	Bands, 90 AF 55 JF, 2 nurseries; UAC, IR/LT.	Descriptive	Annual survival AF = 0.64, seemingly constant. J survival lower 1st yr.	Thompson (1987)

Species, Location, Years	Methods and Sample	Hypotheses Tested	Summary of Findings, Comments	Reference
Pipistrellus pipistrellus, Germany, 1996–2000	Bands, 4,857 AF, 3,263 AM, 4,408 JF, 3,311 JM; one site used as both hibernaculum and "swarming" at night in summer; bats only caught while in flight to minimize disturbance; CJS models following Lebreton et al. (1992), program MARK, AIC_c.	Modeled effects of sex, age, time, season, winter severity on survival, and effort on capture probabilities.	Survival varied by age class and was time-dependent; no effects by sex, and survival was not lower in winter. Annual survival A = 0.80 (SE 0.05), J = 0.53 (SE 0.10). Transience effects noted.	Sendor and Simon (2003)
Pipistrellus subflavus, West Virginia, 1952–1965	Bands, 2 hibernacula, IR/LT each sex. Ad hoc correction to % recaptured using number alive but not recaptured. Justification for the form of the correction not provided, properties unknown.	Descriptive. Age independence of survival.	Survival age-dependent: low yrs 1, 2 (0.41, 0.51 F, 0.46, 0.68 M), increased to yrs 3–4 (0.74, 0.51 AF, 0.98, 0.81 AM), declined to yrs 10–14.	Davis (1966b)
Plecotus auritus, P. austriacus, England, ca. 1960–1969	Bands, year-round colony, 16–20 *P. auritus*, 5–15 *P. austriacus* known alive each yr; IR/LT.	Descriptive	Annual survival AM *P. auritus* 0.54, AF 0.76, *P. austriacus* AM 0.45, AF 0.62.	Stebbings (1966, 1970)
Plecotus auritus, England, 1975–1985	Bands, 401, nurseries in bat boxes. Two methods: KAC IR, CJS.	Descriptive; comparisons of analytical methods. Integrated with reproduction, sex ratio, and population size information.	Geometric mean annual survival estimates 0.78 ± 0.03 SE AF (range 0.68–0.98 over 9 annual estimates), 0.62 ± 0.08 SE AM (range 0.49–0.94). AF survival = 0.86 using regression corrected for rate of increase and age skewness; uncorrected estimate AF = 0.80, AM = 0.60.	Boyd and Stebbings (1989)

Species, Location, Years	Methods and Sample	Hypotheses Tested	Summary of Findings, Comments	Reference
Rhinolophus ferrumequinum, England, 1948–1958	Bands, 1,808, primarily hibernacula. Number known alive each yr after banding.	Descriptive	Simple reporting of number banded and number known alive by yr.	Hooper (1983)
Rhinolophus ferrumequinum, England, ca. 1970–1986	Bands, toe claw clipping J, hibernacula and nurseries, hand captures. IR/LT, UAC and KAC separately.	Descriptive: AM vs. AF, J vs. A, survival, survival by season.	M, F SC similar. Survival to ages 1 (0.65) and 2 (0.62) lowest. Annual survival increases to 0.86 at age 9. Survival lowest winter, % varied by yr from 0.55 to 0.85.	Stebbings and Arnold (1987)
Tadarida brasiliensis, Texas, 1957–1958	Ad hoc extrapolation from relative areas of hypothetical SCs, and best fit by age ratios based on 3 classes of tooth wear in 22,498 Fs.	Descriptive	AF survival of 70–80% "reasonable" (Davis et al. 1962: 337) based on age class ratios.	Davis et al. (1962)
Taphozous georgianus, Australia, 1985–1989	Bands, 229 J, mist nets at caves, subtropical. IR/LT.	Descriptive	Annual survival F 0.42–0.75, M 0.24–0.66.	Jolly (1990)

16

Evaluating Methods for Monitoring Populations of Mexican Spotted Owls: A Case Study

Joseph L. Ganey, Gary C. White,
David C. Bowden, and Alan B. Franklin

Monitoring population status of rare or elusive species presents special challenges. Understanding population trends requires separating signal (true and important changes in abundance) from noise (normal temporal and sampling variation; e.g., Block et al. 2001). This is particularly difficult when small numbers or elusive habits make it difficult to obtain precise estimates of population parameters (Thompson et al. 1998:68). We conducted a pilot study to evaluate a sampling design proposed for monitoring populations of the threatened Mexican spotted owl (*Strix occidentalis lucida*; USDI Fish and Wildlife Service 1995). Here, we discuss this effort as a case study illustrating some pitfalls and issues involved in sampling rare and/or elusive populations. We begin with a brief history of events leading up to the pilot study to provide a context for the study. We then describe study design and implementation and summarize results of the study and conclusions following from those results. Future monitoring efforts may benefit from our efforts and the lessons we learned.

Background and History

The Mexican spotted owl inhabits forested mountains and canyonland terrain throughout the southwestern United States and Mexico (Gutiérrez et al. 1995; Ward et al. 1995). This owl is associated with late-successional forests throughout much of its range (Ganey and Dick 1995; Gutiérrez et al. 1995) and was listed as threatened in 1993 (USDI Fish and Wildlife Service 1993) because of concerns about the effects of timber harvest and wildfire on amount and distribution of late-successional forest.

A recovery plan produced for this owl contained five delisting criteria that must be met to remove the owl from the list of threatened species. One criterion called for monitoring populations of owls in three of the six "Recovery Units" identified in the United States (USDI Fish and Wildlife Service 1995:76–77; see also Figure 16.1). The monitoring criterion specified that delisting should occur only if the populations monitored were shown to be stable or increasing using a monitoring design capable of detecting a 20% decline over 10 yrs with a power of 90% and a Type I error rate of 5%. The specificity in this delisting criterion represented a deliberate attempt to provide a rigorous and unambiguous measure of population status and was a direct response to the history of litigation and administrative appeals surrounding the spotted owl issue (e.g., Marcot and Thomas 1997).

The recovery plan also outlined a general monitoring approach. Briefly, this approach involved (1) surveying for owls on randomly located sample "quadrats" approximately 40–75 km^2 in area (number of quadrats to be determined); (2) capturing and color-banding owls so that individuals could be uniquely identified; (3) estimating abundance of owls on the quadrats using capture-recapture models for closed populations; and (4) monitoring changes in abundance over time. As a first step in evaluating this approach, the USDA Forest Service, Southwestern Region (hereafter the SW Region), provided funding to evaluate the sampling methods on four quadrats in Arizona. Results of this test indicated that the field methodology was feasible at the limited scale tested (May et al. 1996).

The SW Region formally committed to "monitor changes in owl populations and habitat needed for delisting" in an amendment to Land Management Plans covering 11 national forests in the region (USDA Forest Service 1996:88). The SW Region also engaged in formal consultation with the U.S. Fish and Wildlife Service on the effects of this amendment on the owl, under Section 7 of the Endangered Species Act (Stanford Environmental Law Society 2001:Chapter 3). The resulting Biological Opinion stated "Population monitoring will follow the design in the proposed action (i.e., the amendment; USDA Forest Service 1996) and beginning on page 107 of the Recovery Plan" (USDI Fish and Wildlife Service 1996:34). It also called for a pilot study to evaluate the monitoring approach and

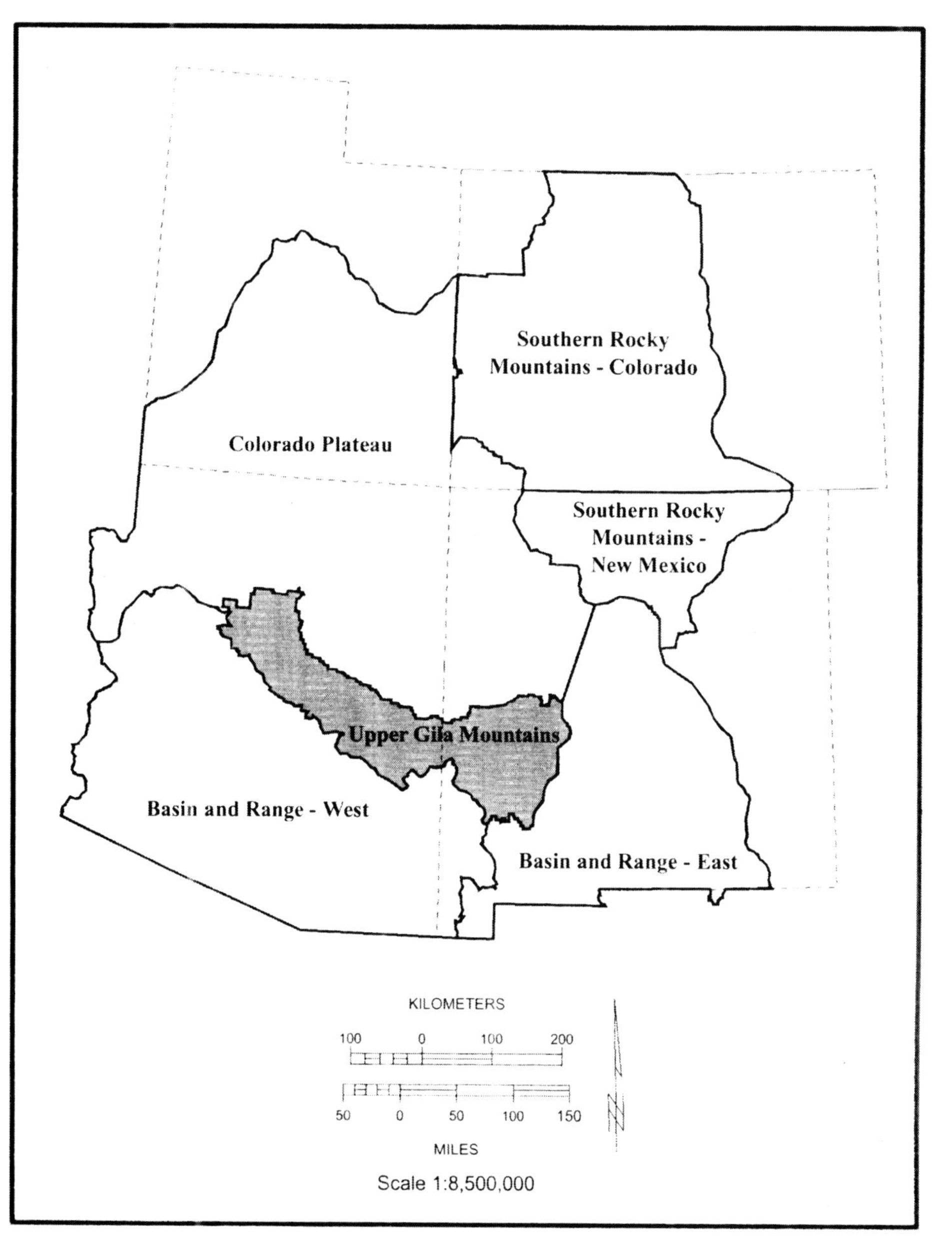

Figure 16.1. Recovery Units recognized for the Mexican spotted owl within the United States (from USDI Fish and Wildlife Service 1995:fig. II.B.1). The Upper Gila Mountains Recovery Unit is highlighted.

specified that the Forest Service should initiate the pilot study within 1 yr and "make timely progress thereafter towards full implementation" (USDI Fish and Wildlife Service 1996:34).

In late 1997, the SW Region requested that the Rocky Mountain Research Station (RMRS) assist in implementing a population monitoring program. RMRS appointed J. L. Ganey to head this effort in January 1998, and Ganey recruited the other authors of this paper to assist in planning, implementing, and evaluating the pilot study. The SW Region provided preliminary funding in 1998 to begin planning for the pilot study and funded implementation of the pilot study in 1999.

Based on the legal and administrative record discussed above, we concluded that our charter was to evaluate the specific monitoring approach described in the recovery plan. Consequently, rather than considering alternative approaches, we designed and conducted a pilot study to further evaluate the methods proposed in the recovery plan. Objectives of this pilot study were to (1) evaluate the feasibility of mark-recapture sampling at the spatial scale specified in the recovery plan; (2) collect preliminary data on owl density, capture probabilities, and spatial variation in those parameters; and (3) use these data to estimate sample-size requirements and costs for conducting a long-term population monitoring program for the Mexican spotted owl.

The Pilot Study

The target population identified in the recovery plan for the long-term monitoring program included the population of territorial Mexican spotted owls in three Recovery Units (Upper Gila Mountains, Basin and Range-West, and Basin and Range-East; Figure 16.1; see Rinkevich et al. 1995 for discussion of Recovery Units). These Recovery Units were chosen because they included what likely are the core populations within the range of the Mexican spotted owl in the United States (Ward et al. 1995). The design focused on territorial owls because no techniques were available to locate nonterritorial owls; hence, the number of nonterritorial owls could not be estimated reliably.

To simplify the pilot study, we restricted the scope of this study to the population of territorial owls in the Upper Gila Mountains Recovery Unit outside of the White Mountain and San Carlos Apache reservations. The Upper Gila Mountains Recovery Unit was chosen because it appeared to contain a large population of Mexican spotted owls (i.e., 424 occupied "sites" [56% of known occupied sites] were documented in this Recovery Unit as of 1993 [Ward et al. 1995]). Tribal lands, which comprised about 18% of the Recovery Unit, were excluded from the pilot study because working on these lands would have required special permissions and access to these lands was not essential to meet the objectives of the pilot study.

Study Area

The Upper Gila Mountains Recovery Unit, approximately 54,860 km^2 in area, corresponds loosely to Bailey's (1980) Upper Gila Mountains Forest Province and William's (1986) Datil-Mogollon Section. This area is transitional between the canyonlands of the Colorado Plateau province to the north and the Basin and Range province to the south. It is characterized by steep mountains and entrenched river drainages dissecting high plateaus. For further information on vegetation patterns, land uses, and other features of this area, see Rinkevich et al. (1995).

Design

Sampling units for the pilot study consisted of randomly located quadrats approximately 40–76 km^2 in area. Quadrat size represented a compromise between the desire for spatial replication and unbiased, precise density estimates. Franklin et al. (1990) found that density estimates for northern spotted owls (*S. o. caurina*) in northwestern California stabilized at a study area size of 90–130 km^2. This was larger than our proposed quadrat size, but there were mitigating circumstances. First, Franklin et al. (1990) did not use ecological boundaries in subsampling their study area, and edge effects consequently may have been exaggerated in their analysis. Second, visual inspection of their figure 4 (a plot of the first derivative of change in density versus study-area size) indicates that the rate of change in density as study area size increased approached zero above 50 km^2. Therefore, we concluded that our quadrat size was both large enough to reduce edge effects and estimate density accurately and small enough to allow spatial replication.

The sampling frame for quadrat selection was developed using spatial modeling in a geographical information systems (GIS) environment. We first developed a GIS coverage that delineated the entire Recovery Unit into separate quadrats of 40–76 km^2 using size and shape constraints, topography (1:250,000 digital elevation models), and other spatial information. To the extent possible, quadrat boundaries followed ecological features such as ridgelines and watersheds in an attempt to reduce edge effects and minimize the probability that a quadrat boundary ran through an owl's activity center. Further details on developing the quadrat coverage are contained in Arundel (1999).

Once the quadrat sampling frame was developed, quadrats were selected using stratified random sampling. Random sampling was used to ensure an unbiased, representative sample so that inferences about owl abundance could be extended to the entire Recovery Unit (exclusive of tribal lands), rather than being restricted to the sample quadrats. Stratification was used to reduce the extent of among-quadrat variability in owl density.

We were unable to locate a useful vegetation-based GIS coverage to aid in stratifying quadrats, so instead stratified quadrats based on elevation. Quadrats with ≥ 33% of their area occurring at elevations ≥ 1,982 m and < 2,745 m were placed in a "high-density" stratum, and all other quadrats were placed in a "low-density" stratum. These represented strata with higher and lower expected owl density, respectively. Most of the quadrat

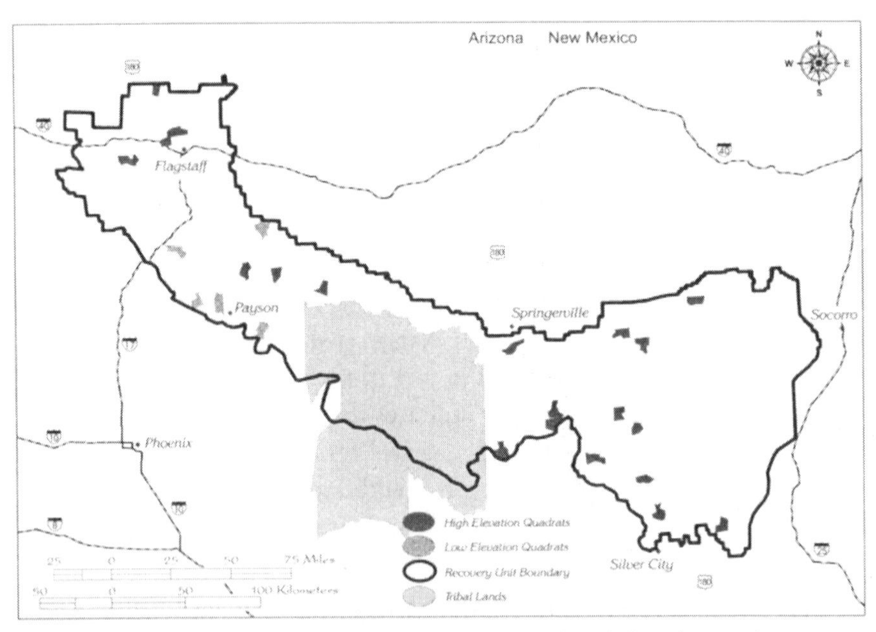

Figure 16.2. Sampling quadrats included in a study of abundance of Mexican spotted owls, Upper Gila Mountains Recovery Unit, Arizona and New Mexico, 1999. Also shown are tribal lands intersecting the Recovery Unit; these lands were not included in the sampling frame.

area within high-density quadrats fell within the elevational range where most Mexican spotted owls occur, whereas low-density quadrats were dominated by areas lower or higher than those where most spotted owls are found (Fletcher and Hollis 1994; Ganey and Dick 1995).

We selected 20 quadrats from the high-density and five from the low-density strata (Figure 16.2). The reduced emphasis on sampling quadrats from the low-density stratum was based on the expectation that few owls would be found in such quadrats and that spatial variation in owl density therefore would be low in that stratum. Conversely, greater expected variation in the high-density stratum resulted in a need for greater sampling intensity. For descriptions of quadrats sampled, see Ganey et al. (1999).

Field Methods

All field sampling for this study was accomplished from April–September 1999. Sampling followed methods established for studies of spotted owl demography (Franklin et al. 1996), with some differences. Fixed call points were established to ensure that the entire area within the quadrat was surveyed effectively. This generally meant that call points were established at intervals of ≤ 0.8 km along roads, trails, ridge tops, or drainage bottoms, depending on terrain. Crew leaders experienced in conducting spotted owl surveys guided establishment of call points.

Nocturnal calling surveys (Forsman 1983; Franklin et al. 1996) were conducted from call points to locate general areas occupied by owls. Surveyors spent 10 min at each call point, alternately calling and listening for a response from territorial owls.

We attempted to complete four survey passes through the quadrats, except for quadrats where no owls were located by the end of the third survey pass. We did not survey those quadrats a fourth time. A survey pass was considered completed when all call points were surveyed and all required daytime follow-up surveys were completed. Daytime follow-up surveys were conducted wherever an owl was detected during nocturnal surveys, to locate nest and/or roost areas, capture and mark owls, resight previously marked owls, and/or determine the reproductive status of resident owls. Two follow-up surveys were allowed per response per survey pass.

Crews attempted to capture and individually mark all territorial adult and subadult owls observed on the quadrat. Owls were captured using noose poles and board traps (Forsman 1983; Johnson and Reynolds 1998) and marked with a numbered locking aluminum band on one leg and a unique color band on the other leg (Forsman et al. 1996). After an initial physical capture, owls were considered to be "recaptured" on subsequent passes if we were able to resight colored leg bands. Thus, each survey pass was treated as a potential capture occasion. Banding records were not reviewed before resighting attempts, to avoid potential bias in recognizing color combinations.

Survey Results

We established 2,345 call points across the 25 sample quadrats. The required number of survey passes was completed on 21 of the 25 quadrats (Table 16.1). Of the remaining quadrats, three survey passes were completed on one, a third pass was largely completed on a second, and two passes were completed on a third. On the fourth quadrat, four passes were completed on the southern half, whereas only two passes were completed on the northern half, where no owls were located during these passes. Failure to complete surveys in these areas was primarily due to limited numbers of surveyors working in roadless areas that featured difficult terrain.

Total survey effort involved calling 8,257 points (including repeat visits) over the course of all passes. This resulted in 434 nocturnal detections of owls, which triggered 400 daytime follow-up surveys. One or more owls were located on 202 of these follow-up surveys, and 57 territorial adult and subadult owls were captured and color-banded.

Spotted owls were heard or seen on 18 quadrats, 16 (80%) in the high-elevation stratum and two (40%) in the low elevation stratum. Of these, we estimated that 14 quadrats in the high-elevation stratum and one in the low-elevation stratum actually supported resident owls whose activity center was located within the quadrat boundary (70% and 20% of total quadrats within strata, respectively). Ganey et al. (1999) provided further information on numbers of territories resolved, reproductive histories, and numbers of owls banded by quadrat.

Initial Estimates of Population Size and Spatial Variation in Owl Density

We used the following steps, each of which is outlined in more detail below, to estimate the total number of Mexican spotted owls (N_{RU}) and its standard error ($\text{SE}[N_{RU}]$) for the Upper Gila Mountains Recovery Unit. First, we used the capture-recapture data to estimate capture probabilities

Table 16.1.

General location, area, number of survey passes completed, and number of call points for 25 quadrats surveyed for Mexican spotted owls, Upper Gila Mountains Recovery Unit, 1999. Quadrat numbers begin with H or L to indicate expected high- and low-density quadrats, respectively.

Quadrat						
No.	Name	Area (km^2)	National forest	District	Survey passes	Call points
H01	Red Mtn.	43.7	Coconino/Kaibab	Peaks/Chalender	3[a]	60
H02	Peaks	61.2	Coconino	Peaks	4	89
H03	Wing Mtn.	46.5	Coconino	Peaks	3[a]	67
H04	White Horse	76.4	Kaibab	Chalender	4	90
H07	Gen. Springs	59.6	Coconino	Blue Ridge	4	66
H08	Chevelon	55.7	Apache/Sitgreaves	Chevelon	4	86
H09	Heber	54.9	Apache/Sitgreaves	Heber	3[a]	96
H11	Datil	57.9	Cibola	Magdalena	3[a]	102
H12	Slaughter Mesa	63.2	Apache[b]	Quemado	4	77
H13	Springerville	66.8	Apache/Sitgreaves	Springerville	4	105
H14	Mangas Mtn.	68.1	Apache[b]	Quemado	4	134
H16	Brushy Mtn.	69.0	Apache[b]	Glenwood	3	153
H17	Telephone Canyon	62.3	Gila	Reserve	4	102
H18	Pueblo Creek	66.9	Apache[b]	Glenwood/Alpine	2	148
H19	Pitchfork Canyon	49.9	Gila	Reserve	4	37
H20	Rose Peak	71.0	Apache/Sitgreaves	Clifton	4	85
H21	Whitewater Creek	72.8	Gila	Glenwood	3[c]	123
H22	Cliff Dwellings	52.5	Gila	Mimbres	4	123
H23	Tadpole Ridge	66.3	Gila	Silver City	4[d]	149
H24	Hillsboro Peak	66.9	Gila	Silver City/Black Range	4	168
L01	Lake Montezuma	40.9	Coconino	Beaver Creek	3[a]	47
L02	Jacks Canyon	66.1	Coconino	Blue Ridge	4	114
L03	Buckhead Mesa	68.5	Tonto	Payson	3[a]	108
L05	The Gorge	45.0	Tonto	Payson	3[a]	47
L06	Spring Creek	51.3	Tonto	Pleasant Valley	4	52

[a] Fourth survey pass not required; no owls found through three survey passes.
[b] Apache National Forest administered by the Gila National Forest.
[c] A third survey pass was largely but not completely finished.
[d] Four survey passes were completed on the southern half of the quadrat, which contained spotted owls. Due to time constraints, only two passes were completed on the northern half, where no spotted owls were located during the first two passes.

(p) for owls on each quadrat. Second, we determined the numbers of banded and unbanded owls observed on each quadrat. Third, we used the estimates of p to correct the observed counts of banded and unbanded owls to estimate the total number of owls ($N_{quadrat}$) for each quadrat. The sampling variances and covariances associated with these estimates formed a variance-covariance (VC) matrix that represented the precision of the $N_{quadrat}$ estimates (sampling variances on the diagonal of the VC matrix) and the degree to which the estimates were spatially dependent (sampling

Table 16.2.

Distribution of quadrats among and characteristics of the two strata used to estimate abundance of Mexican spotted owls in the Upper Gila Mountains Recovery Unit, Arizona and New Mexico, 1999. Areas are in km².

Stratum	Stratum area	Area sampled	Total quadrats	Quadrats sampled
Low	10,635.20	271.81	186	5
High	34,795.66	1,231.4	559	20
Total	45,430.86	1,505.21	745	25

covariances in the off-diagonal elements of the VC matrix) because p's were estimated with a capture-recapture model jointly across quadrats. Fourth, density ($D_{quadrat}$) was estimated for each quadrat by dividing each $N_{quadrat}$ by quadrat area. Again, an associated VC matrix was estimated. Finally, we estimated the total number of owls (N_{RU}) and the standard error in the Recovery Unit by (a) multiplying the density for each stratum by the total area of that stratum within the sampling frame from which the sample quadrats were drawn (Table 16.2), and (b) summing across strata (e.g., Bowden et al. 2003).

We use the following notation (after Thompson et al. 1998) in the discussion that follows (see Table 16.2). We used U_h to denote the number of quadrats in the sampling frame for strata h, and u_h to define the number of quadrats sampled in strata h. The usual notation for sampling theory uses N and n, but we reserved N for the size of the biological population being estimated.

Estimating Capture Probability (p)

We used the capture histories of banded owls on quadrats where owls were banded and subsequently resighted (Table 16.3) to estimate p, the probability of capture on a given trapping occasion (Huggins 1989). We used a closed capture-recapture modeling procedure developed by Huggins (1989, 1991) and implemented in Program MARK (White and Burnham 1999). A closed-capture model was appropriate because (1) the population estimate focused on territorial adults; and (2) the study was conducted over a short time period at a time of year when territorial adults are sedentary and little emigration or immigration occurs in this population (personal observation).

The goal at this stage was to estimate p's as precisely as possible because the sampling variances of the p's contribute to the sampling variances of

Table 16.3.

Sample quadrats used to estimate capture probabilities of color-banded Mexican spotted owls in the Upper Gila Mountains Recovery Unit, Arizona and New Mexico, 1999. Quadrats were included if owls were color-banded and subsequently resighted on the quadrat.

Quadrat	Number of marked owls		Number of survey passes completed	Road access
	Male	Female		
H02	3	3	4	Roaded
H04	1	1	4	Roaded
H08	1	0	4	Roaded
H16	4	3	3	Roaded
H17	1	1	4	Roaded
H18	1	2	2	Unroaded
H20	1	0	4	Roaded
H21	4	4	3	Unroaded
H22	2	2	4	Unroaded
H23	3	2	4	Roaded
H24	5	5	4	Roaded

the estimated N's. In addition to p, the probability of recapture (c) also can be estimated, adding an additional parameter to be modeled. Both of these probabilities actually represent a combination of events, each with its own probability. For example, capture probability (p) is a function of (1) detecting an owl during a nocturnal survey; (2) locating the owl during a subsequent daytime follow-up survey; and (3) physically capturing the owl. Thus, capture probability is a combination of the probabilities of detecting the owl on both a nocturnal and diurnal survey and then physically capturing it. Recapture probability (c) requires largely the same steps, except that the owl does not need to be physically captured. Instead, it is necessary only to resight and correctly read the owl's color-band combination.

In standard closed capture-recapture models, maximum-likelihood estimation is used to estimate both p and N simultaneously (Otis et al. 1978), so the resulting estimates represent the joint maximum-likelihood estimates. The Huggins models used here differ from the standard models in that only p and c are modeled, with N estimated as a derived parameter (N is computed algebraically from p and the number of individual owls observed). Thus, our initial efforts centered on modeling the capture-recapture data to obtain parsimonious estimates of p.

To estimate p, we ran 26 closed-capture models in Program MARK (Table 16.4; notation follows Lebreton et al. 1992). In this set of models, we modeled the effects on p of sex, road access to the quadrat, occasion-specificity, and behavioral response to initial capture (i.e., inclusion of the recapture parameter c in the model). We used a bias-corrected version of Akaike's Information Criterion, AIC_c (Akaike 1973; Hurvich and Tsai 1995; Burnham and Anderson 1998), to rank models. The model best explaining the structure in p and c had the lowest AIC_c. This best model was $p = c_{T+roadless+sex}$ (Table 16.4), which constrained p's equal to c's (capture and recapture probabilities did not differ), changed linearly across occasions (T), had an additive effect of roadless versus roaded quadrats, and had an additive sex effect. The linear occasion, roadless, and sex effects were all negative and different from zero ($\beta_T = -0.350$, 95% CI = –0.637, –0.063; $\beta_{roadless} = -1.614$, 95% CI = –2.742, –0.486; $\beta_{sex} = -0.983$, 95% CI = –1.764, –0.203). This model indicated that capture probabilities declined over occasions in a linear fashion, were lower in roadless than in roaded quadrats, and were lower for females than for males.

We used model averaging to estimate the p's rather than using the p's from just the best model (Stanley and Burnham 1998a,b). This process, which incorporates model uncertainty into the estimates of p, involved estimating Akaike weights (the AIC_c weights in Table 16.4) for each model in Program MARK. These weights represented the likelihood that a given model was the best model to explain this particular data set, relative to the other models examined in our set of models (Buckland et al. 1997; Burnham and Anderson 1998). We then used these weights to derive a weighted mean estimate of capture probabilities (p_i) for each occasion for each sex and within roaded and unroaded quadrats across all models (see Stanley and Burnham1998a,b). These weighted estimates of p_i had estimated standard errors that included a variance component due to model selection uncertainty, that is, uncertainty about which model was best for providing an adequate structure on the p's (Buckland et al. 1997; Burnham and Anderson 1998).

Thus, because the "best" model included occasion, road access, and sex effects, we ended up with 16 unique estimates of p (four occasions × two types of quadrats [roaded vs. unroaded] × two sexes; Table 16.5). These 16 unique values were expanded to 32 estimates of p assigned to the different sex, quadrat type, and pass number combinations. Associated with these 32

Table 16.4.

Models considered in Program MARK for estimating capture probabilities for Mexican spotted owls in quadrats listed in Table 16.3, and resulting summary statistics, including the AIC weights used for model averaging.

Model	AIC_c[a]	ΔAIC_c[b]	AIC_c weight[c]	No. of parameters[d]	Deviance
{p(T+Roadless+Sex) = c(T+Roadless+Sex), 0 passes}	221.62	0.00	0.41	4	231.41
{p(T*Roadless+Sex) = c(T*Roadless+Sex), 0 passes}	222.99	1.37	0.21	5	212.68
{p(T+Roadless+Sex) = c(T+Roadless+Sex)+ constant, 0 passes}	223.62	2.00	0.15	5	213.31
{p(T+Roadless*Sex) = c(T+Roadless*Sex), 0 passes}	223.68	2.06	0.14	5	213.37
{p(T+Roadless) = c(T+Roadless)+constant, 0 passes}	227.75	6.13	0.02	4	219.55
{p(Roadless) c(Roadless), 0 passes}	228.37	6.75	0.01	4	220.16
{p(.) c(Roadless), 0 passes}	228.64	7.02	0.01	3	222.51
{p(T+Roadless+Sex) c(T+Roadless+Sex), 0 passes}	229.49	7.87	0.01	8	212.73
{p(Roadless) = c(Roadless), 0 passes}	229.74	8.12	0.01	2	225.68
{p(T+Sex) = c(T+Sex), 0 passes}	229.75	8.13	0.01	3	223.62
{p(Sex) c(Sex), 0 passes}	230.21	8.59	0.01	4	222.00
{p(T*Sex) = c(T*Sex), 0 passes}	231.41	9.79	<0.01	4	223.20
{p(T+Sex) = c(T+Sex)+constant, 0 passes}	231.62	10.00	<0.01	4	223.42
{p(Roadless continuous) = c(Roadless continuous), 0 passes}	231.76	10.14	<0.01	2	227.70
{p(Sex) = c(Sex), 0 passes}	232.23	10.61	<0.01	2	228.16
{p(T*Sex) = c(T*Sex)+constant, 0 passes}	233.25	11.63	<0.01	5	222.94
{p(t+Sex) = c(t+Sex), 0 passes}	233.50	11.88	<0.01	5	223.18
{p(Roadless) c(.), 0 passes}	233.50	11.88	<0.01	3	227.37
{p(.) c(.), 0 passes}	233.79	12.17	<0.01	2	229.73
{p(T) = c(T)+constant, 0 passes}	235.70	14.08	<0.01	3	229.58
{p(.) = c(.), 0 passes}	236.77	15.15	<0.01	1	234.75
{p(t) = c(t), 0 passes}	238.12	16.50	<0.01	4	229.91
{p(Quadrat) = c(Quadrat), 0 passes}	239.38	17.76	<0.01	11	215.98
{p(t*Sex) = c(t*Sex), 0 passes}	239.41	17.79	<0.01	8	222.66
{p(Quadrat) c(Quadrat), 0 passes}	247.59	25.97	<0.01	22	197.87
{p(.) = c(.)}	268.57	46.95	<0.01	1	266.55

[a] AIC_c = A bias-corrected version of Akaike's Information Criterion for the model currently under consideration (Akaike 1973; Hurvich and Tsai 1989).
[b] ΔAIC_c = The change in AIC_c between the model currently under consideration and the "best" model tested.
[c] AIC_c weight = The weight given to the model currently under consideration in model averaging.
[d] Number of parameters estimated in the current model.

Table 16.5.

Estimates of occasion-specific capture probabilities ($\hat{p}_i$) and indexing of overall capture probability ($\hat{p}^*$) among quadrats surveyed for Mexican spotted owls in the Upper Gila Mountains Recovery Unit, 1999. Estimates and indices were assigned based on combinations of sex, road access, and number of survey passes completed, based on modeling of capture probabilities (Table 16.4).

Gender	Access	Number of survey passes	Number of quadrats[a]	$\hat{p}_i$	$\hat{p}^*$
Male	Roaded	3	1	p_1, p_2, p_3	1
	Roaded	4	20	p_1, p_2, p_3, p_4	2
	Unroaded	2	1	p_5, p_6	3
	Unroaded	3	1	p_5, p_6, p_7	4
	Unroaded	4	2	p_5, p_6, p_7, p_8	5
Female	Roaded	3	1	p_9, p_{10}, p_{11}	6
	Roaded	4	20	$p_9, p_{10}, p_{11}, p_{12}$	7
	Unroaded	2	1	p_{13}, p_{14}	8
	Unroaded	3	1	p_{13}, p_{14}, p_{15}	9
	Unroaded	4	2	$p_{13}, p_{14}, p_{15}, p_{16}$	10

[a] Number of quadrats that were covered by a particular set of capture probabilities.

estimates of p was the estimated VC matrix $\mathbf{V\hat{a}r}(\hat{\mathbf{p}})$ (bold terms represent matrices). The VC matrix of the original 16 estimates from the model averaging procedure in MARK was used to build the 32 × 32 matrix $\mathbf{V\hat{a}r}(\hat{\mathbf{p}})$. The variances for each $\hat{p}$ were placed on the diagonal of this matrix, with the covariances between $\hat{p}$ in the appropriate off-diagonal elements and zeros elsewhere.

From the 32 estimates of p, we estimated an overall probability of seeing an owl (p^*) for each of the 10 sex, quadrat type, and pass-number combinations (rows in Table 16.5), as

$$p^* = 1 - (1 - \hat{p}_1)(1 - \hat{p}_2)(1 - \hat{p}_3)(1 - \hat{p}_4).$$

Subscripts 1 through 4 refer to survey pass, which was treated as a capture occasion. Thus, they correspond to subscripts in Table 16.5 only for the first two rows. For all other rows, sequential subscripts refer to passes 1 through the highest number pass completed. For quadrats with only three passes, $\hat{p}_4$ was set equal to zero, and for quadrats with only two passes, $\hat{p}_3$ and $\hat{p}_4$ were set equal to zero ($\hat{p}_3 = \hat{p}_4 = 0$) when p's were modeled. Ten values of p^* were required because there were 10 combinations of road access × number of passes × sex (Table 16.5).

Using $\hat{\text{Var}}(\hat{p})$ (a 32 × 32 matrix) and a 10 × 32 matrix of partial derivatives of p^* with respect to p_i, we used the delta method (Seber 1982) to compute the VC matrix [$\hat{\text{Var}}(\hat{p}^*)$]:

$$\hat{\text{Var}}(\hat{\mathbf{p}}^*) = \frac{\partial \mathbf{p}^*}{\partial \mathbf{p}} \hat{\text{Var}}(\hat{\mathbf{p}}) \frac{\partial \mathbf{p}^{*T}}{\partial \mathbf{p}},$$

where T indicates the transpose of the matrix of partial derivatives. This resulted in the 10 × 10 VC matrix $\hat{\text{Var}}(\hat{p}^*)$.

Estimating Numbers of Unbanded Owls per Quadrat

We estimated abundance of both banded and unbanded owls observed on quadrats. Numbers of banded owls were based on a direct count. To estimate numbers of unbanded owls observed, we first identified unique owl "territories," then determined how many owls had been conclusively observed on those territories, based on daytime observations or other conservative criteria. We identified unique territories using a combination of nocturnal and/or diurnal observations clustered in a distinct geographic location. We assumed that a territory existed wherever (a) a nesting pair or juveniles were located; (b) ≥ 1 owl was located roosting in the same area (roost locations within 400 m of each other) on two or more passes, with all locations before 1 September; or (c) ≥ 1 owl was heard at night on two or more passes and one diurnal roost was located, with all locations before 1 September (again, locations within 400 m of each other).

Once unique territories were identified, we assessed their occupancy status to estimate numbers of unbanded birds. We counted unbanded birds where they were (a) unbanded mates of banded owls; or (b) unbanded members of nesting pairs (i.e., both pair members unbanded). We also classified an area as occupied by a pair if a male and female were observed within 400 m of each other on ≥ 1 daytime surveys, with all locations before 1 September. We counted only a single owl where we had daytime observations of only one owl in an area. Only daytime observations of known-sex owls were used in estimating numbers of unbanded owls, so that all observations could be adjusted by the correct sex-specific capture probability (see above).

Table 16.6.

Sex-specific estimates of Mexican spotted owl abundance in 25 quadrats surveyed within the Upper Gila Mountains Recovery Unit, Arizona and New Mexico, 1999. Shown are sex-specific numbers of banded and unbanded owls observed, quadrat index (p^*) used in estimating capture probabilities, estimated capture probability ($\hat{p}^*$), corrected estimate of owl abundance ($\hat{N}$), and the estimated standard error of $\hat{N}$.

		Number of owls					
Quadrat	Gender	Banded	Unbanded	p^*	$\hat{p}^*$	$\hat{N}$	$\hat{SE}(\hat{N})$
H01	Male	0	0	2	0.98	0	0
H02	Male	3	0	2	0.98	3.04	0.04
H03	Male	0	0	2	0.98	0	0
H04	Male	1	0	2	0.98	1.02	0.01
H07	Male	0	1	2	0.98	1.02	0.01
H08	Male	1	0	2	0.98	1.02	0.01
H09	Male	0	0	2	0.98	0	0
H11	Male	0	0	2	0.98	0	0
H12	Male	0	1	2	0.98	1.02	0.01
H13	Male	0	1	2	0.98	1.02	0.01
H14	Male	0	0	2	0.98	0	0
H16	Male	4	0	1	0.97	4.13	0.09
H17	Male	1	0	2	0.98	1.02	0.01
H18	Male	1	1	3	0.58	3.45	1.13
H19	Male	0	0	2	0.98	0	0
H20	Male	1	0	2	0.98	1.02	0.01
H21	Male	4	5	4	0.69	13.01	3.64
H22	Male	2	1	5	0.76	3.95	0.99
H23	Male	3	0	2	0.98	3.04	0.04
H24	Male	5	3	2	0.98	8.12	0.11
L01	Male	0	0	2	0.98	0	0
L02	Male	0	0	2	0.98	0	0
L03	Male	0	0	2	0.98	0	0
L05	Male	0	0	5	0.76	0	0
L06	Male	0	0	2	0.98	0	0
H01	Female	0	0	7	0.90	0	0
H02	Female	3	0	7	0.90	3.34	0.28
H03	Female	0	0	7	0.90	0	0
H04	Female	1	0	7	0.90	1.11	0.09
H07	Female	0	0	7	0.90	0	0
H08	Female	0	1	7	0.90	1.11	0.09
H09	Female	0	0	7	0.90	0	0
H11	Female	0	0	7	0.90	0	0
H12	Female	0	1	7	0.90	1.11	0.09
H13	Female	0	1	7	0.90	1.11	0.09
H14	Female	0	0	7	0.90	0	0
H16	Female	3	1	6	0.85	4.69	0.48

Quadrat	Gender	Number of owls		p^*	$\hat{p}^*$	$\hat{N}$	$\hat{SE}(\hat{N})$
		Banded	Unbanded				
H17	Female	1	0	7	0.90	1.11	0.09
H18	Female	2	0	8	0.33	6.61	3.51
H19	Female	0	0	7	0.90	0	0
H20	Female	0	0	7	0.90	0	0
H21	Female	4	4	9	0.42	19.07	10.44
H22	Female	2	2	10	0.48	8.27	4.39
H23	Female	2	0	7	0.90	2.22	0.19
H24	Female	5	2	7	0.90	7.79	0.66
L01	Female	0	0	7	0.90	0	0
L02	Female	0	0	7	0.90	0	0
L03	Female	0	0	7	0.90	0	0
L05	Female	0	0	10	0.48	0	0
L06	Female	0	1	7	0.90	1.11	0.09

Estimating the Number and Density of Owls per Quadrat

It is highly unlikely that all resident owls on the quadrats were observed during surveys. Consequently, we used the estimates of p^* derived earlier to correct the number of owls banded (plus the additional number of unbanded owls known to exist) on each of the 25 quadrats (Table 16.6). Counts were corrected as:

$$\hat{N}_i = \frac{M_{t+1,i}}{p_m^*},$$

where i indexes quadrat, m indexes the combination of sex, quadrat type, and number of passes, and M_{t+1} is the number of known owls on the quadrat (Table 16.7).

On each quadrat, we estimated N_i separately for males and females. Thus, we had 50 estimates (2 estimates × 25 quadrats) of N_i. To compute the VC matrix of the 50 estimates of N_i, that is, $\hat{\text{Var}}(\hat{N})$, $\hat{\text{Var}}(\hat{\mathbf{p}}^*)$ was first expanded to be a 50 × 50 matrix, in which the variances for each estimate of p were placed on the diagonal corresponding to the appropriate N_i and the covariances between respective p's were placed in the appropriate off-diagonal elements, with zeros elsewhere. This matrix then was used in the delta method (Seber 1982) to estimate $\hat{\text{Var}}(\hat{N})$:

$$\hat{\mathbf{Var}}(\hat{N}) = \frac{\partial N}{\partial p^*} \hat{\mathbf{Var}}(\hat{p}^*) \frac{\partial N^T}{\partial p^*}.$$

Table 16.7.

Estimated abundance ($\hat{N}_{hq}$) and density ($\hat{D}_{hq}$) of Mexican spotted owls (males and females combined) in 25 surveyed quadrats. Observed counts (M_{t+1}) of banded and unbanded owls were corrected by capture probabilities to generate these estimates.

Quadrat	Area (km²)	Owls observed	$\hat{N}_{hq}$	$\hat{SE}(\hat{N}_{hq})$	$\hat{D}_{hq}$	$\hat{SE}(\hat{D}_{hq})$
H01	43.7	0	0	0	0	0
H02	61.2	6	6.38	0.29	0.10	0
H03	46.5	0	0	0	0	0
H04	76.4	2	2.13	0.10	0.03	0
H07	59.6	1	1.01	0.01	0.02	0
H08	55.7	2	2.13	0.10	0.04	0
H09	54.9	0	0	0	0	0
H11	57.9	0	0	0	0	0
H12	63.2	2	2.13	0.10	0.03	0
H13	66.8	2	2.13	0.10	0.03	0
H14	68.1	0	0	0	0	0
H16	69.0	8	8.82	0.52	0.13	0
H17	62.3	2	2.13	0.10	0.03	0
H18	66.9	4	9.52	4.35	0.14	0.07
H19	49.9	0	0	0	0	0
H20	71.0	1	1.01	0.01	0.01	0
H21	72.8	17	32.08	13.16	0.44	0.18
H22	52.5	7	12.21	5.10	0.23	0.10
H23	66.3	5	5.27	0.20	0.08	0
H24	66.9	15	15.9	0.69	0.24	0.01
L01	40.9	0	0	0	0	0
L02	66.1	0	0	0	0	0
L03	68.5	0	0	0	0	0
L05	45.0	0	0	0	0	0
L06	51.3	1	1.11	0.10	0.02	0

To compute the estimate of the total number of owls (males plus females) on each of the 25 quadrats, the matrix $\boldsymbol{B}$ was defined as consisting of two 25 × 25 identity matrices (one for each sex) set one on top of the other. Then $\hat{N}_{quadrat} = \boldsymbol{B}^T\hat{N}$, where $\hat{N}$ is a 50 × 1 vector and $\hat{N}_{quadrat}$ is a 25 × 1 vector, with the estimated variance-covariance matrix $\hat{\text{Var}}(\hat{N}_{quadrat}) = \boldsymbol{B}^T\hat{\text{Var}}(\hat{N})\boldsymbol{B}$.

We estimated density (owls/km²) for each of the 25 quadrats as $\hat{D} = \boldsymbol{E}\hat{N}_{quadrat}$, where the matrix $\boldsymbol{E}$ was a 25 × 25 diagonal matrix with diagonal elements consisting of the reciprocal of the area of each quadrat, with $\hat{\text{Var}}(\hat{D}) = \boldsymbol{E}\hat{\text{Var}}(\hat{N}_{quadrat})\boldsymbol{E}^T$ (Table 16.7).

Estimating Total Population Size on the Recovery Unit

To compute the total population size for the recovery unit ($\hat{N}_{RU}$), we used a ratio estimator following Cochran (1977) and Bowden et al. (2003). We used the ratio of estimated population size ($\hat{N}_{hq}$, obtained from the $\hat{N}_{quadrat}$ vector, where h is the strata and q is the quadrat) to quadrat area (a_{hq}, with A_h the area of each strata h), such that:

$$\hat{N}_{RU} = \sum_{h=1}^{2} \hat{N}_h = \sum_{h=1}^{2} \hat{R}_h A_h ,$$

where

$$\hat{R}_h = \frac{\sum_{q=1}^{u_h} \hat{N}_{hq}}{\sum_{q=1}^{u_h} a_{hq}} .$$

In our case, we had only two strata (low- or high-density). Because sampled quadrat population sizes were estimates instead of exact counts, and because of the covariances of the $\hat{N}_{hq}$ induced by estimating common resighting probabilities across quadrats and strata, the estimated variance of $\hat{N}_{RU}$, $\hat{\text{Var}}(\hat{N}_{RU})$, was considerably more complicated than the usual case:

$$\begin{aligned}
\hat{\text{Var}}(\hat{N}_{RU}) = & \sum_{h=1}^{2}\left(\frac{A_h}{\bar{a}_h}\right)^2\left(\frac{1}{u_h}-\frac{1}{U_h}\right)\sum_{q=1}^{u_h}\frac{\left(\hat{N}_{hq}-\hat{R}_h a_{hq}\right)^2}{u_h-1} \\
& + \sum_{h=1}^{2}\frac{\left(\frac{A_h}{\bar{a}_h}\right)^2}{u_h}\left[\sum_{q=1}^{u_h}\frac{\hat{\text{Var}}(\hat{N}_{hq})}{u_h}+\sum_{q=1}^{u_h}\sum_{q'\neq q}^{u_h}\frac{\hat{\text{Cov}}(\hat{N}_{hq},\hat{N}_{hq'})}{u_h}\right] \\
& + \sum_{h=1}^{2}\sum_{h\neq h'}^{2}\frac{\left(\frac{A_h}{\bar{a}_h}\right)\left(\frac{A_{h'}}{\bar{a}_{h'}}\right)}{u_h u_{h'}}\sum_{q=1}^{u_h}\sum_{q'=1}^{u_{h'}}\hat{\text{Cov}}(\hat{N}_{hq},\hat{N}_{h'q'}) \\
& - \sum_{h=1}^{2}\left(\frac{A_h}{\bar{a}_h}\right)^2\left(\frac{1}{u_h}-\frac{1}{U_h}\right)\left[\sum_{q=1}^{u_h}\frac{\hat{\text{Var}}(\hat{N}_{hq})}{u_h}-\sum_{q=1}^{u_h}\sum_{q'\neq q}^{u_h}\frac{\hat{\text{Cov}}(\hat{N}_{hq},\hat{N}_{hq'})}{u_h(u_h-1)}\right],
\end{aligned}$$

where $\bar{a}_h$ is the average area of quadrats in the sample. The first term of this equation is the variation in the $\hat{N}_{hq}$ multiplied by a finite correction factor. The second term incorporates both sampling variation due to estimating the number of owls on the sampled quadrats and the covariances among quadrat estimates within the same stratum induced by model averaging. The third term incorporates the covariances between quadrats, and the final term is a correction term that subtracts out mean sampling variance and covariances adjusted by a finite population correction factor. Finite population correction factors were incorporated into the equation because we sampled from a finite sampling frame. A reasonable approximation to this equation in which the finite population correction is ignored and $u_h^2 \approx u_h(u_h - 1)$ is:

$$\hat{\text{Var}}(\hat{N}_{RU}) = \sum_{h=1}^{2} \frac{\left(\frac{A_h}{\bar{a}_h}\right)^2}{u_h} \left[\sum_{q=1}^{u_h} \frac{\left(\hat{N}_{hq} - \hat{R}_h a_{hq}\right)^2}{u_h - 1} \right]$$

$$+ \sum_{h=1}^{2} \frac{\left(\frac{A_h}{\bar{a}_h}\right)^2}{u_h(u_h - 1)} \sum_{q=1}^{u_h} \sum_{q' \neq q}^{u_h} \hat{\text{Cov}}(\hat{N}_{hq}, \hat{N}_{hq'})$$

$$+ \sum_{h \neq h'}^{2} \frac{\left(\frac{A_h}{\bar{a}_h}\right)\left(\frac{A_{h'}}{\bar{a}_{h'}}\right)}{u_h u_{h'}} \sum_{q=1}^{u_h} \sum_{q'=1}^{u_{h'}} \hat{\text{Cov}}(\hat{N}_{hq}, \hat{N}_{hq'}).$$

In this equation, the first term is the variation in the $\hat{N}_{hq}$, the second term incorporates the covariances induced within strata, and the third term incorporates the covariances induced among quadrats in different strata. These equations are extensions of the formulae given by Skalski (1994) because they incorporate the covariances of the $\hat{N}_{hq}$ across quadrats and strata.

Our estimate of total population for the Recovery Unit (exclusive of tribal lands) was imprecise (Table 16.8). For comparison with this estimate, we also computed a "naive" estimate of population size using raw counts of observed owls without adjusting these counts by p^*. These naive estimates thus represent counts of owls assuming complete detectability (i.e., $p^* = 1$). Resulting estimates for high and low strata were 2,091.0 ($\hat{\text{SE}} = 574.0$) and 39.1 ($\hat{\text{SE}} = 37.4$), respectively, giving a naive estimate of 2,130.1 ($\hat{\text{SE}} = 575.3$) for the recovery unit (95% confidence interval = 930–3,330 owls). The difference in the standard errors for the naive estimate and $\hat{N}_{RU}$ was substantial, illustrating the impact of a few quadrats in greatly increasing the variance of the estimate when capture probability corrections are applied. For example, quadrats H21, H22, and H18 had some of the lowest capture probabilities, and hence much larger variances for $\hat{N}_{hq}$. These three quadrats contributed 99.6% of the variance to the term:

$$\sum_{h=1}^{2} \sum_{q=1}^{u_h} \hat{\text{Var}}(\hat{N}_{hq}).$$

Table 16.8.

Estimates of population size ($\hat{N}$) of Mexican spotted owls for high- and low-density strata, and for the Upper Gila Mountains Recovery Unit (exclusive of tribal lands).

Stratum	$\hat{N}$	$\hat{SE}(\hat{N})$	CV (%)
High density	2,896.6	1,073.6	37.1
Low density	44.0	44.2	100.5
Recovery Unit (without tribal lands)	2,940.6	1,075.4	36.6

Initial Power Analyses

Using the data from the pilot study, we evaluated statistical power using two approaches. In the first approach we regressed annual population estimates (calculated assuming a 20% decline over 10 yrs, see below) versus time and estimated the rate of change of the population (λ) from the slope of the regression. In the second approach, we estimated λ directly from the quadrat counts. If there is less among-quadrat variance in λ than in population size, this approach should remove much of the among-quadrat variance.

Regression Approach to Power Analysis

A population declining by 20% over 10 yrs would decrease by $1–0.8^{1\backslash 9}$ = 0.02449 per year. We assessed the power to detect this rate of decline using the methods of Gerrodette (1987) as described by Thompson et al. (1998). Here, we denote $N_{RU,t}$ as the population estimate of owls within the recovery unit in year t. We conducted a one-sided test with a sample of n = 10 yrs, $\alpha = 0.05$, and $r = -0.02449$ in the multiplicative model $N_{RU,t} = N_{RU,1}(1 + r)^{t-1}\varepsilon_t$ using three assumed error structures. Error structures assessed included (1) (CV_t (the variation [CV] for the estimate of population size in year t) proportional to $1/\sqrt{\hat{N}_{RU,t}}$; (2) CV_t constant; and (3) CV_t proportional to $\sqrt{\hat{N}_{RU,t}}$. Statistical power, modeled as a function of the $CV(\hat{N}_{RU})$, was similar for each error structure (Figure 16.3).

This analysis clearly showed that the current level of effort was inadequate to detect a 20% decline with 10 yrs of surveys and would result in a power of approximately 18% when CV = 36% (Table 16.8). Even if the $\hat{N}_{hq}$ were estimated with no error, this level of effort with the current stratification would be inadequate to meet the power requirements in the Recovery Plan. This can be shown by examining the naive estimator based on

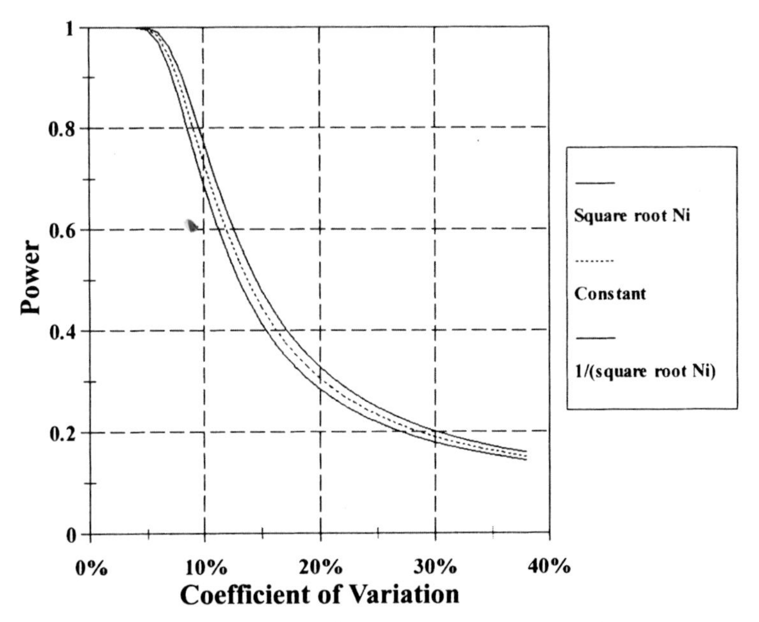

Figure 16.3. Estimated power to detect a 20% decline in Mexican spotted owl abundance over 10 yrs with a power of 90% for $\alpha = 0.05$, shown as a function of the coefficient of variation (CV) of the population estimate. Power was modeled assuming three separate error structures: CV proportional to $\sqrt{N_i}$, CV constant, and CV proportional to $1/\sqrt{N_i}$ (shown from top to bottom curves, respectively).

the minimum number of owls known to be on a quadrat (discussed above). The CV of this naive estimate was still 29%, resulting in power of approximately 20%, depending on the error structure assumed (Figure 16.3).

Meeting the power requirements in the Recovery Plan would necessitate that the CV for the annual estimates of $\hat{N}_{RU}$ be reduced to approximately 7% (Figure 16.3). To assess the feasibility of achieving such a reduction, we identified and evaluated several modifications to the existing sampling plan that could potentially reduce CV. These included (1) raising capture probabilities, especially on quadrats where capture probabilities were low; (2) better a priori stratification of quadrats to reduce among-quadrat variance within strata; and (3) increasing sample size.

As noted earlier, low capture probabilities in three high-density quadrats introduced considerable variance into the estimator of N_{RU}. We identified two changes in field techniques that should allow us to raise these capture probabilities. One would involve simply putting more surveyors in difficult roadless areas. A second would involve requiring surveyors to carry and use the proper equipment. Surveyors working in these backcountry quadrats sometimes carried small binoculars rather than the heavier binoculars they were issued. These smaller binoculars proved inadequate for resighting color bands. These changes likely would result in greater capture probabilities but would be of limited help with respect to statistical power to detect trends. Comparisons with the naive estimator based on minimum number known alive suggest that consistently high capture probabilities would lower $CV(\hat{N}_{RU})$ from approximately 36% to approximately 29%. This reduction by itself is insufficient to meet the specified power requirements.

Better a priori stratification of quadrats could reduce among-quadrat variance in owl abundance within strata, especially in the high-density stratum. The estimate for the low-density stratum ranged from 0–1 owls per quadrat and contributed little variance to the total. However, naive estimates for the high-density stratum ranged from 0–17 owls per quadrat. The large variance in $\hat{N}_{RU}$ comes predominantly from this high variability across the high-density quadrats.

We were unable to determine the magnitude of potential improvements in a priori stratification of quadrats. We stratified by elevation in the pilot study because we were unable to locate a digital coverage of vegetation types that was mapped at a reasonable scale and covered the entire study area. This problem still exists and severely constrains our ability to stratify more precisely. Consequently, the problem here is similar to the problem with capture probabilities: improvements can be made, but those improvements by themselves are unlikely to be great enough to result in adequate statistical power.

The final alternative considered was to increase the sample size. Absent improvements in stratification, roughly a 10-fold increase in sampling effort would be required to obtain the necessary precision to detect a 20% decrease in 10 yrs (i.e., a $CV(\hat{N}_{RU}) = 7\%$) for the naive estimates based on the minimum number known alive. Approximately an 18-fold increase in sampling effort would be required to obtain the necessary precision for the exact estimator of the variance of $\hat{N}_{RU}$ when the current variability of capture probabilities is used. The effect of better stratification cannot be modeled here, as discussed above. Given the already large expense of this sampling plan and the practical difficulties encountered in sampling even 25 quadrats, we concluded that such large increases in sampling effort were not feasible. This conclusion led us to consider a second approach to estimating λ.

Approach to Estimating λ Directly from Quadrat Data

Because the large among-quadrat variance in $\hat{N}_{RU}$ caused a large variance in estimates of λ obtained by taking ratios of consecutive $\hat{N}_{RU}$ values, we considered an alternative approach in which λ was estimated directly from the sample and then extrapolated to the population. To illustrate, consider the 20 values of $\hat{N}_{hq}$ for the high-density strata (i.e., the owls observed on each quadrat, given in Table 16.7). Assuming a second survey 1 yr later, an estimate of λ for the year from t to t+1 for strata h is:

$$\hat{\lambda}_t = \frac{\sum_{q=1}^{u_h} \hat{N}_{t+1,hq}}{\sum_{q=1}^{u_h} \hat{N}_{t,hq}} .$$

If survey techniques were improved sufficiently to eliminate the few large standard errors for $\hat{N}_{hq}$, the standard errors of $\hat{N}_{hq}$ could be ignored. We simulated this by using the M_{t+1} values from Table 16.7 in place of $\hat{N}_{hq}$ in the following power calculation. The sampling variance of λ can be estimated by recognizing that $\hat{\lambda}$ is a simple ratio estimator (e.g., Cochran 1977:150–188). For a given strata h, and finite population correction *fpc*:

$$\hat{\text{Var}}(\hat{\lambda}_t) = \left[\frac{1 - fpc}{\frac{1}{u_h}\left(\sum_{q=1}^{u_h} \hat{N}_{t,hq}\right)^2} \right] \left[\frac{\sum_{q=1}^{u_h} (\hat{N}_{t+1,hq} - \hat{\lambda}\hat{N}_{t,hq})^2}{u_h - 1} \right].$$

The overall estimate of λ across a 10-yr period would consist of the product of the nine $\hat{\lambda}$ values as calculated by the above formula. The variance of the λ estimate for the 10-yr period can be obtained as the variance of the product of the nine estimates by repeatedly applying the formula (Goodman 1960) $\hat{\text{Var}}(xy) = x^2\hat{\text{Var}}(y) + y^2\hat{\text{Var}}(x) - \hat{\text{Var}}(x)\hat{\text{Var}}(y)$. However, the product $\prod_{t=1}^{9} \hat{\lambda}_t$ simplifies to the ratio estimator for the surveys at time t = 1 and t = 10 (the first year and the last year of the monitoring period), that is,

$$\hat{\lambda} = \prod_{t=1}^{9} \hat{\lambda}_t = \frac{\sum_{q=1}^{u_h} \hat{N}_{t+9,hq}}{\sum_{q=1}^{u_h} \hat{N}_{t,hq}} ,$$

because the intermediate terms cancel out of the equation. Thus, the variance for the decline across 10 yrs can be estimated as:

$$\hat{\text{Var}}(\hat{\lambda}_t) = \left[\frac{1 - fpc}{\frac{1}{u_h}\left(\sum_{q=1}^{u_h} \hat{N}_{t,hq} \right)^2} \right] \left[\frac{\sum_{q=1}^{u_h} (\hat{N}_{t+9,hq} - \hat{\lambda}\hat{N}_{t,hq})^2}{u_h - 1} \right].$$

The idea that only estimates of N from the first and last years are needed to estimate λ was not intuitive, but can be demonstrated mathematically with a simple example. Given a 5-yr monitoring period, we obtain

four estimates of λ via $\hat{\lambda}_1 = \frac{\hat{N}_2}{\hat{N}_1}$, $\hat{\lambda}_2 = \frac{\hat{N}_3}{\hat{N}_2}$, $\hat{\lambda}_3 = \frac{\hat{N}_4}{\hat{N}_3}$, and $\hat{\lambda}_4 = \frac{\hat{N}_5}{\hat{N}_4}$. A product of λ over the 5-yr period can be estimated as $\hat{\lambda} = \hat{\lambda}_1 \bullet \hat{\lambda}_2 \bullet \hat{\lambda}_3 \bullet \hat{\lambda}_4$, which can be rewritten as:

$$\hat{\lambda} = \left(\frac{\hat{N}_2}{\hat{N}_1} \right) \bullet \left(\frac{\hat{N}_3}{\hat{N}_2} \right) \bullet \left(\frac{\hat{N}_4}{\hat{N}_3} \right) \bullet \left(\frac{\hat{N}_5}{\hat{N}_4} \right).$$

The numerator of the first term and the denominator of the second term ($\hat{N}_2$) cancel out, as do $\hat{N}_3$ and $\hat{N}_4$ in the second, third, and fourth terms. This leaves us with $\hat{\lambda} = \hat{N}_5 / \hat{N}_1$, which includes only the first and last years of the monitoring period.

For the following power analysis, we again simulated declines of 80% over a 10-yr period (i.e., annual $\lambda = 0.80^{(1/9)}$). We used two sets of population parameters to model annual rate of change of the population. Set 1 had survival – emigration (ϕ) equal to 0.865 and immigration + recruitment (f) equal to 0.110, giving $\lambda = \phi + f = 0.975$. Set 2 had $\phi = 0.845$ and $f = 0.130$, again giving $\lambda = 0.975$. The parameter ϕ is the probability that an owl on the quadrat at time t remains alive and on the quadrat at time $t+1$. The parameter f is the number of new recruits into the population at time t divided by the number of owls present at time $t - 1$. New recruits can be either offspring of individuals on the quadrat or new immigrants. These parameter sets were derived from demographic data from two long-term study areas in Arizona and New Mexico (Seamans et al. 1999) and were biologically reasonable. These study areas yielded mean estimates of 0.845 for ϕ and 0.110 for f. We adjusted these values slightly so that both parameter sets gave a rate of change of exactly 80% across 10 yrs.

For this analysis, the initial number of owls per quadrat was selected at random (with replacement) from the observed distribution of owls on the quadrats sampled. Demographic variation was modeled as a random binomial variable. Thus, the population size on a quadrat at time $t + 1$ was computed as the random binomial variable of the owls surviving from time t plus another random binomial variable of the number of new immigrants and recruits since time t. Once a quadrat reached a zero population level, the recolonization probability was zero, that is, quadrats never experienced immigration unless at least one animal was present. This assumption was biologically unrealistic but made little difference in terms of statistical power because it had little effect on the variance of the estimate. That is, colonization reduced effect size but not variance.

We conducted simulations using sample sizes of 25, 40, 50, 60, 75, 100, 125, 150, 175, and 200 quadrats. For each sample size, 100,000 simulations were conducted to estimate power. Estimates of β (where β = [1–power]) were computed by determining the proportion of confidence intervals on the estimate of λ for the 9-yr interval that included 1. Results indicated that 90% power was achieved when between 100 and 150 quadrats were sampled

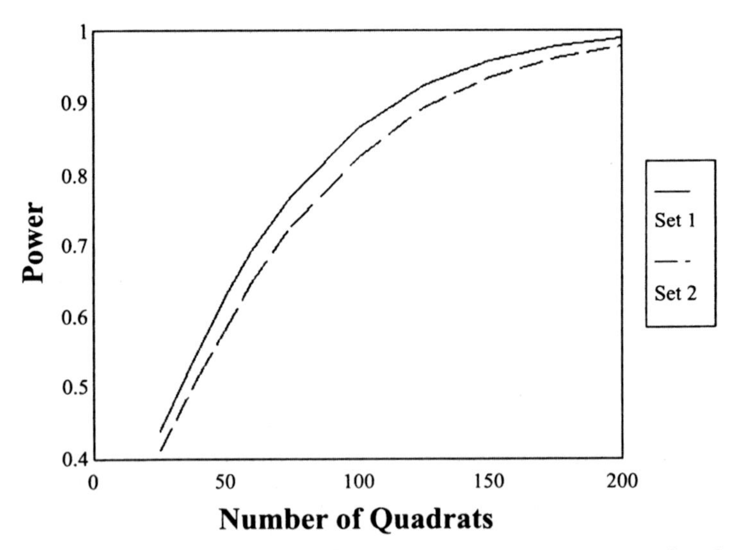

Figure 16.4. Power to detect a decline of 20% in Mexican spotted owl abundance over a 10-year period when randomly selected quadrats were sampled in years 1 and 10. Two sets of population parameters were used to model annual rate of change of the population. Set one (top curve) had survival–emigration (ϕ) = 0.865 and immigration + recruitment (f) = 0.110. Set two (bottom curve) had $\phi = 0.845$ and $f = 0.131$. For both sets, annual $\lambda = 0.975511$ and 10-year $\lambda = 0.8$.

in year 1 and again in year 10 (Figure 16.4). Additional analyses demonstrated that sampling during the intervening years did not increase the power of this design. This was because temporal variation in the population growth process affected the probability that the population would decline by 20% but did not affect the power of the survey to detect that decline.

This method thus appears to offer powerful advantages over the regression approach. Using this method, the number of quadrats required to be sampled was considerably reduced, and sampling was required in only 2 yrs. The second advantage is largely illusory, however, because understanding the magnitude of temporal variation is important in an ecological and interpretive context. For example, we might find that the population regularly fluctuates by as much as 20%, suggesting that a 20% decline may not be cause for alarm and that the delisting criterion should be revised. Conversely, we might note a change of $< 20\%$ at the end of 10 yrs but observe large fluctuations in population size during the intervening years. In this case, delisting might not be indicated, as populations that undergo such large fluctuations may be especially vulnerable to extinction (Goodman 1987). Therefore, we would be reluctant to recommend a monitoring program that did not provide information on the magnitude of temporal variation in population size at intervals shorter than a decade.

Secondary Analyses Aimed at Increasing Precision of Estimates

Our overall conclusion from the previous section mirrored our earlier conclusion: the large increases necessary in sampling effort were not feasible. This led us to conduct a second set of analyses aimed at evaluating ways to improve the precision of estimates without large increases in sampling effort. We focused on two approaches: reducing quadrat size and using covariates to aid in estimating population size.

Effects of Quadrat Size on Mexican Spotted Owl Density Estimates

Smaller quadrats could be sampled for less money than large quadrats, allowing for greater spatial replication for a fixed level of funding. This greater spatial replication should in turn reduce the spatial variation component. Consequently, we explored the potential effects of reducing quadrat size to approximately 35–50 km^2. Parameters evaluated included (1) the magnitude and sampling variance of density estimates, (2) temporal process variation in density estimates, and (3) incidence of movement across quadrat boundaries by owls holding territories near quadrat boundaries (i.e., edge effects). Density was used as the basis for comparison to control for differences in quadrat area.

Briefly, comparisons were made between quadrats of the size used in the pilot study conducted in 1999 and quadrats approximately half the size of those sampled. Two relatively large study areas delineated to investigate the demography of Mexican spotted owls in Arizona and New Mexico (585 and 323 km^2, respectively; Seamans et al. 1999) were used to develop these comparisons. Spotted owls were color-banded and resighted annually on each study area from 1991 through 1999. Personnel conducting these studies provided data on spotted owl locations (roost and nest sites occupied by pairs, single males, or single females), with one location representing a unique spotted owl territory center for each year. Territory locations were delineated based on color-marked birds and geographic locations. These data provided the basis for estimating density within quadrats.

Using GIS, we subdivided the study areas into large quadrats (similar in size to those used in the pilot study) and small quadrats (approximately half the size of those sampled in 1999). This involved first delineating watersheds within the demography areas, then subdividing these into subwatersheds in the appropriate size range, using topographic features. We then overlaid owl locations on the resulting quadrats and computed annual estimates of density (owls/km^2) for both large and small quadrats where (1) the number of individuals was computed for each quadrat and year (including quadrats with no owls); (2) the density of each quadrat was computed based on number of individuals and quadrat area; and (3) mean density, with its standard error, was estimated for each year.

In this analysis, we assumed that the number of individuals within a quadrat was estimated without error, that is, with no sampling variance, and therefore assumed that the arithmetic standard deviation associated with mean annual densities estimated spatial process standard deviation ($\sigma_{spatial}$) among quadrats within each year. Annual estimates of $\sigma^2{}_{spatial}$ were expressed as coefficients of spatial process variation, $\hat{CV}_{spatial} = \frac{\hat{\sigma}^2{}_{spatial}}{\hat{\bar{D}}}$.

We estimated temporal process variation only for small quadrats because our interest centered on whether or not small quadrats could be used. Because the mean annual estimates had sampling variance associated with them, $\hat{Var}(\hat{D}) = [\hat{SE}(\hat{D})]^2$, we used a method-of-moments random-effects model using shrinkage estimators to estimate $\sigma^2_{temporal}$, $\hat{D}$, and SE($\bar{D}$), using an intercepts-only model, and β_1 (trend in estimates over time) and $\sigma^2_{residual}$ using a linear trend model. These models removed the sampling variation from the total variation to estimate temporal process variation. In the linear trend models, $\sigma^2_{residual}$ was the amount of temporal process variation remaining after accounting for a linear trend. We conducted separate analyses using estimates for (1) all 9 yrs and (2) for only the last 7 yrs. The latter analysis reflected the potential for a "learning" effect by which numbers of owls may have been underestimated in early years because the investigators were still becoming familiar with the study areas and the territorial owl populations.

Annual estimates of density were similar between large and small quadrats (Figure 16.5), indicating that reducing quadrat size did not increase

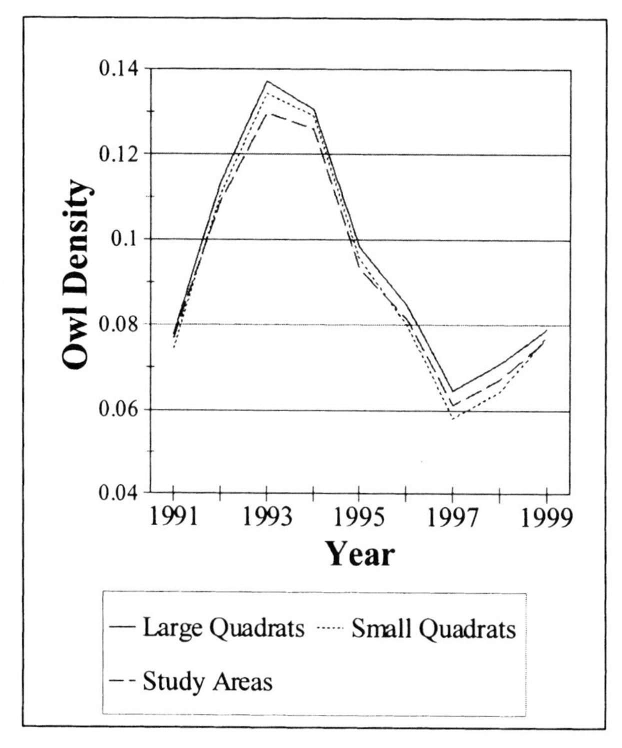

Figure 16.5. Trends in annual density (owls/km^2) of Mexican spotted owls derived from large and small quadrats delineated in two demography study areas in Arizona and New Mexico. Annual density estimates averaged over the two study areas are shown for comparison.

bias in density estimates. In addition, density estimates from the large and small quadrats were similar to annual density estimates averaged across the two demography study areas (Figure 16.5). Mean standard errors of density estimates were 12.5% smaller for small quadrats than for large quadrats. In contrast, annual estimates of spatial process variation were approximately 24% higher for small quadrats than for large quadrats (Figure 16.6), likely due to the patchy distribution of owls across the landscape. We expect that the sample size could be approximately doubled with smaller quadrats, however. This increase in sample size should reduce the variance of the mean by $1/\sqrt{2} = 0.707$. Thus, use of small quadrats should result in a net decrease of approximately 12% in the CV of the density estimate (i.e., 0.707×1.239 [ratio of process variation on small quadrats relative to large quadrats] $= 0.88$).

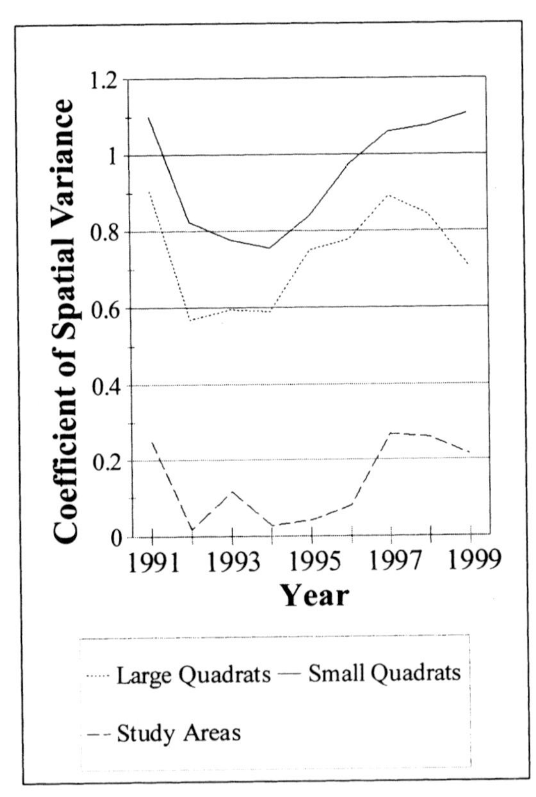

Figure 16.6. Trends in annual estimates of the coefficient of spatial process variation for estimates of density of Mexican spotted owls in large and small quadrats delineated in two demography study areas in Arizona and New Mexico. Annual estimates averaged over the two study areas are shown for comparison.

The estimates of temporal process standard variation (based on the random effects means models) were high for both the 9- and 7-yr periods (Table 16.9). Process CVs were 0.251 and 0.296 for the 9-yr and 7-yr periods, respectively. The linear model reduced temporal process variation substantially for the 7-yr period, but not for the 9-yr period (Table 16.9). This suggests that temporal process variation around the trend line was minimal for small quadrats where there was both a learning effect and a linear trend.

The degree to which territories crossed quadrat boundaries in different years was evaluated in two ways for large and small quadrats separately. First, we computed the total number of territories that crossed quadrat boundaries (e.g., a territory found in one quadrat in one year was found in another quadrat in a different year) across all years. Second, we estimated annual rates of territories crossing boundaries. This involved dividing the number of territories in year $t+1$ that were on quadrats different from those in year t by the total number of territories in year t that were available for reobservation in year $t+1$ (i.e., territories that were observed in one year but not the next were not included).

Annual rates of territories crossing boundaries were almost identical for small and large quadrats (Table 16.10). These rates varied in only 2 yrs (1997 and 1998), with this difference due to one territory crossing small quadrat boundaries in each of these years (Table 16.10). Rates of territories crossing quadrat boundaries appeared to increase over time, which may indicate a general pattern of shifts in territory locations over time.

Table 16.9.

Parameter estimates from means and linear trends models of Mexican spotted owl density estimated from small quadrats over 9- and 7-year periods. Estimates of σ^2 from the means model represent temporal process variation, whereas estimates from the linear trend model represent residual temporal variation remaining from that variation accounted for by the model.

	Estimates from means model		Estimates from trend model	
Parameter	9 yrs	7 yrs	9 yrs	7 yrs
β_0	0.0894	0.0893	0.1168	0.1369
SE(β_0)	0.0088	0.0113	0.0181	0.0159
β_1	–	–	–0.0054	–0.0120
SE(β_1)	–	–	0.0032	0.0034
σ^2	0.0005	0.0007	0.0004	0.0001
(95% CI)	(0.0002, 0.0034)	(0.0001, 0.0030)	(0.0001, 0.0023)	(0, 0.0006)

Table 16.10.

Annual rates at which Mexican spotted owl territories crossed boundaries of large and small quadrats delineated from demography study areas (Seamans et al. 1999) in the Upper Gila Mountains Recovery Unit, Arizona and New Mexico.

		Number of territories found in different quadrat in year $t + 1$		Rate of territories crossing quadrat boundaries	
Year	Number of territories in year t	Large quadrats	Small quadrats	Large quadrats	Small quadrats
1991	42	–	–	–	–
1992	57	1	1	0.024	0.024
1993	62	1	1	0.018	0.018
1994	51	2	2	0.032	0.032
1995	44	1	1	0.020	0.020
1996	32	7	7	0.159	0.159
1997	32	4	5	0.125	0.156
1998	29	2	3	0.063	0.094
1999	–	5	5	0.172	0.172

In summary, reducing quadrat size did not bias estimates of owl density and did not inflate "edge effects" in terms of owls moving among quadrats between years. Further, because more small quadrats could be sampled for a given funding level, use of smaller quadrats should improve the precision of estimates of owl abundance, as well as allow for greater spatial replication. It is impossible to realistically assess the cost per small quadrat, however, as this cost will depend largely on the actual sample of quadrats drawn. For example, this cost is a function of quadrat distribution (e.g., travel time to access quadrats) and logistics (e.g., road access or lack thereof, as well as topography).

Evaluation of Topographic Covariates to Aid in Estimating Population Size

We used a ratio estimator with covariates in our initial estimation of population size to reduce spatial variability relative to a ratio estimator that did not use covariates (Bowden et al. 2003). We used quadrat area as a covariate in this estimate for lack of alternatives. Quadrat area was not tightly linked to owl density, however, and use of covariates that are linked to owl density should reduce spatial variation and improve the precision of the estimate. Consequently, we developed alternative terrain-based covariates and evaluated their use with this ratio estimator. We focused on terrain-based covariates because they could be computed from digital elevation models, which were readily available for the study area. A combination of terrain-based and habitat-based covariates (e.g., forest types) likely would work even better, but development of habitat-based covariates was beyond the scope of this study.

We developed and screened numerous terrain-based covariates but focus here on a covariate that provided an index of topographic roughness. This index (hereafter TIN ratio – 1) represented the ratio of true surface area to planar area for each quadrat, with true surface area derived from a triangulated irregular network (TIN). Because this ratio is always ≥1, we subtracted 1 from the ratio to set the minimum value to zero and force the relationship through the origin.

This covariate was highly correlated with owl abundance on sampled quadrats (Figure 16.7) and including it improved the precision of the estimate considerably. Using the CV as an index of variability, use of this covariate reduced CV from 37% (when quadrat area was used as the covariate; Bowden et al. 2003) to 24%. The relationship between abundance and TIN ratio – 1 appeared to be nonlinear, however, suggesting that further increases in precision might be possible using nonlinear functions

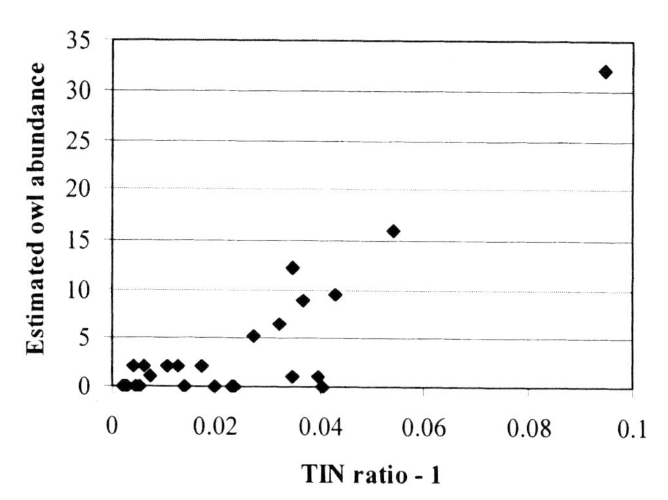

Figure 16.7. The relationship between estimated abundance of Mexican spotted owls and a topographic covariate (TIN ratio–1) for 25 quadrats sampled in Arizona and New Mexico, 1999.

of TIN ratio – 1. Consequently, we explored several nonlinear functions of TIN ratio – 1 as covariates. The best fit (CV = 21%) was obtained using $(\text{TIN ratio} - 1)^{1.52}$. Thus, we reduced variability in the estimate of abundance by 43% through use of an appropriate topographic covariate. We suspect that further reductions would be possible using a multivariate ratio estimator with both habitat- and terrain-based covariates. This analysis thus suggests that considerable reductions in variability are possible, with concurrent increases in the precision of the estimate and in our ability to detect trends in owl abundance.

Evaluating the Effect of Temporal Variation in Abundance

As a final step in evaluating the proposed monitoring program, we conducted a set of analyses to evaluate the effect of temporal process variation on (1) the probability of observing a 20% decline over 10 yrs; and (2) the estimated power to detect population trend. These analyses used estimates of temporal variation derived from the two demography study areas (Seamans et al. 1999). No such estimates were available when the recovery plan was produced or when the pilot study was initiated.

Estimation of the Probability of a 20% Decline

Because of temporal process variation in owl abundance, the Mexican spotted owl population could decline by 20% over a 10-yr period even if the population was stable over longer time horizons. To determine the probability of such a scenario, we estimated the process variance of λ, σ^2_λ, using the capture-recapture data from the demographic study areas in Arizona and New Mexico (Seamans et al. 1999). We used two approaches to estimate σ^2_λ. The first compared consecutive annual counts of owls observed on the study areas (i.e., $\lambda = N_{t+1}/N_t$, where N_t was the combined number of owls observed on both study areas in year t). The first two estimates of λ (for the intervals 1991–1992 and 1992–1993) were discarded because of the potential learning effect by observers during the first years of the study, leaving six estimates of λ for the combined study areas. The standard deviation of these six estimates ($\text{SD} = \sigma_\lambda = \sqrt{\sigma^2_\lambda} = 0.175$) represents the process standard deviation of the population growth rate ($\bar{\lambda} = 0.925$) assuming that annual "censuses" were accomplished. This estimate should be biased high because the counts used to estimate λ likely contained some sampling variation.

The second approach used estimates of λ obtained using capture-recapture data with the Pradel (1996) model in Program MARK (White and Burnham 1999). Five estimates of λ from the model $\{\phi$ (t), p(area*t), λ(t)$\}$ for the two study areas were used to estimate σ_λ. The first two estimates again were discarded because of potential learning effects by observers, and λ cannot be estimated for the first and last intervals with this model because it is confounded with estimates of p. The variance components estimator of Program MARK (White et al. 2002) estimated process standard deviation for the remaining five estimates as $\hat{\sigma}_\lambda = 0.114$ (95% confidence interval = 0.040–0.377; $\lambda = 0.905$; $\hat{SE} = 0.0557$). This estimate should be less biased than the value based on consecutive counts because it incorporated owl detection probabilities and, hence, sampling variances for the estimates.

Each of these estimates of temporal variability was used to generate 1,000,000 values of the product $\sum_{i=1}^{9} \lambda_i$, where λ_i was drawn from a normal distribution with a mean of 1 and a standard deviation equal to the appropriate estimate of σ_λ. The proportion of the simulations with the product less than 0.9, 0.8, and 0.5 then represent the probability of the owl population declining by 10, 20, and 50%, respectively, over a 10-yr period.

Results of these simulations indicated that the likelihood of observing a 20% decline in owl abundance over a 10-yr period was high given current estimates of temporal variation in abundance (Table 16.11). To assess the sensitivity of these results to the assumption of a normal distribution, two beta distributions also were examined. To set the bounds on the distribution of λ, a projection matrix was developed:

$$L = \begin{bmatrix} 0 & s_{SA} fR & s_A fR \\ s_J & 0 & 0 \\ 0 & s_{SA} & s_A \end{bmatrix},$$

where s_A = adult survival, s_{SA}= subadult survival, S_J = juvenile survival, f = fecundity, and R = proportion of young that are female. To estimate an upper bound on λ, we set $s_A = 0.95$, $s_{SA} = 0.95$, $s_J = 0.60$, $f = 2.8$, and R = 0.50. These values were very liberal compared to observed vital rates (Seamans et al. 1999). We obtained an asymptotic value of $\lambda = 1.49$ based on the dominant eigenvalue of the matrix:

Table 16.11.

Probability of the Mexican spotted owl population declining by 10, 20, or 50%, assuming that the population rate of growth (λ) is distributed with process standard deviation $\sigma_\lambda = 0.175$ or 0.114. We simulated both a normal and a beta distribution. Probabilities shown were based on 1,000,000 simulations, so standard errors of the estimates were $\sqrt{\hat{p}(1-\hat{p})/1{,}000{,}000}$.

			Probability of decline	
Statistical distribution	Bounds	Decline (%)	$\sigma_\lambda = 0.175$	$\sigma_\lambda = 0.114$
Normal	$-\infty,\infty$	10	0.515	0.439
		20	0.429	0.312
		50	0.156	0.037
Beta	0.60, 1.49	10	0.521	0.441
		20	0.434	0.312
		50	0.151	0.033
Beta	0.75, 1.49	10	0.527	0.446
		20	0.436	0.313
		50	0.138	0.027

$$\lambda = \frac{s_A + \sqrt{s_A^2 + 4Rs_J s_{SA} f}}{2}.$$

To estimate a lower bound on λ, we set $s_A = 0.60$, $s_{SA} = 0.60$, $s_J = 0$, $f = 0$, and $R = 0.50$. These values, which were very conservative (Seamans et al. 1999), yielded an asymptotic value of $\lambda = 0.60$. As can be seen from the eigenvalue equation, if $f = 0$, $\lambda = s_A$. The value $s_{SA} = 0.60$ was taken from Franklin et al. (2000) for northern spotted owls. A lower bound on $\lambda = 0.75$ also was considered.

Values of λ were simulated from a beta distribution as $B(\alpha,\beta)(\lambda_{upper} - \lambda_{lower}) + \lambda_{lower}$, where $B(\alpha, \beta)$ had a mean of $\mu = \dfrac{\bar{\lambda} - \lambda_{lower}}{\lambda_{upper} - \lambda_{lower}}$ and standard deviation of $\sigma = \dfrac{SD(\lambda)}{\lambda_{upper} - \lambda_{lower}}$, yielding $\alpha = \dfrac{-\mu(\mu^2 - \mu + \sigma^2)}{\sigma^2}$ and $\beta = \dfrac{(\mu - 1)(\mu^2 - \mu + \sigma^2)}{\sigma^2}$. Use of these alternate distributions did not greatly affect the results of the simulations (Table 16.11).

To verify that these simulations were approximately correct, we derived analytical results (as discussed earlier) to provide approximate estimates.

This analysis documented that the expected value of the product $\prod_{t=1}^{9} \lambda_i$ was 1, as we found in the simulations. Further, the approximate variance of this product based on a Taylor series expansion (delta method) was $\hat{\text{Var}}(\prod_{i=1}^{9} \lambda_i) = \prod_{i=1}^{9} \lambda_i \sum_{i=1}^{9} \text{CV}(\lambda_i)^2 = 9\sigma_\lambda^2$, although we know that this estimate is biased low. Again, this analytical result was consistent with simulation results.

The delisting criterion in the Mexican spotted owl Recovery Plan (USDI Fish and Wildlife Service 1995) specified monitoring for a 20% decline in owl abundance over a 10-yr period. Our results demonstrated that the likelihood of observing such a decline was high (e.g., Tables 16.11, 16.12) even when the overall population trend was stable (i.e., $\bar{\lambda} = 1$), given the level of process variation observed in the demography studies (Seamans et al. 1999). This suggests that observing a 20% decline in owl abundance over a 10-yr period does not provide unambiguous evidence that the owl population is in trouble. In other words, such a decline could occur with relatively high probability even though the owl population was stable over longer time horizons. Consequently, we considered alternative approaches to deriving a less ambiguous delisting criterion.

One such approach would involve basing the criterion on the probability of observing a decline of specified magnitude. To illustrate this approach, we modeled the percent decline in abundance expected as a func-

Table 16.12.

The percent decline expected in the Mexican spotted owl population over 10 yrs, given the specified probability levels and levels of temporal process variation (σ_λ).

			Expected percent decline	
Statistical distribution	Bounds	Probability	$\sigma_\lambda = 0.175$	$\sigma_\lambda = 0.114$
Normal	$-\infty,\infty$	0.05	65.7	47.3
		0.10	57.4	39.8
		0.25	39.4	25.1
Beta	0.60, 1.49	0.05	64.3	46.7
		0.10	56.4	39.4
		0.25	39.3	25.1
Beta	0.75, 1.49	0.05	61.8	45.5
		0.10	54.5	38.7
		0.25	38.5	25.0

tion of various probability levels, using the three distributions and two estimates of process standard deviation discussed above. Results suggested that neither the particular distribution used nor the lower bound on the beta distribution had much effect on the predictions (Table 16.12). In contrast, the predictions were extremely sensitive to the process standard deviation. Thus, one of the biggest uncertainties in devising an appropriate delisting criterion for the Mexican spotted owl population is the true magnitude of the process standard deviation. The available estimates may overestimate overall temporal variation because localized temporal variation may influence the populations on the demography areas. However, no other estimates of σ_λ were available at this time.

Estimation of Power to Detect a Decline

We also evaluated how process variation affected the power to detect trends in the population, because (1) the process variance of λ (i.e., temporal variability in owl abundance) strongly influences predicted probabilities of decline, and (2) considerable uncertainty exists concerning the true magnitude of this process variance (see above). Most research evaluating the power of monitoring programs to detect a trend has not considered process variance of λ directly. For example, Gerrodette (1987) developed analytical formulae for determining power but assumed that process variance was zero. Hatfield et al. (1996) considered serial correlation of population estimates but did not consider process variance explicitly. Gibbs (1999, 2000) provided a program (MONITOR) to simulate power of a monitoring program that "can accommodate random trend variation among plots or sites if estimates of its magnitude are available" (Gibbs 2000:228), but he assumed "deterministic linear trends are projected from the initial abundance index on each plot over the series of survey years" (Gibbs 2000:225) for analyses presented. Gibbs also presented estimated CVs to aid in designing population monitoring programs but did not separate sampling variance from temporal process variance in these estimates. Thompson et al. (1998) recognized that temporal process variation influences the appropriate level of decline for which to monitor but did not directly incorporate process variation into monitoring programs. Rather, Thompson et al. (1998) suggested that the analytical results presented by Gerrodette (1987) could be used with nonzero process variance by adding the process variance to the sampling variance when estimating power.

We evaluated the usefulness of Gerrodette's (1987) formulae when process variance was added to the sampling variance (Thompson et al.

1998) by comparing power estimates obtained from those formulae with power estimates from Monte Carlo simulations. Ten years of population estimates were considered, with a process standard deviation for λ of $\hat{\sigma}_\lambda = 0.114$, obtained from the Pradel (1996) model λ estimates from the two demographic study areas. We considered seven values of the sampling CV, ranging from 0 to 30% in increments of 5%. Sampling CV was assumed constant for population size, and the exponential model of decline, $\hat{N}_i = \hat{N}_1\lambda^{i-1}$, $i = 1, \ldots, 10$, was used to detect trend. Taking the natural logarithm of both sides of this equation recast it in the form of a linear regression ($y = a + bx$), such that $\ln(\hat{N}_i) = \ln(\hat{N}_1) + \ln(\lambda)(i - 1)$ and $\hat{N}_1 = \exp(\hat{a})$, $\hat{\lambda} = \exp(\hat{b})$, $SE(\hat{N}_1) \cong \exp(\hat{a})SE(\hat{a})$, and $SE(\hat{\lambda}) \cong \exp(\hat{b})SE(\hat{b})$.

We conducted Monte Carlo simulations with 10,000 replications, where λ_{i-1} was generated as a random normal deviate with mean = 1 and standard deviation = 0.114. Then, $\hat{N}_i = N + \sigma_i$, where σ_i was generated as a random normal deviate with mean zero and standard deviation CV_1N_i. Resulting power estimates differed considerably between the analytical approach and Monte Carlo simulations, where power estimates determined via simulation were considerably lower than estimates from the analytical procedure (Figures 16.8, 16.9). Thus, we concluded that simulation procedures were required to determine power for monitoring populations with nonzero temporal process variation, rather than using the approach in Thompson et al. (1998).

We also evaluated the percent relative bias (PRB) of estimates of N_1 and λ, $\hat{PRB}(\hat{\theta}) = (\hat{E}(\hat{\theta}) - \theta)/\theta$. Response surface models were fit for both $\hat{\lambda}$ and $\hat{N}_1$ with independent variables of process standard deviation (σ_λ), decline, decline squared, sampling CV, and sampling CV squared. The squared terms for decline and sampling CV were included to examine potential nonlinear, quadratic relationships. Based on AIC model selection, the best models were $P\hat{R}B(\hat{\lambda}) = 0.0528 - 10.404\,\sigma_\lambda + 0.00008182\,CV^2 - 0.00003257$ decline^2 and $P\hat{R}B(\hat{N}_1) = -0.47877 + 9.97573\,\sigma_\lambda + 0.00796\,CV - 0.00380\,CV^2 - 0.00002171$ decline^2. Relative bias was < 4% and < 1.5% for $\hat{N}_1$ and $\hat{\lambda}$, respectively, suggesting that the regression estimators used in simulations were useful for evaluating trends over the range of process standard deviations, sampling CVs, and declines considered.

To illustrate how temporal variation simultaneously influenced the power to detect a trend in abundance and the probability of occurrence of declines of specific magnitude, we again modeled power to detect a linear trend versus specified probabilities of decline. Unlike our earlier analysis, here we explicitly included temporal process variation. We

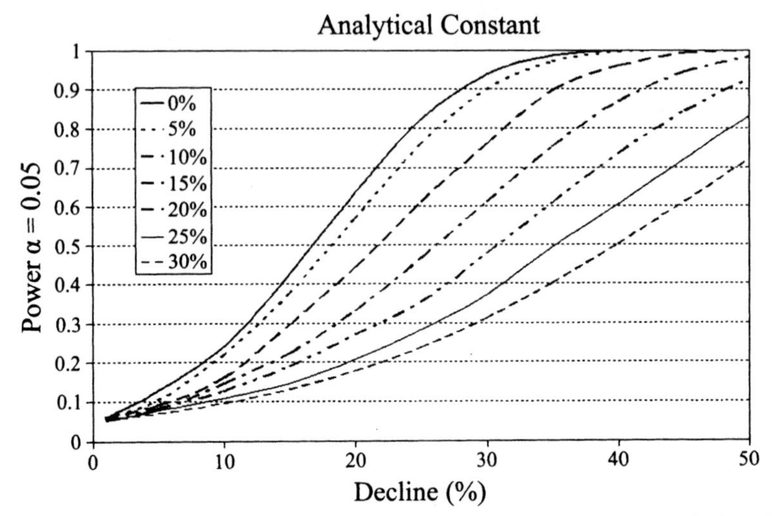

Figure 16.8. Power to detect a decline in Mexican spotted owl abundance modeled for declines of varying magnitude. Power curves were generated with the analytical formulae of Gerrodette (1987) for seven values of the coefficient of sampling variation assuming process standard deviation was 0.114. The coefficient of sampling variation was assumed to be constant and the exponential trend model was used with 10 yrs of population estimates. Curves shown represent CVs of 0–30% (from top to bottom, respectively).

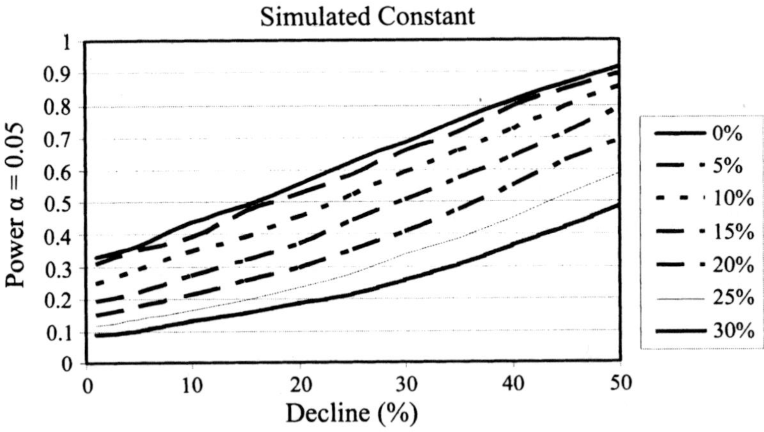

Figure 16.9. Simulation results for models relating power to percent decline in abundance of Mexican spotted owls, using seven values of the sampling coefficient of variation assuming process standard deviation was 0.114. An exponential model (Gerrodette 1987) was used to detect a trend over 10 yrs of population estimates. Curves represent CVs of 0–30% (from top to bottom, respectively).

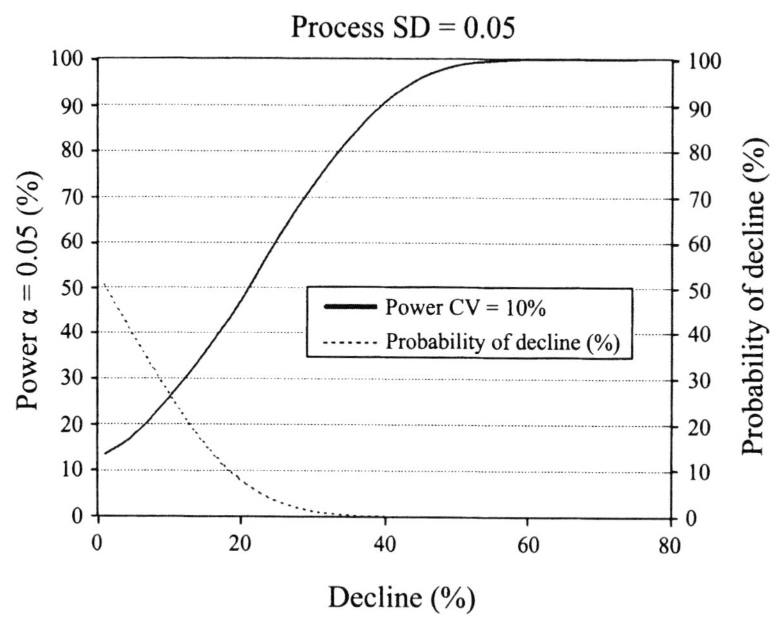

Figure 16.10. Probability of observing a decline in abundance of Mexican spotted owls of specific magnitude (descending curve) when mean $\lambda = 1$ and process standard deviation = 0.05. This probability is plotted at the same scale as the power to detect a decline (ascending curve) with a coefficient of sampling variation = 10%. Power was estimated from 10,000 simulations using 10 yrs of population estimates and the exponential model (Gerrodette 1987).

assumed CV = 10% for sampling variation and considered three levels of process variation: 0.05, 0.10, and 0.15 (Figures 16.10–16.12). We discuss these results relative to a decline in abundance of 20% (the magnitude specified in the current delisting criterion for the Mexican spotted owl). The expected probability of observing a 20% decline was approximately 10% for $\sigma_\lambda = 0.05$, 28% for $\sigma_\lambda = 0.10$, and 35% for $\sigma_\lambda = 0.15$. Although power curves varied among the three levels of process variance simulated, power to detect a 20% decline was approximately 40% in all cases (Figures 16.10–16.12). Thus, our results suggest that reasonable levels of temporal variation compound problems with both power to detect a trend and the probability of occurrence of declines of specified magnitude. Both factors

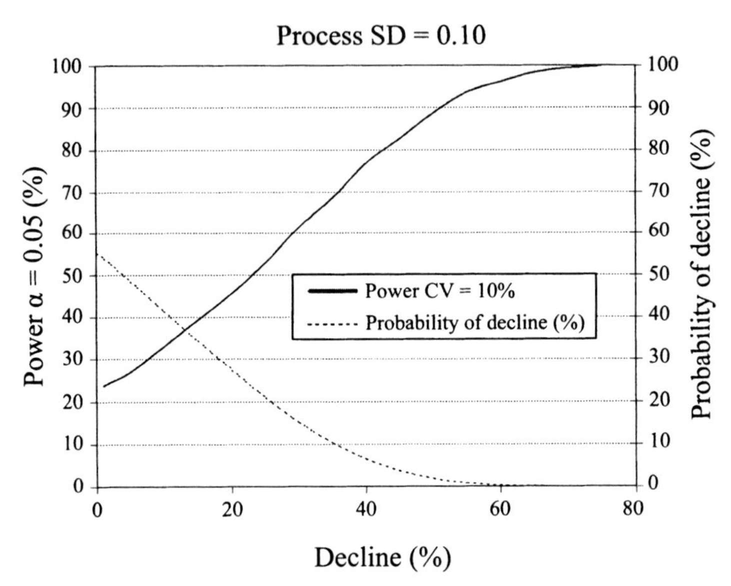

Figure 16.11. Probability of observing a decline in abundance of Mexican spotted owls of specific magnitude (descending curve) when mean $\lambda = 1$ and process standard deviation = 0.10. This probability is plotted at the same scale as the power to detect a decline (ascending curve) with a coefficient of sampling variation = 10%. Power was estimated from 10,000 simulations using 10 yrs of population estimates and the exponential model (Gerrodette 1987).

thus complicate devising a reasonable and unambiguous delisting criterion for the Mexican spotted owl.

Conclusion

Our experience in this pilot study suggests that mark-recapture sampling of Mexican spotted owls can be conducted and coordinated at large spatial scales, but that such sampling is relatively impractical. It requires large numbers of well-qualified and highly motivated field assistants, and finding and training these numbers of such individuals consistently on a seasonal basis will be difficult. The approach evaluated also is highly labor-intensive and therefore requires high funding levels. Both of these factors present serious practical hurdles to implementing such a program.

We also found that sample quadrats were too large to survey effectively in roadless areas containing large numbers of owls given limited person-

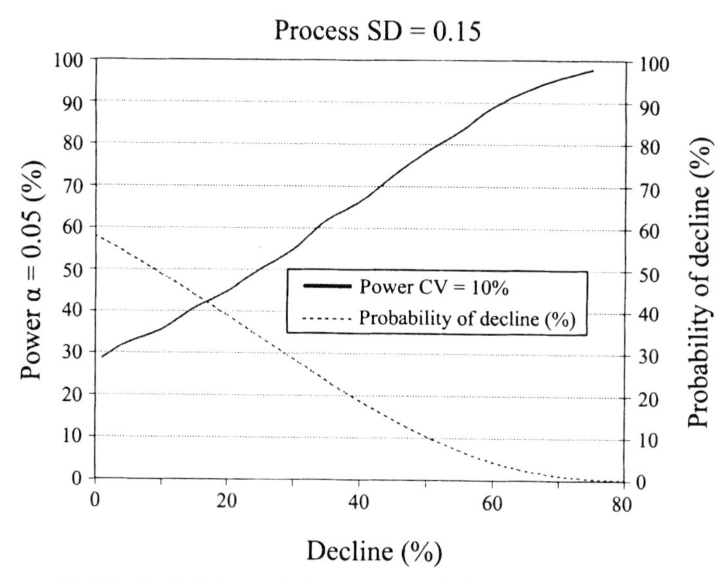

Figure 16.12. Probability of observing a decline in abundance of Mexican spotted owls of specific magnitude (descending curve) when mean $\lambda = 1$ and process standard deviation = 0.15. This probability is plotted at the same scale as the power to detect a decline (ascending curve) with a coefficient of sampling variation = 10%. Power was estimated from 10,000 simulations using 10 yrs of population estimates and the exponential model (Gerrodette 1987).

nel. Survey passes in such areas typically were not completed in a short time period, which is undesirable given the sampling design, and the required number of passes was never completed on some of these areas. Reducing quadrat size could alleviate both of these problems. Our analysis suggested that sampling smaller quadrats would not result in biased estimates of owl density or exaggerated edge effects. Thus, reducing quadrat size appears to be a feasible option. This should allow for an increase in spatial replication of quadrats, a decrease in overall variance, and an increase in the number of quadrats sampled for a given funding level.

We identified several practical ways to improve sampling efficiency, including placing more surveyors in roadless areas and ensuring that surveyors carry proper equipment. Despite such improvements, however, the mark-recapture sampling called for remains impractical at the required scale.

The use of a ratio estimator with quadrat area as a covariate increased precision in estimates of abundance over a simple estimator without covariates (Bowden et al. 2003), and use of appropriate topographical covariates resulted in further improvements in precision. It should be possible to increase precision further by using a combination of habitat-based and topographic covariates. Appropriate habitat-based covariates were not available to us but could be developed in future efforts.

Our analyses suggest that the approach of estimating a population trend directly from quadrat data holds more promise than the more typical regression approach. Despite proposed and actual improvements in efficiency of field sampling and estimation procedures, however, power to detect a 20% decline over a 10-yr period (as required by the delisting criterion in USDI Fish and Wildlife Service [1995]) remained low given current estimates of process variation. Further, there was a high probability of observing a 20% decline by chance alone, given current estimates of temporal process variation. This suggests that the delisting criterion does not provide an unambiguous monitoring target. We present results that may aid in devising a more appropriate delisting criterion, but recommending a specific criterion represents a decision-theory problem beyond the scope of this study.

Given the high economic costs involved in large-scale mark-recapture sampling, the serious logistical hurdles involved in such sampling, and uncertainty about an appropriate delisting criterion, we cannot recommend full-scale implementation of the monitoring design proposed in USDI Fish and Wildlife Service (1995). Instead, we recommend revising the delisting criterion to incorporate current data on temporal variation. This may require monitoring for a larger decline, an idea likely to prove unpopular with segments of the American public. Given the difficulty inherent in implementing the approach evaluated here, we also recommend considering practical aspects of implementing a monitoring program. This likely will require evaluating alternative approaches to monitoring that require less labor-intensive sampling. For example, we are currently working on a design based on simple presence-absence monitoring of a large number of small (1 km^2) quadrats.

General Lessons

This effort provides insight into several potential problems that may be encountered when sampling rare and elusive populations. The first is the potential magnitude of spatial variation. In this case study, spatial variation in abundance was the most significant source of variation encountered. This may not be a significant problem in all populations, but it will undoubtedly prove problematic in some, particularly populations that are widely but patchily distributed.

A second significant problem encountered in this study was temporal variation. Again, the magnitude of this problem likely will vary among populations, but our results suggest that it should not be ignored. In particular, choosing an appropriate monitoring target becomes difficult in the face of significant temporal variation. In our case, monitoring for a decline that would occur with low probability (e.g., 5–10% of the time) would have required us to monitor for a very large decline (roughly 30–60%, depending on the extent of temporal variation; Figure 16.12) within the 10-yr period. Monitoring for a decline of such magnitude is likely to prove socially or politically unacceptable when dealing with a threatened species. Monitoring for longer time periods may alleviate this problem somewhat, a possibility that we have not yet explored. The obvious drawbacks to that approach are delayed decisions on listing status and greater funding requirements.

A related problem concerns the power to detect a trend. In this case study, power was low due to spatial variation when temporal variation was ignored, and lower still when both sources of variation were considered explicitly. Our results also suggested that traditional analytical analyses of power to detect trends did not provide valid results when temporal variation was significant and that simulations were required to accurately estimate power in such situations.

It is unclear at this time whether the Mexican spotted owl presents a best- or worst-case scenario relative to these points. This owl likely is both more abundant and more easily detected than many rare and elusive species. All else being equal, both factors should aid in generating precise estimates and adequate power to detect trends. However, both spatial and temporal variation may be higher for the Mexican spotted owl than for some rare and elusive species. Both sources of variation strongly affect precision of estimates and power to detect trends. Thus, although we can note general problems likely to be encountered by similar efforts, we cannot at this time estimate the magnitude of problems likely to be encountered by such efforts.

Our results clearly demonstrate that monitoring population trends in a variable world is a difficult and expensive task. Many federal agencies operate under mandates to monitor multiple species of plants and animals, however. Our experience suggests that fulfilling this mandate will present great practical difficulties. Recent advances in analysis of presence-absence data (e.g., MacKenzie et al. 2003; Royle and Nichols 2003) may help alleviate this problem. Nevertheless, we recommend that agencies proceed extremely carefully when identifying particular species to monitor and when designing monitoring criteria and protocols. Particular attention should be paid to the appropriate level of decline for which to monitor, power to detect a decline of that magnitude, sample sizes required to achieve that power, and associated costs and feasibility issues. Failure to consider these aspects could result in poorly designed monitoring schemes and unattainable goals, which in turn could place agencies in legal jeopardy.

Acknowledgments

This study was made possible by the efforts of many dedicated individuals. Special thanks to field crew leaders S. Ackers, P. Fonken, J. Jenness, C. Kessler, P. Shaklee, and E. Swarthout and crew members D. April, J. Bennett, S. Bird, B. Coyle, J. Cummings, K. Damm, R. Davis, C. Dodge, G. Dunn, B. Gill, K. Gilligan, J. Gulbransen, J. Hagerty, M. Holley, T. Holt, R. Hunt, J. Justus, K. Krause, A. Lavoie, M. Layes, M. Lucid, D. McCallum, M. McCarty, H. McClure, S. Meares, B. Millay, J. Moser, M. Moyer, J. Palmer, A. Ringia, M. Schenk, S. Schmitz, T. Shewan, A. Shovlain, A. Stein, J. Trent, A. Trombley, E. Vincent, S. Vuturo, R. Wagner, P. Wenninger, T. Williams, A. Winters, and J. York. Special thanks also to K. Nodal, administrative assistant extraordinaire, and to T. Garito for data entry. Funding was provided by the Southwestern Region, U.S. Forest Service, and numerous Forest Service district personnel helped in innumerable ways. S. Arundel created the quadrat sampling frame, and J. S. Jenness computed the topographic covariates. J. P. Ward, Jr., helped considerably during the initial design phase. Numerous individuals from the Rocky Mountain Research Station also assisted in numerous ways, including B. Block, J. Dwyer, S. Elefant, K. Gurley-Davis, S. Kyle, D. Parker, D. Prince, B. Strohmeyer, and B. Whiteman, Flagstaff; and L. Bemis, R. King, and T. Pues, Fort Collins. C. May, M. Petersburg, M. Seamans, and R. J. Gutiérrez kindly allowed us to use unpublished data from their demography studies to evaluate the effects of quadrat size and temporal variability. W. L. Thompson and an anonymous reviewer reviewed earlier drafts of this chapter.

References

Akaike, H. 1973. Information theory and an extension of the maximum likelihood principle. pp. 267–281 in B. N. Petran and F. Csaki, eds., *Second International Symposium on Information Theory*. Akademiai Kiado, Budapest, Hungary.

Arundel, S. 1999. *Developing a Sampling Frame to Guide Quadrat Selection for a*

Program to Monitor Population Trend of the Mexican Spotted Owl: Methodology. Unpublished report. USDA Forest Service, Rocky Mountain Research Station, Flagstaff, Arizona.

Bailey, R. G. 1980. *Descriptions of the Ecoregions of the United States.* USDA Forest Service Miscellaneous Publication 1391, Intermountain Region, Ogden, Utah.

Block, W. M., A. B. Franklin, J. P. Ward Jr., J. L. Ganey, and G. C. White. 2001. Design and implementation of monitoring studies to evaluate the success of ecological restoration on wildlife. *Restoration Ecology* 9:293–303.

Bowden, D. C., G. C. White, A. B. Franklin, and J. L. Ganey. 2003. Estimating population size with correlated sampling unit estimates. *Journal of Wildlife Management* 67:1–10.

Buckland, S. T., K. P. Burnham, and N. H. Augustin. 1997. Model selection: An integral part of inference. *Biometrics* 53:603–618.

Burnham, K. P., and D. R. Anderson. 1998. *Model Selection and Inference: A Practical Information-Theoretic Approach.* Springer-Verlag, New York.

Cochran, W. G. 1977. *Sampling Techniques,* 3rd ed. Wiley, New York.

Fletcher, K. W., and H. E. Hollis. 1994. *Habitats Used, Abundance, and Distribution of the Mexican Spotted Owl* (Strix occidentalis lucida) *on National Forest System Lands in the Southwestern Region.* Unpublished report. USDA Forest Service, Southwestern Region, Albuquerque, New Mexico.

Forsman, E. D. 1983. *Methods and Materials for Locating and Studying Spotted Owls.* USDA Forest Service General Technical Report PNW-162, Pacific Northwest Research Station, Portland, Oregon.

Forsman, E. D., A. B. Franklin, F. M. Oliver, and J. P. Ward Jr. 1996. A color band for spotted owls. *Journal of Field Ornithology* 67:507–510.

Franklin, A. B., D. R. Anderson, E. D. Forsman, K. P. Burnham, and F. W. Wagner. 1996. Methods for collecting and analyzing demographic data on the Northern spotted owl. *Studies in Avian Biology* 17:12–20.

Franklin, A. B., D. R. Anderson, R. J. Gutiérrez, and K. P. Burnham. 2000. Climate, habitat quality, and fitness in Northern spotted owl populations in northwestern California. *Ecological Monographs* 70:69–120.

Franklin, A. B., J. P. Ward, R. J. Gutiérrez, and G. I. Gould Jr. 1990. Density of northern spotted owls in northwest California. *Journal of Wildlife Management* 54:1–10.

Ganey, J. L., S. Ackers, P. Fonken, J. S. Jenness, C. Kessler, K. Nodal, P. Shaklee, and E. Swarthout. 1999. *Monitoring Populations of Mexican Spotted Owls in Arizona and New Mexico: 1999 Progress Report.* USDA Forest Service Rocky Mountain Research Station, Flagstaff, Arizona. Available from http://www.rms.nau.edu/lab/4251/spowmonitoring.html.

Ganey, J. L., and J. A. Dick. 1995. Habitat relationships of Mexican spotted owls: Current knowledge. Chapter 4:1–42 in *Recovery Plan for the Mexican Spotted Owl. Vol. II - Technical Supporting Information.* USDI Fish and Wildlife Service, Albuquerque, New Mexico.

Gerrodette, T. 1987. A power analysis for detecting trends. *Ecology* 68:1364–1372.

Gibbs, J. P. 1999. Effective monitoring for adaptive wildlife management: Lessons from the Galápagos Islands. *Journal of Wildlife Management* 63:1055–1065.

_____. 2000. Monitoring populations. pp. 213–252 in L. Boitani and T. K. Fuller, eds., *Research Techniques in Animal Ecology: Controversies and Consequences.* Columbia University Press, New York.

Goodman, D. 1987. The demography of chance extinction. pp. 11–34 in M. E. Soulé, ed., *Viable Populations for Conservation*. Cambridge University Press, Cambridge.

Goodman, L. 1960. On the exact variance of products. *Journal of the American Statistical Association* 55:708–713.

Gutiérrez, R. J., A. B. Franklin, and W. S. LaHaye. 1995. Spotted owl (*Strix occidentalis*). No. 179 in A. Poole and F. Gill, eds., *The Birds of North America*. The Academy of Natural Sciences, Philadelphia, and The American Ornithologists Union, Washington, D.C.

Hatfield, J. S., W. R. Gould IV, B. A. Hoover, M. R. Fuller, and E. L. Lindquist. 1996. Detecting trends in raptor counts: Power and Type I error rates of various statistical tests. *Wildlife Society Bulletin* 24:505–515.

Huggins, R. M. 1989. On the statistical analysis of capture-recapture experiments. *Biometrika* 76:133–140.

_____. 1991. Some practical aspects of a conditional likelihood approach to capture experiments. *Biometrics* 47:725–732.

Hurvich, C. M., and C.-L. Tsai. 1995. Model selection for extended quasi-likelihood models in small samples. *Biometrics* 51:1077–1084.

Johnson, C. L., and R. T. Reynolds. 1998. A new trap design for capturing spotted owls. *Journal of Raptor Research* 32:181–182.

Lebreton, J.-D., K. P. Burnham, J. Clobert, and D. R. Anderson. 1992. Modeling survival and testing biological hypotheses using marked animals: A unified approach with case studies. *Ecological Monographs* 62:67–118.

MacKenzie, D. I., J. D. Nichols, J. E. Hines, M. G. Knutson, and A. B. Franklin. 2003. Estimating site occupancy, colonization, and local extinction when a species is detected imperfectly. *Ecology* 84:2200–2207.

Marcot, B. G., and J. W. Thomas. 1997. *Of Spotted Owls, Old Growth, and New Policies: A History Since the Interagency Scientific Committee Report*. USDA Forest Service General Technical Report PNW-GTR-408, Pacific Northwest Research Station, Portland, Oregon.

May, C. A., M. Z. Peery, R. J. Gutiérrez, M. E. Seamans, and D. R. Olson. 1996. *Feasibility of a Random Quadrat Study Design to Estimate Changes in Density of Mexican Spotted Owls*. USDA Forest Service Research Paper RM-RP-322, Rocky Mountain Forest and Range Experiment Station, Fort Collins, Colorado.

Otis, D. L., K. P. Burnham, G. C. White, and D. R. Anderson. 1978. Statistical inference from capture data on closed animal populations. *Wildlife Monographs* 62:1–135.

Pradel, R. 1996. Utilization of capture-mark-recapture for the study of recruitment and population growth rate. *Biometrics* 52:703–709.

Rinkevich, S. E., J. L. Ganey, W. H. Moir, F. P. Howe, F. Clemente, and J. F. Martinez-Montoya. 1995. Recovery Units. pp. 36–51 in *Recovery Plan for the Mexican Spotted Owl*. USDI Fish and Wildlife Service, Albuquerque, New Mexico.

Royle, J. A., and J. D. Nichols. 2003. Estimating abundance from repeated presence-absence data or point counts. *Ecology* 84:777–790.

Seamans, M. E., R. J. Gutiérrez, C. A. May, and M. Z. Peery. 1999. Demography of two Mexican spotted owl populations. *Conservation Biology* 13:744–754.

Seber, G. A. F. 1982. *The Estimation of Animal Abundance and Related Parameters*, 2nd ed. Griffin, London.

Skalski, J. R. 1994. Estimating wildlife populations based on incomplete area surveys. *Wildlife Society Bulletin* 22:192–203.

Stanford Environmental Law Society. 2001. *The Endangered Species Act.* Stanford University Press, Stanford, California.

Stanley, T. R., and K. P. Burnham. 1998a. Estimator selection for closed-population capture-recapture. *Journal of Agricultural, Biological, and Environmental Statistics* 3:131–150.

_____. 1998b. Information-theoretic model selection and model averaging for closed-population capture-recapture studies. *Biometrical Journal* 40:475–494.

Thompson, W. L., G. C. White, and C. Gowan. 1998. *Monitoring Vertebrate Populations.* Academic Press, San Diego.

USDA Forest Service. 1996. *Record of Decision for Amendment of Forest Plans. Arizona and New Mexico.* USDA Forest Service, Southwestern Region, Albuquerque, New Mexico.

USDI Fish and Wildlife Service. 1993. Endangered and threatened wildlife and plants: Final rule to list the Mexican spotted owl as a threatened species. *Federal Register* 58:14248–14271.

_____. 1995. *Recovery Plan for the Mexican Spotted Owl* (Strix occidentalis lucida), *Vol. I.* USDI Fish and Wildlife Service, Albuquerque. Available from http://mso.fws.gov/recovery_plan.htm.

_____. 1996. *Final Biological Opinion. Mexican Spotted Owl and Critical Habitat and Forest Plan Amendments: U. S. Forest Service, Southwestern Region.* USDI Fish and Wildlife Service, Albuquerque.

Ward, J. P., Jr., A. B. Franklin, S. E. Rinkevich, and F. Clemente. 1995. Distribution and abundance of Mexican spotted owls. Chap. 1:1–14 in *Recovery Plan for the Mexican Spotted Owl. Vol. II - Technical Supporting Information.* USDI Fish and Wildlife Service, Albuquerque, New Mexico.

White, G. C., and K. P. Burnham. 1999. Program MARK: Survival estimation from populations of marked animals. *Bird Study* 46 Supplement:S120–S138.

White, G. C., A. B. Franklin, and T. M. Shenk. 2002. Estimating parameters of PVA models from data on marked animals. pp. 169–190 in S. R. Beissinger and D. R. McCullough, eds., *Population Viability Analysis.* University of Chicago Press, Chicago.

Williams, J. L. 1986. *New Mexico in Maps.* University of New Mexico Press, Albuquerque, New Mexico.

Part V

THE FUTURE

17

Future Directions in Estimating Abundance of Rare or Elusive Species

William L. Thompson

Natural resource professionals often are charged with gathering information about population status and trend of rare, elusive or otherwise difficult-to-detect plants and animals. Obtaining reliable information of this kind is of paramount importance because these species typically are the ones of greatest management concern. However, rare or elusive species pose a difficult sampling challenge at all stages of a survey (see Chapter 2). For example, consider a two-stage sample to estimate abundance wherein a random sample of plots is selected from a sampling frame during the first stage and counts are conducted on selected plots during the second stage (see Chapter 1). Further assume that the species of interest is spatially rare; hence, there are a relatively large number of plots containing no individuals. The number of individuals within occupied plots may vary from few (truly rare) to many (rare at a larger spatial scale but locally abundant). Sampling challenges at both stages relate to detection probability, that is, detection of which plots are occupied and detection of individuals within these plots. This chapter discusses potential avenues of future research and methodological development for estimating abundance of rare or elusive species within this two-stage design context.

Plot Selection

A primary goal of a survey is to obtain a sample of plots in which the relevant characteristics of the study species, such as abundance, approximate those of the sampled population. This often is referred to as a representative sample. Because biological populations are usually clustered in the environment (Cole 1946), sampling designs that provide good spatial coverage of plots are preferred. This is especially true when sampling a species that is spatially rare. Here, the goal is to maximize the detection probability of occupied plots to avoid expending too much effort sampling unoccupied ones. Sampling across multiple time periods demands additional design considerations. In this case, a sampled population can vary both spatially and temporally. This section discusses potential future applications and directions in developing designs to select plots for sampling rare or elusive species within single and multiple time periods.

Single Time Period

Traditional designs for sampling rare human populations, such as snowball sampling, screening, and sequential sampling (Kalton and Anderson 1986; Sudman et al. 1988), usually do not translate well to surveying plant and animal populations (but see Christman and Lan 2001; Chapter 7, this volume). Thompson (1990) developed adaptive cluster sampling designs to address this inadequacy, with further developments by Thompson and Seber (1994, 1996), Brown and Manly (1998), and Chao and Thompson (2001) (see also Chapter 5, this volume). However, given that there have only been a few new sampling designs of any type developed within the recent past (e.g., adaptive cluster sampling [Thompson 1990], simple Latin square sampling +1 [Munholland and Borkowski 1996], generalized random-tessellation stratified [GRTS] design [Stevens and Olsen 2004]), future advances will probably be manifested in some combined form of existing designs. For instance, a combined adaptive cluster–inverse sampling design may prove to be more efficient, both statistically and logistically, than either design alone (see Christman and Lan 2001). In general, greater use of auxiliary variables in model-assisted survey sampling (Sarndal et al. 1992) may increase efficiency of estimators when sampling rare or elusive species. Such variables could include habitat or environmental measures that are related to population abundance or density. Ranked set sampling (McIntyre 1952; Johnson et al. 1996; see also Volume 6, Issue 1 of *Environmental and Ecological Statistics*) is an example of such a design that may be useful for sampling these species, especially if applied adaptively.

Another use of auxiliary or prior information is through placement of plots or sampling units such as in directed sampling designs. Gradsect sampling is a directed sampling approach that chooses sampling units along the largest environmental gradients, such as elevation and rainfall, in the area of interest to maximize coverage of habitat types and, hence, increase the probability of detecting the maximum number of species present (Gillison and Brewer 1985). Johnson and Patil's (1995) covariate-directed sampling design for estimating species richness is in a similar vein. Future developments may include combining classic probabilistic designs (Cochran 1977) with directed sampling designs. Stahl et al. (2000) adopted this approach in developing their guided transect sampling design for assessing rare populations.

Stable isotopes are a source of auxiliary information that could be used to focus more sampling effort within specific areas to maximize detection of rare or elusive species. These chemical elements, which occur in the tissues of organisms and in the environment (Hobson 1999), possess different numbers of neutrons and their quantities (ratios) are influenced by those occurring naturally in the environment. Hence, isotopic signatures can be related to geographical location, although the spatial resolution upon which this can be performed depends on the organism and geochemical context. Nonetheless, ecologists have increasingly applied stable isotope technology to address ecological questions, such as identifying source or seasonal use areas of highly mobile or migratory organisms (Hobson 1999; Rubenstein et al. 2002). Thus, there is potential to sample individuals from a rare or elusive species and backtrack them to where they came from or previously visited, depending on the spatial resolution of the isotopic signature. This information then could be used to increase efficiency of subsequent surveys by pinpointing higher probability areas of species occupancy.

Multiple Time Periods

Population monitoring is a common activity for many natural resource agencies and often is necessary for attainment of legislative mandates (e.g., Endangered Species Act of 1973, National Forest Management Act of 1976, and National Parks Omnibus Management Act of 1998 in the United States). One of the key design objectives is to ensure that the initial sample of plots remains representative. A sample may lose this quality if there are shifts in population numbers and/or spatial distribution during later time periods that are no longer reflected in the original sampled units (Overton and Stehman 1996; Wikle and Royle 1999). These population shifts across time could be generated by a number of factors, such as changes in habitat due to anthropomorphic influences (Schweiger et al. 2000; Coppedge et al. 2001) or ecological succession (Ballinger and Watts

1995; Skelly et al. 1999). This situation is exacerbated when monitoring rare or elusive species because the number of detections is small and population estimates are highly variable.

Probability sampling of units over time falls into three general categories. Sampling units may be repeatedly sampled across time (permanent plots), be replaced with a new random sample during each time period (temporary plots), or have only a portion repeatedly sampled with the rest chosen anew during each time period (partial replacement plots) (Schreuder et al. 1993; Scott 1998). As long as the initial sample remains representative of the population of interest, permanent plots yield the most precise trend estimator of these three approaches (Patterson 1950; Cochran 1977).

Sampling with partial replacement designs combines gains in precision from permanent plots with increases in representativeness from temporary plots. These designs have been used most often in ecological applications by foresters to estimate current values and change in forest resources (Ware and Cunia 1962; Scott 1984; Schreuder et al. 1993). However, their widespread use is curtailed because of their complexity, especially of their variance estimators (Scott 1998), which are typically based on regression estimators. Further, estimators have not been developed for partial replacement designs for sampling rare or elusive species, such as adaptive cluster sampling, although bootstrapping (Efron and Tibshirani 1993) may be a feasible approach to generating variance estimates for these designs (Schreuder et al. 2004). Future work may be focused in this direction as well as in applying nonlinear variance estimators, such as negative binomial regression, when many sampling zeros occur.

Simplicity of design is an important component of an effective monitoring protocol (Schreuder and Czaplewski 1993; Schreuder 1994; Overton and Stehman 1995). Thus, a simpler design that provides comparable levels of precision would be more attractive to practitioners. As mentioned previously, partial replacement designs are complex, so investigations into alternatives based on simpler designs would be welcomed. For instance, in situations in which the population is expected to shift or change during the course of the study, one may incorporate Overton and Stehman's (1996) poststratification scheme with simple random sampling of permanent plots. Note that stratification is typically an important aspect of sampling designs for single time periods but may be counterproductive under multiple time periods and changing populations. In fact, Overton and Stehman (1996:355) suggested, "stratification will often be a liability in long-term monitoring." The challenge will be to develop a design that balances simplification with realities of sampling rare or elusive species. Nonetheless, this is a ripe area for research.

Counts within Plots

When a complete or nearly complete count is not possible within a selected plot, the objective is to reliably estimate the detection probability of the species of interest, which then will be used to correct the observed count. Note that this applies to sampling either individuals (occurrence, abundance) or species (species richness) within plots. Capture-recapture models normally do not function well for populations that are sparse, have low capture probabilities, or both (White et al. 1982; Menkens and Anderson 1988; but see Chao 1989 and Rosenberg et al. 1995). Distance sampling estimators (Buckland et al. 2001) also suffer when numbers of detections are low (but see Ramsey et al. 1987, Fancy 1997). Note, however, that Pollard et al. (2002) combined adaptive sampling with line-transect sampling to survey spatially clustered populations. This section discusses potential future directions in development of statistical models for estimating abundance and spatial distribution as well as the counting techniques used to gather data to fit these models.

Statistical Models

Capture-recapture models have a long history of use in sampling biological populations (see recent reviews by Schwarz and Seber 1999; Borchers et al. 2002; and Williams et al. 2002). Although mainly applied to fish and wildlife populations, there have been recent applications to plants (Alexander et al. 1997). However, nearly all work to date has been on discrete-time, capture-recapture models (White et al. 1982) wherein capture data are collected at specific sampling occasions and previous capture information is recorded. Scant attention has been paid to continuous-time models (Becker 1984), that is, capture data are collected and recorded for each marked individual over continuous time periods and each capture is a separate occasion (these may be grouped under the discrete-time model; Wilson and Anderson 1995). In fact, Wilson and Anderson (1995) suggested that continuous-time, capture-recapture models might be appropriate for species with low capture probabilities and/or densities. Hwang and Chao (2002) developed a class of continuous-time, capture-recapture models that incorporated covariates and allowed individual capture probabilities to vary with time, behavioral response, and heterogeneity. Although novel applications of discrete-time, capture-recapture models will continue to be important (see Farnsworth et al. 2002), more focus could be placed on development and application of continuous time models, especially applied to sparse or elusive species.

Mixture models represent a semiparametric approach to fitting data whose distributional shape is unknown. Use of these models has greatly expanded in the statistical literature since 1995 (McLachlan and Peel 2000) and will likely continue. Pledger's (2000) class of discrete-time, capture-recapture models represents a notable application within ecology. This general class of models offers great potential for estimating abundance or species richness of rare or elusive species, with the Poisson gamma (negative binomial) mixture being particularly suited to modeling rare and clustered populations.

Counting Techniques

As technology has improved and new analytical methods are developed and refined, the number of possible counting techniques available to biologists and ecologists has grown. Noninvasive sampling methods, in particular, have increased in both popularity and use. These approaches avoid potential problems related to handling or marking individuals, are usually more effective at detecting presence of rare or elusive species, and tend to be much more cost-effective than their invasive counterparts. Three types of noninvasive sampling methods will be briefly discussed in this section because they appear to hold the broadest potential for estimating abundance of rare or elusive species: (1) genetic analyses of collected feces or hair; (2) bioacoustical sampling; and (3) thermal infrared imaging in aerial surveys.

DNA microsatellite analysis of collected feces (Kohn and Wayne 1997) or hair (Woods et al. 1999) to identify individuals has been used to fit mark-recapture and similar models to estimate abundance of rare or elusive species (see Chapter 11). Genetic analysis is not without potential problems (Taberlet and Waits 1998; Mills et al. 2000), but continued improvements in laboratory techniques and technology should address these shortcomings (Taberlet et al. 1999). Further, increased automation in processing genetic samples should make this approach much more cost-effective in the future. Genetic information might be useful for answering demographic questions such as degree of exchange among populations (metapopulations) and rates of dispersal.

Use of bioacoustics offers a way to sample and "mark" individuals or species based on their vocal signatures. This is a growing area of research and technological development that has been successfully applied to identifying individuals of certain species, such as frogs (Bee et al. 2001) and owls (Delport et al. 2002). As with DNA microsatellite analysis, data collected from bioacoustical sampling could be used with mark-recapture models to estimate abundance and species richness as well as demographic parameters such as survival rate.

Thermal infrared imagery has been occasionally used to count animals, particularly deer (*Odocoileus* spp.; Havens and Sharp 1998). However, present technology does not allow for complete counts of animals over the size of area typically surveyed. Consequently, bias due to incomplete and variable detection rates remains a concern, particularly in habitats where thermal signals are difficult to identify. Combining thermal imagery with methods that correct for incomplete detection, such as distance sampling, might be a fruitful avenue of research. For instance, Gill et al. (1997) used a thermal imager in conjunction with an internal range finder in ground surveys to estimate densities of various species of deer. Aerial surveys would allow much larger areas to be sampled, so even if a species occurred in low densities, one may obtain a suitable number of detections through use of longer transects.

Summary

Future directions in estimating abundance of rare or elusive species can be summarized in three words: (1) innovation, (2) technology, and (3) software. Innovation plays a role in using existing designs, models, or counting techniques, singly or in combination, in new ways or contexts. For instance, Farnsworth et al. (2002) combined mark-recapture theory (removal models) with traditional point counts to estimate abundance of birds. This type of innovative thinking needs to be applied to the more challenging issue of sampling rare or elusive species. Biologists and ecologists should not limit themselves to considering applications in their own discipline; more cross-fertilization among scientific disciplines can offer new perspectives, insights, and perhaps tools (models). Thinking outside the box, by definition, requires broadening one's horizons.

Technological advances in tools for counting plants and animals will be the grist for the future research mill. Much theory currently exists for estimating abundance; the problem often is having the means to gather data in a way that will meet key assumptions in a cost-efficient manner. For example, recent advances in genetic techniques and equipment (e.g., gene sequencers) have allowed biologists to feasibly use noninvasive methods to sample rare or elusive species (see Chapter 11, this volume). As counting tools improve, our ability to properly account for nuisance parameters, such as detection probability, also should improve.

Software development must go hand-in-hand with innovation and technological advances to be accessible and practical for end-users. The most innovative designs, models, or counting techniques are not going to be useful to ecologists and biologists in general if they are not available in a user-friendly computing program. For instance, Program MARK (White and Burnham 1999) has allowed users to fit complex models in a readily available freeware package. The recent proliferation of studies incorporating the latest mark-recapture and related models is due in no small part to this program. Similarly, freeware Program DISTANCE (Thomas 1999) also has enabled biologists and ecologists to analyze distance sampling data; the latest version 4.1 also incorporates components for designing studies in conjunction with GIS data. Nonetheless, there currently exists a void in software available for designing studies to estimate abundance or density in a two-stage context across multiple time periods. In particular, more emphasis should be placed on providing software programs to both design and analyze surveys based on partial replacement of plots during each time period. This is especially true of designs for sampling rare or elusive species. As designs and models become more complex, the need for software increases.

Much of the traditional sampling designs and counting techniques were developed for surveying moderately abundant to abundant species. With increasing human pressures, many plants and animals will continue to decline in numbers, contract in range, or both. Thus, more focus needs to be placed on development of a larger toolbox of sampling designs and counting techniques for surveying rare or elusive species. For many imperiled species, the future is now.

References

Alexander, H. M., N. A. Slade, and W. D. Kettle. 1997. Application of mark-recapture models to estimation of the population size of plants. *Ecology* 78:1230–1237.

Ballinger, R. E., and K. S. Watts. 1995. Path to extinction: Impact of vegetation change on lizard populations on Arapahoe Prairie in the Nebraska Sandhills. *American Midland Naturalist* 134:413–417.

Becker, N. G. 1984. Estimating population size from capture-recapture experiments in continuous time. *Australian Journal of Statistics* 26:1–7.

Bee, M. A., C. E. Kozich, K. J. Blackwell, and H. C. Gerhardt. 2001. Individual variation in advertisement calls of territorial male green frogs, *Rana clamitans*: Implications for individual discrimination. *Ethology* 107:65–84.

Borchers, D. L., S. T. Buckland, and W. Zucchini. 2002. *Estimating Animal Abundance: Closed Populations*. Springer-Verlag, London.

Brown, J. A., and B. F. J. Manly. 1998. Restricted adaptive cluster sampling. *Environmental and Ecological Statistics* 5:49–63.

Buckland, S. T., D. R. Anderson, K. P. Burnham, J. L. Laake, D. L. Borchers, and L. Thomas. 2001. *Introduction to Distance Sampling*. Oxford University Press, Oxford.

Chao, A. 1989. Estimating population size for sparse data in capture-recapture experiments. *Biometrics* 45:427–438.

Chao, C.-T., and S. K. Thompson. 2001. Optimal adaptive selection of sampling sites. *Environmetrics* 12:517–538.

Christman, M. C., and F. Lan. 2001. Inverse adaptive cluster sampling. *Biometrics* 57:1096–1105.

Cochran, W. G. 1977. *Sampling Techniques*, 3rd ed. Wiley, New York.

Cole, L. C. 1946. A study of the cryptozoa of an Illinois woodland. *Ecological Monographs* 16:50–86.

Coppedge, B. R., D. M. Engle, R. E. Masters, and M. S. Gregory. 2001. Avian response to landscape change in fragmented southern Great Plains grasslands. *Ecological Applications* 11:47–59.

Delport, W., A. C. Kemp, and J. W. H. Ferguson. 2002. Vocal identification of individual African wood owls *Strix woodfordii*: A technique to monitor long-term adult turnover and residency. *Ibis* 144:30–39.

Efron, B., and R. J. Tibshirani. 1993. *An Introduction to the Bootstrap*. Chapman and Hall, New York.

Fancy, S. G. 1997. A new approach for analyzing bird densities from variable circular-plot counts. *Pacific Science* 51:107–114.

Farnsworth, G. L., K. H. Pollock, J. D. Nichols, T. R. Simons, J. E. Hines, and J. R. Sauer. 2002. A removal model for estimating detection probabilities from bird counts. *Auk* 119:414–425.

Gill, R. M. A., M. L. Thomas, and D. Stocker. 1997. The use of portable thermal imaging for estimating deer population density in forest habitats. *Journal of Applied Ecology* 34:1273–1286.

Gillison, A. N., and K. R. W. Brewer. 1985. The use of gradient directed transects or gradsects in natural resource survey. *Journal of Environmental Management* 20:103–127.

Havens, K. J., and E. J. Sharp. 1998. Using thermal imagery in the aerial survey of animals. *Wildlife Society Bulletin* 26:17–23.

Hobson, K. A. 1999. Tracing origins and migration of wildlife using stable isotopes: A review. *Oecologia* 120:314–326.

Hwang, W.-H., and A. Chao. 2002. Continuous-time capture-recapture models with covariates. *Statistica Sinica* 12:1115–1131.

Johnson, G. D., B. D. Nussman, and G. P. Patil. 1996. Designing cost-effective environmental sampling using concomitant information. *Chance* 9:4–11.

Johnson, G. D., and G. P. Patil. 1995. Estimating statewide species richness of breeding birds in Pennsylvania. *Coenoses* 10:81–87.

Kalton, G., and D. W. Anderson. 1986. Sampling rare populations. *Journal of the Royal Statistical Society, Series A* 149:65–82.

Kohn, M. H., and R. K. Wayne. 1997. Facts from feces revisited. *Trends in Ecology and Evolution* 12:223–227.

McIntyre, G. A. 1952. A method of unbiased selective sampling, using ranked sets. *Australian Journal of Agricultural Research* 3:385–390.

McLachlan G., and D. Peel. 2000. *Finite Mixture Models*. Wiley, New York.

Menkens, G. E., Jr., and S. H. Anderson. 1988. Estimation of small-mammal population size. *Ecology* 69:1952–1959.

Mills, L. S., J. J. Citta, K. P. Lair, M. K. Schwartz, and D. A. Tallmon. 2000. Estimating animal abundance using noninvasive DNA sampling: Promise and pitfalls. *Ecological Applications* 10:283–294.

Munholland, P. L., and J. J. Borkowski. 1996. Simple Latin square sampling +1: A spatial design using quadrats. *Biometrics* 52:125–136.

Overton, W. S., and S. V. Stehman. 1995. Design implications of anticipated data uses for comprehensive environmental monitoring programmes. *Environmental and Ecological Statistics* 2:287–303.

_____. 1996. Desirable design characteristics for long-term monitoring of ecological variables. *Environmental and Ecological Statistics* 3:349–361.

Patterson, H. D. 1950. Sampling on successive occasions with partial replacement of units. *Journal of the Royal Statistical Society, Series B* 12:241–255.

Pledger, S. 2000. Unified maximum likelihood estimates for closed capture-recapture models using mixtures. *Biometrics* 56:434–442.

Pollard, J. H., D. Palka, and S. T. Buckland. 2002. Adaptive line transect sampling. *Biometrics* 58:862–870.

Ramsey, F. L., V. Wildman, and J. Engbring. 1987. Covariate adjustments to effective area in variable-area wildlife surveys. *Biometrics* 43:1–11.

Rosenberg, D. K., W. S. Overton, and R. G. Anthony. 1995. Estimation of animal abundance when capture probabilities are low and heterogeneous. *Journal of Wildlife Management* 59:252–261.

Rubenstein, D. R., C. P. Chamberlain, R. T. Holmes, M. P. Ayres, J. R. Waldbauer, G. R. Graves, and N. C. Tuross. 2002. Linking breeding and wintering ranges of a migratory songbird using stable isotopes. *Science* 295:1062–1065.

Sarndal, C.-E., B. Swensson, and J. Wretman. 1992. *Model Assisted Survey Sampling*. Springer-Verlag, New York.

Schreuder, H. T. 1994. Simplicity versus efficiency in sampling designs and estimation. *Environmental Monitoring and Assessment* 33:237–245.

Schreuder, H. T., R. Ernst, and H. Ramirez-Maldonado. 2004. *Statistical Techniques for Sampling and Monitoring Natural Resources*. USDA Forest Service General Technical Report RMRS-GTR-126, Rocky Mountain Research Station, Fort Collins, Colorado.

Schreuder, H. T., and R. L. Czaplewski. 1993. Long-term strategy for the statistical design of a forest health monitoring system. *Environmental Monitoring and Assessment* 27:81–94.

Schreuder, H. T., T. G. Gregoire, and G. B. Wood. 1993. *Sampling Methods for Multiresource Forest Inventory*. Wiley, New York.

Schwarz, C. J. and G. A. F. Seber. 1999. A review of estimating animal abundance III. *Statistical Science* 14:427–456.

Schweiger, E. W., J. E. Diffendorfer, R. D. Holt, R. Pierotti, and M. S. Gaines. 2000. The interaction of habitat fragmentation, plant, and small mammal succession in an old field. *Ecological Monographs* 70:383–400.

Scott, C. T. 1984. A new look at sampling with partial replacement. *Forest Science* 30:157–166.

_____. 1998. Sampling methods for estimating change in forest resources. *Ecological Applications* 8:228–233.

Skelly, D. K., E. E. Werner, and S. A. Cortwright. 1999. Long-term distributional dynamics of a Michigan amphibian assemblage. *Ecology* 80:2326–2337.

Stahl, G., A. Ringvall, and T. Lamas. 2000. Guided transect sampling for assessing sparse populations. *Forest Science* 46:108–115.

Stevens, D. L., Jr., and A. R. Olsen. 2004. Spatially balanced sampling of natural resources. *Journal of the American Statistical Association* 99:262–278.

Sudman, S., M. G. Sirken, and C. D. Cowan. 1988. Sampling rare and elusive populations. *Science* 240:991–996.

Taberlet, P., and L. Waits. 1998. Non-invasive genetic sampling. *Trends in Evolution and Ecology* 13:26–27.

Taberlet, P., L. Waits, and G. Luikart. 1999. Noninvasive genetic sampling: Look before you leap. *Trends in Ecology and Evolution* 14:323–327.

Thomas, L. 1999. DISTANCE 3.5: Software for analysis of distance sampling data. *Bulletin of the Ecological Society of America* 80:114–115.

Thompson, S. K. 1990. Adaptive cluster sampling. *Journal of the American Statistical Association* 85:1050–1059.

Thompson, S. K., and G. A. F. Seber. 1994. Detectability in conventional and adaptive sampling. *Biometrics* 50:712–724.

Thompson, S. K., and G. A. F. Seber. 1996. *Adaptive Sampling*. Wiley, New York.

Ware, K. D., and T. Cunia. 1962. Continuous forest inventory with partial replacement of samples. *Forest Science Monograph* 3:1–40.

White, G. C., D. R. Anderson, K. P. Burnham, and D. L. Otis. 1982. *Capture-Recapture and Removal Methods for Sampling Closed Populations*. USDOE Los Alamos National Laboratory Report LA-8787-NERP, Los Alamos, New Mexico.

White, G. C., and K. P. Burnham. 1999. Program MARK: Survival estimation from populations of marked animals. *Bird Study* 46:S120–S139.

Wikle, C. K., and J. A. Royle. 1999. Space-time dynamic design of environmental monitoring networks. *Journal of Agricultural, Biological, and Environmental Statistics* 4:489–507.

Williams, B. K., J. D. Nichols, and M. J. Conroy. 2002. *Analysis and Management of Animal Populations*. Academic Press, San Diego.

Wilson, K. R., and D. R. Anderson. 1995. Continuous-time capture-recapture population estimation when capture probabilities vary over time. *Environmental and Ecological Statistics* 2:55–69.

Woods, J. G., D. Paetkau, D. Lewis, B. N. McLellan, M. Proctor, and C. Strobeck. 1999. Genetic tagging of free-ranging black and brown bears. *Wildlife Society Bulletin* 27:616–627.

Contributors

MATHEW W. ALLDREDGE
Biomathematics Program, Statistics Department
North Carolina State University
Raleigh, NC 27695
USA

LARISSA L. BAILEY
USGS North Carolina State University Cooperative Fish & Wildlife Research Unit &
USGS Patuxent Wildlife Research Center
12100 Beech Forest Road
Laurel, MD 20708
USA

PETER B. BAYLEY
Department of Fisheries and Wildlife
104 Nash Hall
Oregon State University
Corvallis, OR 97331
USA

EARL F. BECKER
Alaska Department of Fish and Game
Division of Wildlife Conservation
333 Raspberry Road
Anchorage, AK 99518
USA

DAVID C. BOWDEN
Department of Statistics
Colorado State University
Fort Collins, CO 80523
USA

JENNIFER A. BROWN
Biomathematics Research Center
Department of Mathematics and Statistics
University of Canterbury
Private Bag 4800
Christchurch
New Zealand

KENNETH P. BURNHAM
USGS Colorado Cooperative Fish & Wildlife Research Unit
Department of Fishery and Wildlife Biology
Colorado State University
Fort Collins, CO 80523
USA

MARY C. CHRISTMAN
Department of Animal and Avian Sciences
University of Maryland
College Park, MD 20742
USA

PAUL B. CONN
Department of Fishery and Wildlife Biology
Colorado State University
Fort Collins, CO 80523
USA

LAURA E. ELLISON
USGS Fort Collins Science Center
2150 Centre Avenue, Bldg C
Fort Collins, CO 80526
USA

George L. Farnsworth
Department of Biology
Xavier University
3800 Victory Parkway
Cincinnati, OH 45207
USA

Alan B. Franklin
USGS Colorado Cooperative Fish & Wildlife Research Unit
Department of Fishery and Wildlife Biology
Colorado State University
Fort Collins, CO 80523
USA

Joseph L. Ganey
USDA Forest Service-Rocky Mountain Research Station
2500 S. Pine Knoll
Flagstaff, AZ 86001
USA

Craig L. Gardner
Alaska Department of Fish and Game
Division of Wildlife Conservation
1300 College Road
Fairbanks, AK 99701
USA

Howard N. Golden
Alaska Department of Fish and Game
Division of Wildlife Conservation
333 Raspberry Road
Anchorage, AK 99518
USA

Dana H. Hanselman
University of Alaska-Fairbanks, SFOS Fisheries Division
11220 Glacier Hwy.
Juneau, AK 99801
USA

K. Ullas Karanth
Wildlife Conservation Society-India Program
26-2, Aga Abbas Ali Road (Apt. 403)
Bangalore
Karnataka 560 042
India

N. Samba Kumar
Centre for Wildlife Studies
823, 13th Cross, Jayanagar 7th Block West
Bangalore 560082
India

Nancy C. H. Lo
National Oceanic and Atmospheric Administration
National Marine Fisheries Service-Southwest Fisheries Science Center
8604 La Jolla Shores Drive
La Jolla, CA 92037
USA

Darryl I. MacKenzie
Proteus Research & Consulting Ltd.
P.O. Box 5193
Dunedin
New Zealand

Bryan F. J. Manly
Western EcoSystems Technology, Inc.
2003 Central Avenue
Cheyenne, WY 82001
USA

Chris R. Margules
CSIRO Sustainable Ecosystems
Tropical Forest Research Centre &
The Rainforest Cooperative Research Center
P.O. Box 780
Atherton, Queensland 4883
Australia

HELENE MARSH
Department of Tropical Environment Studies and Geography
James Cook University &
CRC Reef Research Centre
Townsville 4811
Australia

LYMAN L. MCDONALD
Western EcoSystems Technology, Inc.
2003 Central Avenue
Cheyenne, WY 82001
USA

JAMES D. NICHOLS
USGS Patuxent Wildlife Research Center
12100 Beech Forest Road
Laurel, MD 40708
USA

THOMAS J. O'SHEA
USGS Fort Collins Science Center
2150 Centre Avenue, Bldg C
Fort Collins, CO 80526
USA

JAMES T. PETERSON
USGS Georgia Cooperative Fish & Wildlife Research Unit
Warnell School of Forest Resources
University of Georgia
Athens, GA 30602
USA

KENNETH H. POLLOCK
Department of Zoology
North Carolina State University
Raleigh, NC 27695
USA

ELIZABETH L. POON
Department of Tropical Environment Studies and Geography
James Cook University
P.O. Box 6811
Cairns, Queensland 4870
Australia

TERRANCE J. QUINN II
University of Alaska-Fairbanks, SFOS Fisheries Division
11220 Glacier Hwy.
Juneau, AK 99801
USA

J. ANDREW ROYLE
U.S. Fish and Wildlife Service
Division of Migratory Bird Management
11510 American Holly Drive
Laurel, MD 20708
USA

JOHN R. SAUER
USGS Patuxent Wildlife Research Center
12100 Beech Forest Road
Laurel, MD 20708
USA

THEODORE R. SIMONS
USGS North Carolina Cooperative Fisheries & Wildlife Research Unit
Zoology Department
North Carolina State University
Raleigh, NC 27695
USA

DAVID R. SMITH
USGS Leetown Science Center
Aquatic Ecology Laboratory
11700 Leetown Road
Kearneysville, WV 25430
USA

THOMAS R. STANLEY
USGS Fort Collins Science Center
2150 Centre Avenue, Bldg C
Fort Collins, CO 80526
USA

WILLIAM L. THOMPSON
National Park Service
Southwest Alaska Network
240 West 5th Avenue
Anchorage, AK 99501
USA

LISETTE P. WAITS
Department of Fish and Wildlife Resources
University of Idaho
Moscow, ID 83844
USA

GARY C. WHITE
Department of Fishery and Wildlife Biology
Colorado State University
Fort Collins, CO 80523
USA

Reviewers

LARISSA L. BAILEY
USGS North Carolina State University
Cooperative Fish & Wildlife Research Unit &
USGS Patuxent Wildlife Research Center
12100 Beech Forest Road
Laurel, MD 20708

RICHARD J. BARKER
Department of Mathematics and Statistics
University of Otago
P.O. Box 56
Dunedin, New Zealand

KENNETH P. BURNHAM
USGS Colorado Cooperative Fish & Wildlife Research Unit
Department of Fishery and Wildlife Biology
Colorado State University
Fort Collins, CO 80523

JOHN CLARKSON
Queensland Parks and Wildlife Service
Centre for Tropical Agriculture
P.O. Box 1054
Mareeba, Queensland, Australia

MICHAEL J. CONROY
USGS Georgia Cooperative Fish & Wildlife Research Unit
Warnell School of Forest Resources
University of Georgia
Athens, GA 30602

R. I. C. C. (CHRIS) FRANCIS
National Institute of Water & Atmospheric Research
Private Bag 14901
Wellington, New Zealand

MARY C. FREEMAN
USGS Patuxent Wildlife Research Center
Institute of Ecology
University of Georgia
Athens, GA 30602

WILLIAM R. GOULD
Department of Economics and International Business &
University Statistics Center
New Mexico State University
Las Cruces, NM 88003

TIMOTHY G. GREGOIRE
School of Forestry and Environmental Studies
Yale University
360 Prospect Street
New Haven, CT 06511

ROWAN HAIGH
Pacific Biological Station
Nanaimo, British Columbia V9T 6N7 Canada

WILLIAM L. KENDALL
USGS Patuxent Wildlife Research Center
12100 Beech Forest Road
Laurel, MD 20708

THOMAS H. KUNZ
Center for Ecology and Conservation Biology
Department of Biology
Boston University
5 Cummington Street
Boston, MA 02215

PAUL M. LUKACS
Department of Fishery and Wildlife Biology
Colorado State University
Fort Collins, CO 80523

L. SCOTT MILLS
Wildlife Biology Program
College of Forestry and Conservation
University of Montana
Missoula, MT 59812

TIMOTHY G. O'BRIEN
Wildlife Conservation Society
2300 Southern Blvd.
Bronx, New York 10460

MARK O'DONOGHUE
Dept. of Environment, Fish and Wildlife Branch
P.O. Box 310
Mayo, Yukon Territory Y0B 1M0 Canada

BRENT PATTERSON
Ontario Ministry of Natural Resources
Wildlife Research and Development Section
300 Water Street, 3rd Floor N.
Peterborough, ON K9J 8M5 Canada

JEFFREY S. PONTIUS
Department of Statistics
Kansas State University
Manhattan, KS 66506

MARTIN G. RAPHAEL
USDA Forest Service-Pacific Northwest Research Station
3625 93rd Ave. SW
Olympia, WA 98512

Mohammad Salehi M.
School of Mathematical Sciences
Isfahan University of Technology
Isfahan, Iran

Philip Turk
Department of Mathematical Sciences
Montana State University
Bozeman, MT 59717

Stephen M. Turton
Rainforest Cooperative Research Centre
James Cook University
P.O. Box 6811
Cairns, Queensland 4870, Australia

About the Editor

WILLIAM L. THOMPSON is an ecologist/biometrician with the National Park Service in Anchorage, Alaska, where he oversees the design of long-term monitoring programs for plants and animals in five national parks in southwestern Alaska. He also provides technical training and statistical assistance to biologists and resource managers. He is senior author of *Monitoring Vertebrate Populations.*

Index

f refers to pages with figures.
t refers to pages with tables.

Island Press Board of Directors